Commentary on IET Wiring Regulations 17th Edition

Commentary on IET Wiring Regulations 17th Edition

BS 7671:2008+A3:2015 Requirements for Electrical Installations

Paul Cook

The Institution of Engineering and Technology

Published by The Institution of Engineering and Technology, London, United Kingdom

Commentary on 15th Edition of IEE Wiring Regulations, 1st Edn 1981 (0 90604 851 6)
Commentary on 15th Edition of IEE Wiring Regulations, 2nd Edn 1985 (0 86341 040 5)
Commentary on IEE Wiring Regulations 16th Edition (BS 7671:1992 Amdt No.1) 1997 (0 86341 316 1)
Commentary on IEE Wiring Regulations 16th Edition (BS 7671:1992 Amdt No.2) 1998 (0 86341 318 8)
Commentary on IEE Wiring Regulations 16th Edition (BS 7671:2001 Amdt No.1) 2002 (0 85296 237 1)
Commentary on IEE Wiring Regulations 17th Edition (BS 7671:2008) 2010 (978-0-86341-966-9)
Commentary on IET Wiring Regulations 17th Edition (BS 7671:2008 inc Amd 3) 2016 (978-1-84919-765-6)

The Institution of Engineering and Technology
Michael Faraday House
Six Hills Way, Stevenage
Herts, SG1 2AY, United Kingdom

www.theiet.org

British Library Cataloguing in Publication Data
A catalogue record for this product is available from the British Library

ISBN 978-1-84919-765-6 (hardback)
ISBN 978-1-84919-766-3 (PDF)

Typeset in India by S4Carlisle Publishing Services
Printed in the UK by CPI Antony Rowe, Chippenham

In memoriam

In remembrance of two engineers who had much influence on the modern version of the IET Wiring Regulations: Brian Jenkins and R.G. ('Dick') Parr.

Brian worked for the secretariat of the Wiring Regulations Committee in that special period when the IEC standard was adopted as the 15th Edition. He also wrote the first version of the Commentary. I was and remain flattered that he asked me to take over from him the writing of the Commentary.

Dick Parr worked for the ERA and carried out much of the development work on cable theory that is the backbone of the IEC cable standards; you will see many references to papers of Dick Parr in the Bibliography.

Dick was a calm man, always willing to help and very tolerant of engineers such as myself who did not have a full understanding of his subject.

Brian was outgoing and was always amused by committee work; he would have laughed long at the quote in Brian Bower's article (Appendix A) with respect to the first edition. 'The English of the rules, although in some cases awkward, is not obscure, and its awkwardness is only what one might expect from rules...originally drafted in good English but afterwards amended in committee.'

Contents

Tables

Acknowledgements

I would like to record my thanks to Mark Coates (ERA) (Mark.Coates@era.co.uk), Colin Reed (Prysmian Cables & Systems Limited), Ken Morton (HSE), Bill Rogers (consulting electrical power, Rogers028@aol.com) and Tony Haggis (Central Networks) for their help in preparing this book. I have recognized their contribution to particular chapters in the chapter headings. My thanks go also to Peter Donnachie for reading the draft and providing editorial comment.

If you want help from someone who knows, contact these engineers.

Chapter 1
Introduction

1.1 Some history

There are two very significant dates in the development of BS 7671 *Requirements for Electrical Installations;* they are:

(a) 11 May 1882, when the Society of Telegraph Engineers and of Electricians appointed a committee to consider and report on the rules they would recommend for the prevention of fire risks arising from the use of electric light; and
(b) 2 October 1992, when the British Standards Institution (BSI) and the IEE signed an agreement, making the 16th Edition of the IEE Wiring Regulations into British Standard 7671.

The early history of the IEE Wiring Regulations is described with great flair by Dr Brian Bowers, a historian from the Science Museum, in Appendix A, and Appendix B includes a copy of the first set of Rules and Regulations of the Society of Telegraph Engineers and of Electricians.

For over 100 years, the IEE published what became known to every electrician, designer, consultant and architect as the Wiring Regulations, the current edition being the 17th (full title BS 7671:2008 *Requirements for Electrical Installations*). The IEE (now the IET) has for all this time encouraged all professionals involved in electrical installations to make recommendations and provide advice, and has taken note of agreements reached by the International Electrotechnical Commission (IEC) and by the European Committee for Electrotechnical Standardization (CENELEC), but ultimately decided for itself what requirements would and would not be included in its Rules and Regulations.

This changed on 2 October 1992, when the IEE and the BSI agreed that the 16th Edition of the Wiring Regulations would become a British Standard (BS 7671:1992 *Requirements for Electrical Installations*). It was also agreed that this new standard would be treated in a similar manner to other standards, that is, in accordance with the by-laws of the BSI.

This means that the IEE must honour the agreements the BSI has with the IEC and with CENELEC. The IEE no longer has complete freedom to write Regulations as it sees fit, but must work through the BSI with other members of CENELEC, and the IEC, to agree what Regulations (requirements) are appropriate.

The IEE has participated in international wiring standards since the IEC established its Technical Committee (TC64), known as 'Electrical Installations and Protection against Electric Shock'; and the IEE was a founder member of the British Standards Institution and has always participated in BSI committees on wiring standards. The 15th Edition of the Wiring Regulations was notable for the adoption of the International Electrotechnical Commission structure of wiring standards. Indeed, the wiring regulations from the 15th Edition onwards have taken account of the technical substance of the parts of the IEC publication so far issued, and of the corresponding agreements reached in the European standards-making body, CENELEC. However, the words in the Preface to the Wiring Regulations were carefully chosen: 'account has been taken of the technical substance of the parts of IEC publication 364 so far published and of the corresponding agreements reached in CENELEC'. In the past, where the IEE Wiring Regulations committee considered that the IEC and CENELEC standards were not appropriate, perhaps incomplete, or poorly worded, the rules were not adopted. This changed with the adoption of the Wiring Regulations as a British Standard; it has been agreed that in the preparation of the British Wiring Regulations, BS 7671, the rules of CENELEC will be adopted, and this work continues.

1.2 CENELEC

The Preface to BS 7671 makes reference to CENELEC harmonization documents (HD) in the HD 384 series Electrical Installations of Buildings. A harmonization document is a CENELEC standard that carries with it the obligation to be implemented at a national level by all members of CENELEC at least by the public announcement of the HD number and title, and by withdrawal of any conflicting national standards. Having fulfilled these obligations, a member country is free to maintain or issue a national standard dealing with the subject within the scope of the HD, provided that it does not conflict. There is a requirement that the number, title and date of each such national standard shall be notified to the Central Secretariat. This is the status of the standards associated with electrical installations. A second and more common type of CENELEC standard is an EN (European Norm). These are published by the BSI as BS ENs. An EN, or European Norm, must be implemented at national level by member countries by being given the status of a national standard, and by withdrawal of any conflicting national standards. An EN is implemented identically in technical content and presentation. The implementation of an EN differs from an HD, in that it must be published exactly as prepared by the CENELEC committee. An HD is required to be implemented in technical content, but it need not be identical word for word.

Adoption of standards in CENELEC is determined by a weighted voting system, with members having the numbers of votes as set out in Table 1.1.

Counting of weighted votes

Votes from all members are counted first, and the proposal is adopted if 71 per cent or more of the weighted votes cast (abstentions not counted) are in favour. If the proposal is not adopted by counting all members' votes, the votes of the members from EEA countries only are counted separately. The proposal is adopted if 71 per cent or more of the weighted votes cast by the European Economic Area (EEA) countries (abstentions not counted) are in favour.

Table 1.1 Weightings allocated to CEN/CENELEC national members in case of weighted voting

Situation as of January 2007		
Member country	Weighting	EEA country
France	29	X
Germany	29	X
Italy	29	X
United Kingdom	29	X
Poland	27	X
Spain	27	X
Romania	14	X
Netherlands	13	X
Belgium	12	X
Czech Republic	12	X
Greece	12	X
Hungary	12	X
Portugal	12	X
Austria	10	X
Bulgaria	10	X
Sweden	10	X
Switzerland	10	–
Denmark	7	X
Finland	7	X
Ireland	7	X
Lithuania	7	X
Norway	7	X
Slovakia	7	X
Cyprus	4	X
Estonia	4	X
Latvia	4	X
Luxembourg	4	X
Slovenia	4	X
Iceland	3	X
Malta	3	X

Source: Annex D to CEN/CENELEC Internal Regulations – Part 2: Common Rules for Standards Work (2006) (© CEN/CENELEC, reproduced with permission)

Consultants and designers may now have to refer to the standards for electrical installations of other countries. It is worth noting that not all countries publish their wiring rules in one bound volume. Some publish in many parts, in a similar manner to CENELEC and the IEC. The German standards are published in this way, and English-language versions are available from VDE. However, there are many parts, and a complete set is expensive. The French standard is published in one volume, revised about every ten years. The electrical installation standards in use in all countries in the European Union are harmonized, in that the technical intent of the harmonization documents is incorporated. However, there remain many differences.

Users of BS 7671 will have noted that it is often not specific in its requirements, but is structured on the basis of general principles for the provision of protection against electric shock, fire, overcurrent etc. This approach was adopted in the IEC standard (IEC 60364), as it did allow agreement to be reached on standards that may not have been possible otherwise. Each country has its own installation practices and thinks its own practices are the best. The use of reduced-section protective conductors in general wiring cables and steel wire armoured cables are UK practices not adopted by most other countries.

Countries may continue with their national practices while adopting the same basic principles for protection as all other countries. This fundamental improvement has led the CENELEC and IEC committees that are responsible for installation rules also to be responsible for setting the basic requirements for protection against electric shock to be adopted by all equipment committees.

Historically, all the development work on international and European standards has been carried out under the auspices of the International Electrotechnical Commission (IEC); this procedure generally continues today, with standards initially being prepared and voted on a worldwide basis, and these are then voted on for adoption by CENELEC for European countries.

To speed up the procedures CENELEC has agreed a parallel voting procedure whereby voting in IEC and CENELEC is carried out in parallel (at the same time) at the Committee Draft for Voting stage (CDV). Additionally, if there are no negative votes, the standard will proceed straight to publication, missing out the final FDIS (Final Draft International Standard) stage.

1.3 Deviations

Countries are allowed three types of departure from the European standard (HD or EN):

(a) A special national condition where a national characteristic or practice cannot be changed even over a long period, e.g. climatic conditions, distribution system electrical earthing;

(b) An A-deviation if made necessary by a national legal requirement for the time being outside the competence of CENELEC;

(c) A B-deviation (for HDs only) to reflect a technical difference found for the time being to be really necessary, with a definite date of withdrawal.

Note: Where standards fall under EC Directives, it is the view of the Commission of the European Communities (OJ No. C 59, 1982-03-09) that the effect of the decision of the Court of Justice in Case 815/79 Cremonini/Vrankovich (European Court Reports 1980, p. 3583) is that compliance with A-deviations is no longer mandatory and that the free movement of products complying with such a standard should not be restricted within the EC except under the safeguard procedure provided for in the relevant Directive.

1.4 Note by the Health and Safety Executive

There is an important note by the Health and Safety Executive (HSE) in BS 7671 preceding Part 1. This states that installations which conform to BS 7671:2008 'are regarded by HSE as likely to achieve conformity with the relevant parts of the Electricity at Work Regulations 1989'. However, it goes on to make a further comment that existing installations may have been designed and installed to conform to the standards set by earlier editions of BS 7671 or the IEE Wiring Regulations. While they comply with earlier editions and not the current one, this does not necessarily mean that they will not comply with the relevant parts of the Electricity at Work Regulations. This guidance needs to be considered when periodic inspection report forms in Appendix 6 of BS 7671 are being completed. These require the persons carrying out the inspection and test to make recommendations as to what remedial work is necessary, and also provide guidance as to whether:

- the work is required urgently;
- some improvement is required;
- the inspector needs to make further investigations

This means that the inspector must provide advice. Part of the installation may not comply with BS 7671 and require urgent attention; or, conversely, it may not comply with BS 7671, but the inspector does not consider that any further action is required.

1.5 Plan of BS 7671

The requirements for any particular risk, e.g. electric shock, were to be found in four locations in the 15th and 16th Editions of the Wiring Regulations. This followed the layout of the IEC standard IEC 60364 Electrical installations of buildings. For example, requirements for shock protection were to be found in:

(a) Chapter 13 'Fundamental principles';
(b) Chapter 41 'Protection against shock';
(c) Section 471 'Application of protective measures'; and
(d) Section 531 'Shock protection devices'.

See Figure 1.1.

Subsequently, the IEC plenary committee meeting accepted a UK proposal that all the requirements for any particular risk be brought together. The IEC standard has been restructured on this basis, allowing the UK standard to adopt the same

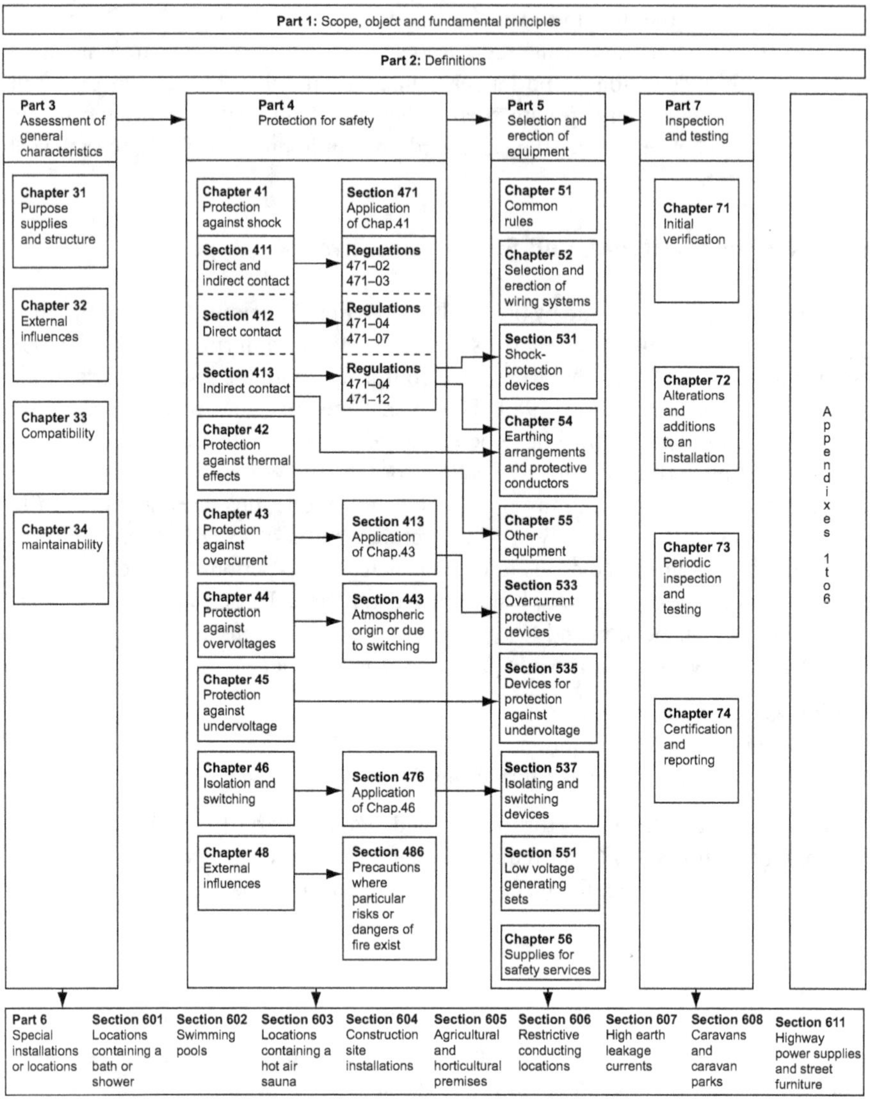

Figure 1.1 Plan of 15th and 16th Editions

simplified structure. In the 17th Edition, requirements for shock are to be found in only two locations:

(a) Chapter 13 ‘Fundamental principles’;
(b) Chapter 41 ‘Protection against electric shock’.

Similarly, the requirements for isolation and switching are found in:

(a) Chapter 13 ‘Fundamental principles’;
(b) Chapter 53 ‘Switchgear (for protection, isolation and switching)’.

See Figure 1.2.

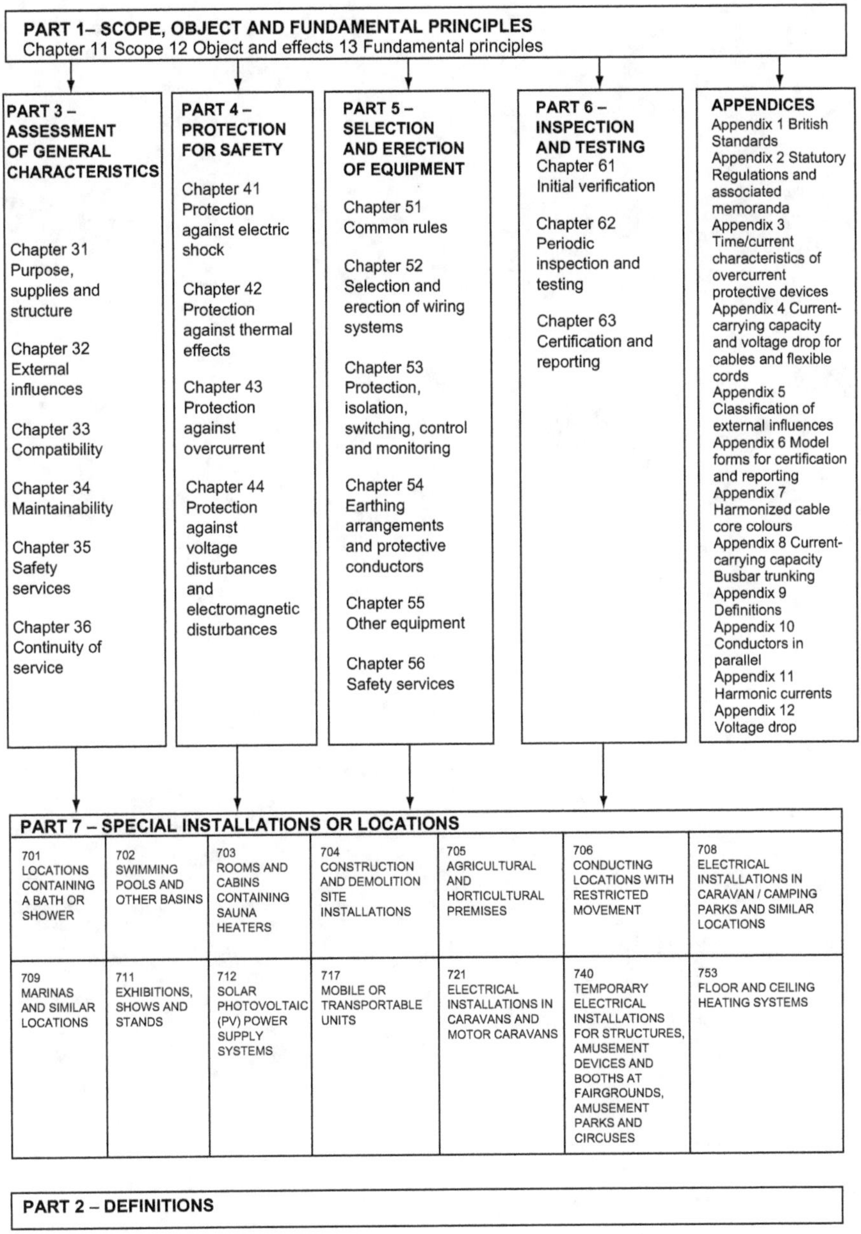

Figure 1.2 Plan of 17th Edition

The 17th Edition now also adopts the numbering of the IEC and CENELEC standards. This will assist in the general adoption worldwide of the UK standard. It also adopts the IEC structure of Inspection and Testing as being Part 6 and Special Locations as Part 7.

Chapter 2
Scope, object and fundamental principles

(Adviser K. J. Morton BSc, CEng, MIET)

2.1 Scope
Chapter 11

BS 7671 is intended to be applied to electrical installations generally but, as stated in Regulation 110.1, it must be supplemented by the requirements or recommendations of certain other British Standards (for example BS 5839: *Fire detection and fire alarm systems for buildings*) and these are listed in Regulation 110.1. There are also installations and equipment that are excluded from the scope of BS 7671, such as ships and aircraft, and these are listed in Regulation 110.2.

BS 7671 is limited to installations within the following nominal voltage ranges:

(a) extra-low voltage – normally not exceeding 50 V a.c. or 120 V ripple-free d.c.; and
(b) low voltage – normally exceeding extra-low voltage, but not exceeding 1000 V a.c. or 1500 V d.c. between conductors, or 600 V a.c. or 900 V d.c. between conductors and earth.

This definition of low voltage is sometimes confusing to electronic engineers, who would consider 230 V a.c. to be a high voltage. These voltage ranges (bands) are as standardized in IEC and CENELEC.

Regulation 113.1 advises that the Regulations are not intended to cover the requirements for the construction of electrical equipment. Reference needs to be made to the appropriate equipment standard. The advice is given that assemblies of electrical equipment are required to comply with appropriate standards, where much fuller information on requirements and tests etc. will be found than in BS 7671, and of course will be much more specific to the equipment. It is quite common for specifiers to require equipment to comply with BS 7671, and this reference is not appropriate; the equipment must comply with its own standard.

2.2 Relationship with statutory regulations
Regulation 114.1

The IET Wiring Regulations (BS 7671) are not regulations in the sense of an Act of Parliament or a Statutory Instrument, which have the force of law. They are simply rules or principles. The statutory regulations with which the Wiring Regulations are most closely associated are the Electricity Safety, Quality and Continuity Regulations 2002 (as amended), and the Electricity at Work Regulations 1989.

Appendix 2 of BS 7671 lists the principal statutory regulations with which certain electrical installations are required to comply. There is much legislation of which installation designers need to be aware, some fairly recently enacted.

2.2.1 The Electricity Safety, Quality and Continuity Regulations 2002 as amended

There are a number of references to BS 7671 in the Electricity Safety, Quality and Continuity Regulations (ESQCR) and in the guidance published by the Department of Trade and Industry (now the Department for Business, Innovation and Skills, or BIS). Table 2.1 summarizes the references. A requirement is that, prior to the connection of new properties, the distributor should seek confirmation that the installation complies with BS 7671. Unless this is inappropriate for reasons of safety, the distributor is required to make available an earth connection. The requirements for embedded generation in Part VI require for low voltage installations that the installations comply with BS 7671, and this would be a particular reference to Section 551 of the latter.

2.2.2 The Health and Safety at Work etc. Act 1974

The Health and Safety at Work etc. Act 1974 is the primary health-and-safety legislation in Great Britain. It imposes general duties on all persons associated with places of work. Among its requirements, Section 6 states that it shall be the duty of any person who designs, manufactures, imports or supplies any article for use at work to ensure, so far as is reasonably practicable, that the article is so designed and constructed that it will be safe and without risk to health at all times when it is being used, cleaned or maintained by a person at work. It is the duty of any person who undertakes the design or manufacture of an article for use at work to carry out or arrange to carry out any necessary research with a view to the discovery and, so far as is reasonably practicable, the elimination or minimization of any risk to health or safety to which the design or article might give rise. It is the duty of any person who erects or installs any article for use at work in any premises to ensure that nothing about the way in which it is erected or installed makes it unsafe or a risk to health. This clearly applies to electrical installations.

2.2.3 Electricity at Work Regulations 1989

The Electricity at Work Regulations 1989 were made under the Health and Safety at Work etc. Act 1974. These Regulations are broad in their scope, imposing requirements for the safe design, installation and maintenance of electrical equipment.

The *Memorandum of guidance on the Electricity at Work Regulations 1989* (HSR25), published by the Health and Safety Executive, states that ‘BS 7671 is

Table 2.1 References to BS 7671 in the Electricity Safety, Quality and Continuity Regulations and DTI guidance

Regulation	Reference	Requirements/guidelines
1(2)	Guidance	Where reference to BS 7671 is made other equivalent CENELEC standards are acceptable.
1(5)	Regulations	British Standard Requirements means BS 7671:2008.
7(1)	Guidance	Reference to danger that can arise in installations not complying with the PME requirements of BS 7671 if the continuity of PEN conductors is lost (Section 543.4).
8(4)	Regulations	The consumer shall not connect the neutral to exposed-conductive-parts, i.e. TN-C systems are not allowed in consumers' installations.
9(4)	Guidance	Distributors are to be satisfied that consumers' installations comply with BS 7671 at time of connection (Section 543.4).
21	Regulations and guidance	Switched alternative sources for low voltage installations to comply with BS 7671 (Section 551).
22(1)	Regulations and guidance	Parallel operation LV installations not allowed unless they comply with BS 7671 (Section 551.7).
22(2)	Regulations	The exclusion for persons with generators of less than 5 kW from the requirements of Regulation 22 does not exclude them from the obligation to comply with the requirements of BS 7671 (Section 551).
24(1)	Guidance	Protective devices to be close to supply terminals. See BS 7671. Regulations 434.3 and 434.2.1 seem appropriate.
25(2)	Regulations and guidance	Before a distributor commences a supply he should be provided with evidence that the installation complies with BS 7671 or the ESQCR as appropriate. For compliance with BS 7671 an Electrical Installation Certificate would normally be acceptable.

a code of practice which is widely recognised and accepted in the UK and compliance with it is likely to achieve compliance with relevant aspects of the 1989 Regulations'. The IET Wiring Regulations (BS 7671) are not as broad in their scope as the Electricity at Work Regulations. The IET Wiring Regulations basically cover the design, installation and testing of fixed installations up to 1000 V a.c. They do not cover safety during construction, safety during use, competence issues, safe systems of work, maintenance or equipment other than the fixed installation, or installations over 1000 V a.c. or 1500 V d.c.

2.2.4 The Building Regulations, Scotland

The Building (Scotland) Act 2003 (2003 asp 8)

This Act gives Scottish Ministers the power to make building regulations (the Building (Scotland) Regulations 2004 as amended) to secure the health, safety, welfare and convenience of persons in or about buildings and of others who may be affected by buildings or matters connected with buildings; to further the conservation of fuel and power and to further the achievement of sustainable development.

Requirements for electrical installations in Scotland are addressed by standard 4.5 – electrical safety for all buildings; and standard 4.6 – electrical fixtures for domestic buildings only (see 12.5.1 and 12.5.2 of the *Electrician's Guide to the Building Regulations* published by the IET); and, as in England and Wales, persons carrying out electrical installations must ensure that the work they carry out complies with both building regulations and the relevant functional standards.

There are no significant differences in general installation requirements for electrical work, with both Scotland and England & Wales citing BS 7671 (as amended) as the recommended means of satisfying building standards requirements. However, the requirements of Part P of the building regulations of England and Wales do not apply to work in Scotland.

2.2.5 The Building Regulations, England

The Building Regulations 2000, Statutory Instrument No. 2531

The Building Regulations 2000 (as amended) are made under the Building Act 1984 and apply in England and Wales. The aim of the Regulations is to ensure the health and safety of building users, promote energy efficiency, facilitate sustainable development and contribute to meeting the access needs of people in and around all types of buildings (that is, domestic, commercial and industrial).

The Building Regulations 2000 do not apply in Scotland or Northern Ireland. In Scotland the requirements of the Building (Scotland) Regulations 2004 (as amended) apply, in particular Regulation 9, and in Northern Ireland the Building Regulations (Northern Ireland) 2000 (as amended) apply.

The technical 'Parts' of the Building Regulations of particular interest to persons designing and installing electrical installations are:

- Part A (Structure): depth of chases in walls, and size and position of holes and notches in floor and roof joists;
- Part B (Fire safety): fire safety of certain electrical installations; provision of fire alarm and fire detection systems; fire resistance of penetrations through floors and walls;
- Part C (Site preparation and resistance to moisture): moisture resistance of cable penetrations through external walls;
- Part E (Resistance to the passage of sound): penetrations through floors and walls;
- Part F (Ventilation): ventilation rates for dwellings;

- Part L (Conservation of fuel and power): energy-efficient lighting;
- Part M (Access to and use of buildings): heights of switches, socket-outlets and other equipment; and
- Part P (Electrical safety – Dwellings); Part P applies only to dwellings.

Electrical designers and installers in all buildings, including buildings other than dwellings, may also be responsible for ensuring compliance with other Parts of the Building Regulations where relevant, particularly if there are no other parties involved with the work.

The Building Regulations (Regulation 4(2)) require that, on completion of work in an existing building, the building should be no worse in terms of the level of compliance with the other applicable Parts of Schedule 1 to the Building Regulations, including Parts A, B, C, E, F, L and M.

2.2.6 The Building Regulations, Wales

On 31 December 2011 the power to make Buildings Regulations for Wales was transferred to Welsh Ministers. This means:

- Welsh Ministers will make any new building regulations or publish any new building regulations guidance applicable in Wales from that date
- The Building Regulations 2010 and related guidance, including Approved Documents as at that date, will continue to apply in Wales until Welsh Ministers make changes to them. For example, the 2006 version of Approved Document P applies at the date of publication of this guide.

As guidance is reviewed and changes made, Welsh Ministers will publish separate Approved Documents.

2.2.7 The Building Regulations, Northern Ireland

In Northern Ireland the Building Regulations (Northern Ireland) 2000 (as amended) apply. The Department of Finance and Personnel has produced Technical Booklet E: 2005 and also refers to other deemed-to-satisfy publications, such as British Standards, to support compliance with the Building Regulations (Northern Ireland) 2000.

2.2.8 The Construction (Design and Management) Regulations 2015

The Construction (Design and Management) Regulations 2015 (CDM 2015) came into force on 6 April 2015, replacing CDM 2007. The Health and Safety Executive in publication L153 provide guidance on the legal requirements for CDM 2015 and is available to help anyone with duties under the Regulations. It describes:

(1) the law that applies to the whole construction process on all construction projects, from concept to completion; and
(2) what each dutyholder must or should do to comply with the law to ensure projects are carried out in a way that secures health and safety.

Regulation 9 imposes duties on designers and is reproduced below.

Regulation 9 Duties of designers

(1) A designer must not commence work in relation to a project unless satisfied that the client is aware of the duties owed by the client under these Regulations.

(2) When preparing or modifying a design the designer must take into account the general principles of prevention and any pre-construction information to eliminate, so far as is reasonably practicable, foreseeable risks to the health or safety of any person—
  (a) carrying out or liable to be affected by construction work;
  (b) maintaining or cleaning a structure; or
  (c) using a structure designed as a workplace.

(3) If it is not possible to eliminate these risks, the designer must, so far as is reasonably practicable—
  (a) take steps to reduce or, if that is not possible, control the risks through the subsequent design process;
  (b) provide information about those risks to the principal designer; and
  (c) ensure appropriate information is included in the health and safety file.

(4) A designer must take all reasonable steps to provide, with the design, sufficient information about the design, construction or maintenance of the structure, to adequately assist the client, other designers and contractors to comply with their duties under these Regulations.

2.2.9 The Electrical Equipment (Safety) Regulations 1994

The Electrical Equipment (Safety) Regulations 1994 implement the Low Voltage Directive in the United Kingdom. The latter requires that electrical equipment shall be safe and constructed in accordance with what is generally regarded as being good practice within the member states of the EU. These Regulations are notable in that equipment that satisfies the safety provisions of harmonized standards is taken to comply with the safety requirements of the Regulations. As discussed in section 8.1 of this publication, a harmonized standard is given the prefix BS EN, e.g. BS EN 60947 *Specification for Low-Voltage Switchgear and Controlgear.* These Regulations require complying equipment to be CE marked; the latest implementation date for marking was 1 January 1997.

2.2.10 The Electromagnetic Compatibility Regulations 2006

Directive 2004/108/EC was transposed into UK law by the Electromagnetic Compatibility (EMC) Regulations 2006 (SI 2006 No. 3418), which came into force on 20 July 2007. These Regulations replaced and repealed the Electromagnetic Compatibility Regulations 2005.

The Regulations require that all electrical and electronic apparatus marketed in the UK, including imports, satisfy the requirements of the EMC Directive. The Regulations introduce a new regime for fixed installations.

The Regulations 2006 as amended implement the EMC Directive in the United Kingdom. This directive requires the electromagnetic emission of equipment to be limited, and requires equipment to be immune to a certain level of interference. This is a New Approach Directive, in that compliance with the directive is achieved by meeting the requirements of mandated standards. These again are ENs. There is a requirement that equipment complying with the EMC Regulations be CE marked in a similar manner to that complying with the Electrical Equipment (Safety) Regulations.

2.2.11 The Supply of Machinery (Safety) Regulations 2008

The EU Machinery Directive has been implemented in the UK by the Supply of Machinery (Safety) Regulations 1992 (as amended). It is the general view, at the time of writing, that a fixed electrical installation is deemed not to be a machine. However, within the machine there may be an electrical installation; such an installation is excluded from the scope of BS 7671. In such instances the machine should comply with the appropriate product standard and the Supply of Machinery (Safety) Regulations, or the Electrical Equipment (Safety) Regulations if the machine is predominantly electrical. The Machinery Directive is also a New Approach Directive, and will generally be met by the use of equipment complying with an appropriate European Standard. There is also a requirement that such equipment be CE marked.

2.2.12 The Construction Products Directive

The Construction Products Directive has been implemented in the UK through a mix of legislation and administrative action. The legislation takes the form of the Construction Products Regulations (Statutory Instrument 1991 No. 1620), which came into force on 27 December 1991. The Construction Products Directive is another New Approach Directive, which is implemented by virtue of mandated standards, which would be BS ENs.

2.3 Object and effects

Chapter 12

Chapter 12 in BS 7671 is a short chapter but it does explain quite simply the construction of the standard.

Regulation 120.3 allows intended departures from Parts 3 to 7 following special consideration by the designer. Any intended departure is to be noted on the Electrical Installation Certificate, however the noting of such departures on the Certificate does not indicate non-compliance with BS 7671.

2.4 Fundamental principles
Chapter 13

Chapter 13 is arguably the most important part of BS 7671 and, regrettably, sometimes one of the most underused parts. It is the part that *must* be complied with. The chapter was expanded in the 2008 edition and has undergone further changes in subsequent amendments to the standard. It sets out the fundamental principles, i.e. protection for safety, for the design, selection of electrical equipment, erection, verification and periodic inspection and testing. The following parts, i.e. Parts 3 to 7, set out technical requirements intended to ensure compliance with the fundamental principles of Chapter 13.

2.4.1 Protection for safety
Section 131

The fundamental principles of protection for safety are detailed, that is, provision for the safety of persons and livestock and the protection of property. Persons and livestock are to be protected against shock currents, excessive temperatures, ignition of explosive atmospheres, under- and overvoltage, EMC, mechanical movement, supply interruptions and arcing and burning.

2.4.2 Design
Section 132

This section describes only the fundamentals of design. It is necessary to determine:

(a) the characteristics of the available supply or supplies;
(b) the nature of the demand, including emergency services, and to select cable sizes, types of wiring, protective equipment, etc., to provide for the protection of persons, livestock and property; and
(c) the proper functioning of the electrical installation for the intended use, so that electrical installations not only must be safe, they must also work.

2.4.3 Selection of electrical equipment
Section 133

The section states the fundamental requirements as to the suitability of equipment for the voltage, current, frequency, power, installation conditions and prevention of harmful effects, expanded on in Part 5 of BS 7671 and Chapter 8 of this Commentary.

Regulation 133.5 is of particular importance as it allows the use of new materials or inventions that do not comply with Parts 4, 5 or 7 of the Regulations, provided the resulting degree of safety is not less than that obtained by compliance with those parts. Such use is to be noted on the Electrical Installations Certificate.

2.4.4 Erection, initial verification and periodic inspection and testing Sections 134 and 135

Section 134 provides a brief summary of procedures to be followed both for erection and initial verification:

- good workmanship by skilled or instructed persons and proper materials are to be used;
- the equipment should not be damaged during erection;
- all joints should be properly constructed;
- the electrical equipment should be suitably selected; and
- on completion of an installation or addition or alteration, inspection and testing should be carried out to verify that the requirements of the standard have been met.

Regulation 134.2.2 states that the designer shall make a recommendation for the interval to the first periodic inspection and test, while Regulation 135.1 makes a recommendation that every electrical installation is subjected to (regular) periodic inspection and testing.

2.5 Equipment marking (CE and approval body)

Designers will generally be able to select products that comply with the directives by looking for CE marking on the product. The CE mark is applied by the manufacturer and is an indication that the product complies with the essential requirements of the relevant directives. There is normally no third-party verification of compliance and the CE mark therefore should not be regarded as a quality-control or quality-assurance mark.

BSI kitemark
BSI has tested the product and has confirmed that the product conforms to the relevant British Standard

British Approvals Service for Cables mark

ASTA diamond
Tested and conforms to standard and factory quality management to ISO 9001

The BEAB Approved mark is an electrical safety mark, for household and similar appliances

HAR mark
European third party certification mark for cables and cords complying with relevant European safety standards (ENs/HDs)

ENEC mark
European third party certification mark for electrical equipment complying with the European safety standards, for luminaires, transformers, power supply units and switches

CEN/CENELEC mark
European third party certification mark for household and similar electrical appliances, complying with relevant European safety standards

Figure 2.1 Equipment approval marks

Where greater assurance that a product complies with a relevant equipment standard is required, a designer will probably look for equipment that has an approval body mark, indicating that there has been independent assessment of compliance with the standard. Some typical approval marks are shown in Figure 2.1. There are moves within Europe to agree a common approval mark, which is to be called the 'key mark'. This has not been implemented at the time of writing but may well be common in the years to come.

However, under the aegis of the European Organization for Testing and Certification (EOTC), voluntary approval schemes are being encouraged. Two of the first of these are the Low voltage agreement (LOVAG) scheme for installation equipment and the European Norms Electrical Certification (ENEC) mark for luminaires. These schemes are voluntary in that each country opts to take part. Certification is to published standards, usually IEC or CENELEC, and is carried out by test houses in each participating country, the results of a test being accepted in any participating country. The UK test house participating in the Lovag scheme is ASTA Certification Services, and the BSI is the test house for the ENEC scheme. It is likely that further schemes of this type will be set up.

Equipment standards are discussed in section 8.1.

2.6 Premises subject to licensing
Regulation 115.1

Readers are reminded that guidance must be sought from local authority planning departments for any particular requirements for licensing, as such requirements may place restrictions on the designer that are more onerous than those of BS 7671.

Chapter 3
Assessment of general characteristics

Part 3

3.1 General
Section 301

Section 301 of BS 7671 is short; however, it is fundamental. The designer is required to give consideration to the fundamentals of the design. All are important, but it is worthwhile drawing attention to the presumptions to be made with respect to maintenance (section 3.11) and construction (section 3.12). Attention is drawn to the selection of the protective measures for electric shock and the suitability of the equipment for the particular general characteristics.

3.2 Purpose, supplies and structure
Chapter 31

Chapter 31 requires the designer to determine the maximum demand and, in consideration of the electricity supply, make decisions on the basic system to be adopted within the installation, and in particular the earthing system; and also to make some fundamental decisions on the circuit arrangements.

3.3 Maximum demand

3.3.1 The Building Regulations

The requirements of the Building Regulations in particular impact on demand. Part L of Schedule 1 to the Building Regulations 2010 is reproduced below.

Requirement	Limits on application
Part L Conservation of fuel and power **L1.** Reasonable provision shall be made for the conservation of fuel and power in buildings by: (a) limiting heat gains and losses: (i) through thermal elements and other parts of the building fabric; and	

> (ii) from pipes, ducts and vessels used for space heating, space cooling and hot water services;
>
> (b) providing fixed building services which are
>
> (a) energy efficient;
>
> (b) have effective controls; and
>
> (c) Are commissioned by testing and adjusting as necessary to ensure they use no more fuel and power than is reasonable in the circumstances
>
> (c) providing to the owner sufficient information about the building, the fixed building services and their maintenance requirements so that the building can be operated in such a manner as to use no more fuel than is reasonable in the circumstances.

3.3.2 BS 7671
Section 311

Regulation 311.1 states that, for economic and reliable design of an installation within thermal limits and admissible voltage drop, the maximum demand shall be determined. In determining the maximum demand of an installation or part thereof, diversity may be taken into account.

In the 15th Edition, Appendix 4 gave the designer guidance on estimating demand. This has since been omitted from BS 7671 but is reproduced in this publication as Appendix D. An International Electrotechnical Commission draft, 'Estimation of maximum demand', document 64 (Secretariat) 254, was circulated in January 1979. There was no agreement within IEC TC 64 to publish the information, but it is summarized below as an aid in the estimation of maximum demand. If demand is overestimated, the cost of an installation is increased unnecessarily and resources are generally wasted. Designers who are responsible for these estimates may wish to compare their own estimates with demands prepared on the basis of the IEC draft, and with those prepared using the guidance in Appendix D.

3.3.3 Estimation of maximum demand

3.3.3.1 IEC Document 64 (Secretariat) 254

Estimates of maximum demand can rarely be made accurately. The guidance given here indicates very approximate values with wide tolerances, and must be subject to many reservations. The designer will need to decide:

(a) whether he can use values known to him personally or from reliable sources;
(b) whether the values given in Tables 3.1 and 3.2, and Figure 3.1 are applicable; and

(c) whether the values given in the tables and figure will need to be adjusted, taking into account:
   (i) the time profiles of the loads;
   (ii) the coincidence of individual loads with other loads – a chart may be helpful in this respect;
   (iii) the relationship of the electrical loading of motors to the mechanical load; the mechanical load is a more accurate guide to the electrical load than motor rating – motor ratings are often conservatively selected (overrated);
   (iv) heating and cooling loads, the seasonal demands and how these might coincide with production demands;
   (v) the availability of other sources of supply;
   (vi) the allowances, if any, for spare capacity or load growth – this must be discussed with the client; and
   (vii) special considerations that apply to the particular job in hand.

Table 3.1 Demand factors for complete installations

	Installation building/premises	Demand factor g for main supply intake	Remarks
1	**Dwellings**		
1.1	Individual	0.4	
1.2	Blocks of flats		The demand factor has to be chosen from the graphs given in Figure 3.1 according to the mean value of the loads connected with each flat.
1.2.1	without electric heating (as main form of heating) or air conditioning		
1.2.1.1	with lighting and some small appliances only	Figure 3.1 curves a, b and c	
1.2.1.2	fully 'electrified' but without electric heating and air conditioning	Figure 3.1 curves a, b and c	
1.2.2	with electric heating (as main form of heating) or air conditioning		The total supply demand will result from the sum of the demand for heating and air conditioning, and all other power demands, see Table 3.2. A reduced demand is to be expected when the heating or other loads are controlled so as not to coincide with other applications.
1.2.2.1	general demand	Figure 3.1 curves a, b and c	
1.2.2.2	heating and cooling demand	0.8–1.0	
2	**Public buildings**		
2.1	Hotels, boarding houses, furnished apartments	0.6–0.8	
2.2	Small offices	0.5–0.7	

Continues

Table 3.1 Continued

Installation building/premises		Demand factor g for main supply intake	Remarks
2.3	Large offices (banks, insurance companies, public administration)	0.7–0.8	
2.4	Shops	0.5–0.7	
2.5	Department stores	0.7–0.9	
2.6	Schools	0.6–0.7	
2.7	Hospitals	0.5–0.75	
2.8	Assembly rooms (sports grounds, theatres, restaurants, churches)	0.6–0.8	
2.9	Terminal buildings (railway stations, airports)	requires investigation	
3	**Mechanical engineering industry**		
3.1	Metal workings	0.25	In general the motor drives are overrated for the mechanical load.
3.2	Car plants	0.25	
4	**Pulp and paper mills**	0.5–0.7	The number of rolling reserve drives considerably affects the demand factor.
5	**Textile industry**		
5.1	Spinning mills	0.75	
5.2	Weaving mills and mixed process installations	0.6–0.7	
6	**Raw-material industry**		
6.1	Wood industry	data not available	
6.2	Rubber industry	0.6–0.7	
6.3	Leather industry	data not available	
7	**Chemical industry petroleum industry**	0.5–0.7	Because the processes in chemical industries are very sensitive to supply* failures, the supply must be secure or backup provided.

* The capacity of a supply is no guarantee of its security, unless I_B is greater than I_n.

Table 3.1 Continued

Installation building/premises		Demand factor g for main supply intake	Remarks
8	**Cement mills**	0.8–0.9	Reference: Production level about 3500 tons a day with about 500 motors (large mills are driven by HV motors).
9	**Food industry**		
9.1	General (including process engineering)	0.7–0.9	
9.2	Silos	0.8–0.9	
10	**Coal mining**		
10.1	Hard coal preparation	0.8–1.0	
	Underground	1.0	
10.2	Lignite general	0.7	
	Excavation	0.8	
11	**Iron and steel mills** (Blast-furnaces, converters)		
11.1	Blowers	0.8–0.9	
11.2	Auxiliary drives	0.5	
12	**Rolling mills**–general	0.5–0.8+	
12.1	Water supply ventilation	0.8–0.9+	
12.2	Auxiliaries for:	0.8–0.9+	
	– rolling mills with cooling bed	0.5–0.7+	
	– rolling mills with loopers	0.6–0.8+	+g depends on the number of standby drives.
	– rolling mills with cooling bed and loopers	0.3–0.5	
12.3	Finishing lines	0.2–0.6+	
13	**Floating docks**		
13.1	Pumping operation during lifting	0.9	Pumping and repair work do not occur simultaneously.
13.2	Repair work without pumping	0.5	
14	**Lighting of street tunnels**	1.0	
15	**Traffic installations**	1.0	Escalators, tunnel (ventilation), traffic lights.

Continues

Table 3.1 Continued

Installation building/premises		Demand factor g for main supply intake	Remarks
16	**Power generation**		
16.1	Power stations, general	requires investigation	
16.1.1	auxiliary power for low voltage circuits	requires investigation	
16.1.2	emergency supply	1.0	
16.2	Nuclear power stations special power demand, e.g. for trace heating for sodium pipes	1.0	
17	**Cranes**	0.7 per crane	Crane work with intermittent duty: power demand depends on the kind of premises where they are used (e.g. harbours, steel mills, dockyards).
18	**Lifts**	0.5 (highly variable depending on the time of the day)	For simultaneous starting of several cranes or lifts the voltage drop has to be considered.

3.3.3.2 Estimation method

The maximum demand of an installation P_{max} is the sum of the loads installed P_i multiplied by a demand factor *g*:

$$P_{max} = gP_i$$

where:

P_i = total installed load for the installation considered, being the sum of all the loads directly connected, generally on the basis of continuous duty

g = demand factor that is the ratio of the maximum demand of an installation to the corresponding total installed load.

Table 3.1 provides g factors for some typical complete installations.

Table 3.1 is not suitable for intermediate distribution boards supplying mainly or wholly one type of load, e.g. lighting or heating. The factors for particular loads are given in Table 3.2.

Table 3.1 g factors may be used for sub-distribution boards where there is a typical mix of load, but care needs to be taken: if a particular type of load predominates it will invalidate the use of the table.

Table 3.2 Estimated values for demand factors g for certain loads intended for use in estimating the demands on intermediate and sub-distribution boards

	Type of premises	Lighting (1)	Socket-outlets (2)	Water heating not central (3)	Water heating centralized (4)	Cooking, canteens (5)	Refrigeration (6)	Domestic appliances (fixed) (7)	Signalling and address system (8)	Lifts, escalators (9)	Heating and air conditioning (10)	Air conditioning not central (11)	Electronic data processing (12)	Experimental and demonstration units (13)	Floodlight installations (14)
1	Individual dwellings	0.6	0.2	0.5	1	0.75	–	0.7	–	0.5	1	0.8	–	–	–
2	Hotels, etc.	0.7	0.1	0.5	1	0.80	0.8	–	0.5	0.5	1	–	requires investigation	–	1
3	Small offices	0.8	0.1	0.3	1	0.50	0.4	–	–	0.7	1	–		–	–
4	Large offices	0.8	0.1	0.3	1	0.80	0.4	–	0.5	0.7	1	–		–	1
5	Shops	0.9	0.3	0.6	1	0.50	0.6	–	–	0.7	1	–		0.2	–
6	Department stores	0.9	0.2	0.3	1	0.80	0.6	–	0.5	0.7	1	–		0.2	1
7	Schools	0.9	0.1	0.3	1	0.80	0.4	–	–	–	1	–		0.4	–
8	Universities and colleges	0.8	0.1	0.3	1	0.80	0.4	–	0.5	0.2	1	–		0.4	–
9	Hospitals	0.7	0.1	0.7	1	0.80	0.8	–	0.5	0.5	1	–		–	–
10	Assembly rooms, public halls	0.9	0.1	0.3	1	0.80	0.6	–	0.5	0.5	1	–		0.4	1

3.3.3.3 Sub-distribution point estimation

For intermediate distribution boards where one type of load may predominate, Table 3.2 provides demand factors for each type of load. The maximum demand on the distribution board (P_{DB}) is then given by:

$$P_{DB} = P_{1g1} + P_{2g2} + P_{3g3} + P_{ngn}$$

where:

P_n = sum of the particular type of load, e.g. lighting
g_n = factor to be applied for that load in the particular type of premises.

3.3.3.4 Tolerance

The difficulty of estimating demand accurately is mentioned in earlier paragraphs.

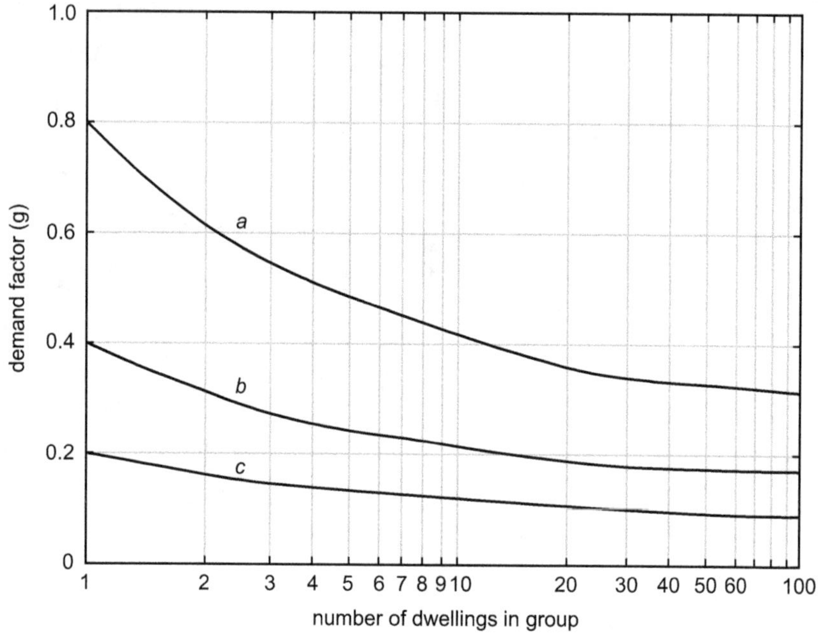

Figure 3.1 Estimated values for demand factors g *for the calculation of the infeed (supply) power demand depending on the number of dwelling units which are connected together*

It must be remembered that the factors given in Tables 3.1 and 3.2 are purely guidance; particular information relevant to a specific installation will always override this guidance.

3.3.3.5 Socket-outlet circuits

The estimation of demand on socket-outlet circuits presents obvious difficulty. The values given in Table 3.2 apply only to circuits comprising a number of outlets that are not expected to be loaded fully and simultaneously. For commercial and industrial installations, a specific estimate of the demand based on the predicted usage of the sockets needs to be made. For dwellings, the following provides guidance:

Demand factor (g) for socket-outlet circuits in dwellings

Number of circuits	g
1	1.0
2	0.6
4	0.3
8	0.15

When fixed equipment is fed by socket-outlets, e.g. water heaters or space heating, specific allowance for this must be made. It is to be noted that the g factors above for socket-outlets are for estimation of the effect of the socket-outlets on the maximum demand. Each socket circuit must be designed for its own maximum demand.

3.3.4 Other methods of demand estimation

The method of estimation of demands provided in the appendices of earlier editions of the Wiring Regulations is included in this Commentary as Appendix D. Designers may find it useful to prepare estimates based on a number of methods in order to give themselves some confidence in the loads they are estimating.

3.3.5 Protective devices in an enclosure

A protective device grouped with other loaded devices inside an enclosure will be derated, as will the other devices, as a result of the increase in ambient temperature and as a result of conducted (e.g. via busbars) and radiated heat from the other devices.

Furthermore, a device installed in a particular enclosure may have a significantly reduced rating from that marked on the device.

A designer must make allowance for these effects by the application of what is known as the rated diversity factor RDF or loading factor.

The required current rating of a device (I_{nc}) in an assembly is given by:

$$I_{nc} \geq \frac{I_b}{RDF}$$

where:

I_b is the design current of the circuit.

RDF is the rated diversity factor, see Table 10.2 in section 10.6.

I_{nc} is the value of current that can be carried by the device in the assembly alone, that is without any other loaded devices.

3.4 Conductor arrangement of live conductors and type of earthing
Section 312

The International standard (IEC 364) and the European harmonization document (HD 384) recognize the following systems of live conductors:

(a) ac systems:
- single-phase – two-wire or three-wire;
- two-phase – three-wire or five-wire;
- three-phase – three-wire or four-wire.

(b) dc systems:
- two-wire or three-wire.

Obviously, the type of system adopted will necessarily depend on the supply offered by the electricity distributor; in the UK the likely systems are:

- single-phase two-wire (line and neutral); and
- three-phase four-wire (three lines and neutral).

However, in rural areas, where supplies are from overhead systems, single-phase three-wire (two-line and neutral) may be offered.

Figure 3.2 shows the usual supply systems in the UK.

3.5 Types of system earthing
Regulation 312.2

The earthing system adopted by a designer will generally be determined by the earthing system of the electricity supply. Typical arrangements are shown in Figure 3.3.

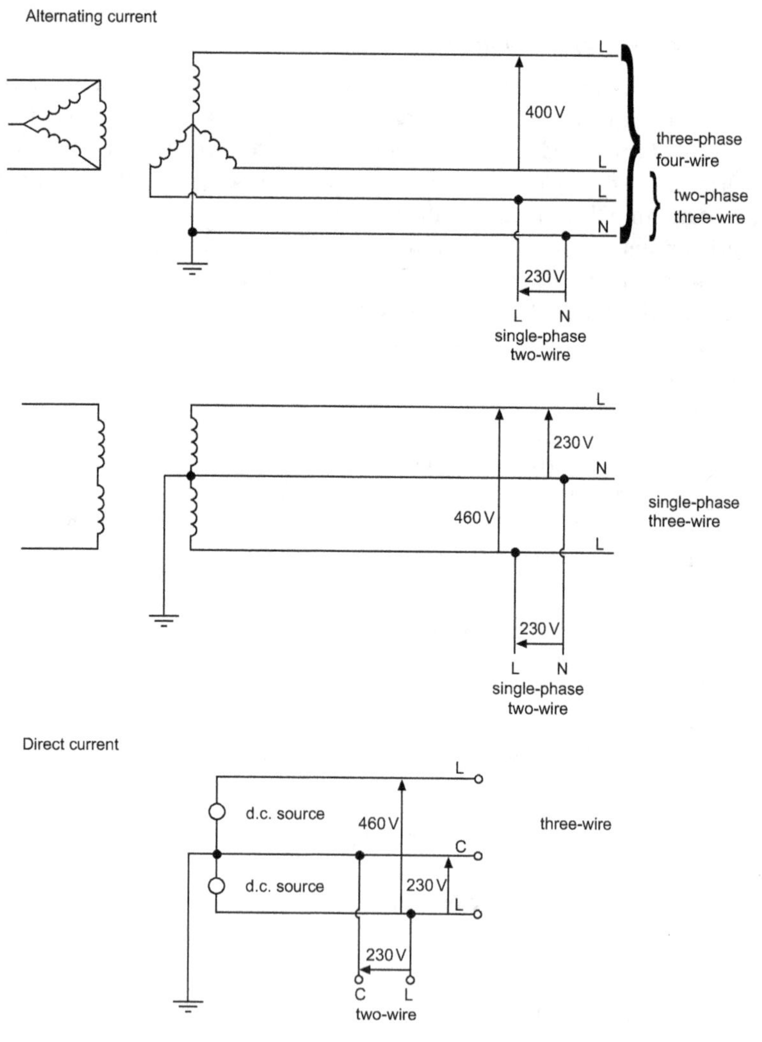

Figure 3.2 Supply systems in the UK

The system of coding uses the letters I, T, N, C and S. The code for each system type comprises two, three or four letters. The letters have their origin in French terms. The first letter indicates the relationship of the supply system to Earth: *T* = direct connection of supply with Earth (T for *terre,* the French word for earth). *I* = all live parts isolated from Earth or supply connected to Earth through an impedance for isolation (I for isolation).

The second letter indicates the relationship of the exposed-conductive-parts of the installation to Earth: *T* = direct electrical connection of exposed-conductive-parts to Earth, independent of the earthing of any point of the power system (T for *terre*). *N* = the direct electrical connection of the exposed-conductive-parts to the earthed point of the power system (in a.c. systems, the earth point is normally the neutral point) (*N* for *neutre*).

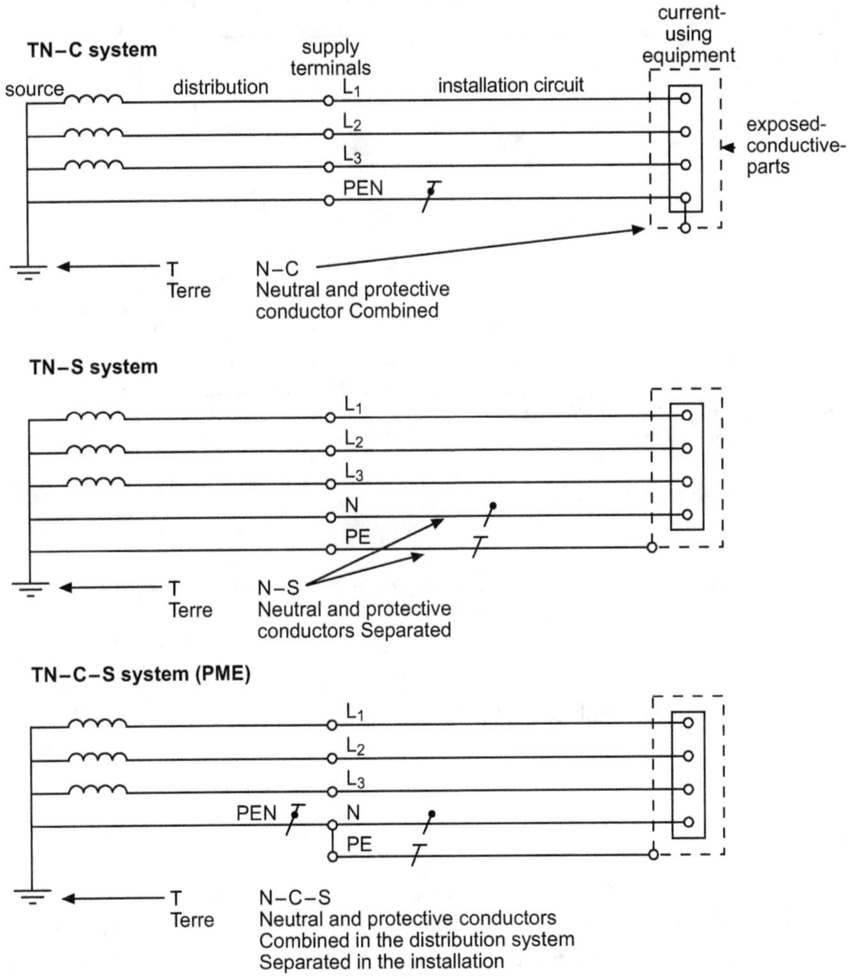

Figure 3.3 Typical earthing arrangements

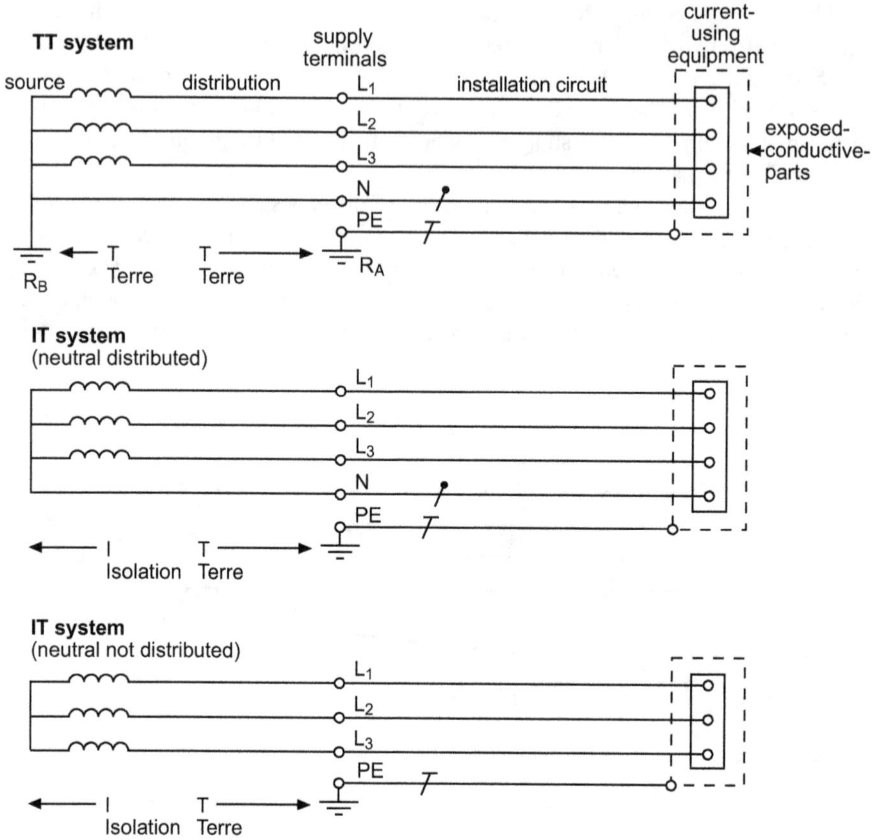

Figure 3.3 Continued

Symbols for conductors are given in IEC 617:

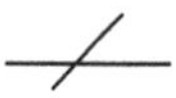 live conductor

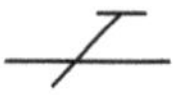 protective conductor (PE)

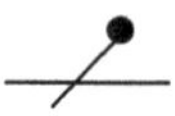 neutral conductor (N)

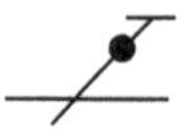 PEN conductor (PEN)

Subsequent letters, if any, indicate the arrangement of neutral and protective conductors. *S* = neutral and protective functions provided by separate conductors (S for *separé*). *C* = neutral and protective functions combined in a single conductor (PEN conductor), (C for *combiné*).

Hyphens are used to link the groups of letters e.g. TN-S, TN-C-S. The alphanumeric notation for particular conductors is as follows:

L1, L2, L3	line conductors 1, 2, 3.
PE	protective conductor.
N	neutral conductor.
PEN	PEN conductor (combined protective and neutral conductor).
FPE	combined functional earth and protective conductor.

TN systems

A TN system has the source of energy directly earthed, hence the initial T, and the exposed-conductive-parts of the installation are connected to that earthed point by a PEN conductor in a TN-C system, by a separate protective conductor in a TN-S system or partly by a PEN conductor and partly by a separate protective conductor in a TN-C-S system.

TT systems

A TT system has the source of energy directly earthed, and the exposed-conductive-parts of the installation are connected to Earth by an independent earth electrode.

IT systems

In an IT system the source of energy has no direct connection with Earth, although it may be earthed via an impedance, and the exposed-conductive-parts of the electrical installation have their own separate earth, as for a TT system. IT systems are not used for public supply in the United Kingdom (prohibited by the ESQCR 2002 (as amended)).

The earthing system designations apply to single-phase as well as three-phase supplies, and also to d.c.

3.6 Supplies
Section 313

BS 7671 requires the designer to determine the characteristics of the supply. This information will normally be provided by the electricity distributor for supplies provided by them, and will include:

- the nominal voltage(s);
- the nature of the current (a.c. or d.c.);
- the frequency;
- the prospective short-circuit current at the origin of the installation (see subsection 6.3.3);
- the earth fault loop impedance Z_e of the part of the system external to the installation (subsection 6.3.3);

- the type and rating of the overcurrent device(s), installed at the origin of the installation (if any); and
- the suitability of the supply for the proposed installation maximum demand and nature of the proposed load.

Electricity suppliers and distributors in the UK are required to comply with the ESQCR 2002 (as amended). These Regulations impose requirements on the distributor, some of which are listed below.

Regulation 27 requires that before commencing a supply to a consumer's installation, the supplier shall ascertain from the distribution and then declare to the consumer:

(a) the number of phases;
(b) the frequency; and
(c) the voltage.

Unless otherwise agreed in writing between the distributor, the supplier and the consumer, the frequency shall be 50 Hz ±1% and for low voltage 400/230 V + 10%–6%. Regulation 28 requires that the distributor shall provide on request a written statement of:

(a) the maximum prospective short-circuit current at the supply terminals;
(b) for LV connections, the maximum earth loop impedance of the earth fault path outside the consumer's installation;
(c) the type and rating of the distributor's protective device nearest the supply terminals;
(d) the type of earthing system applicable to the connection; and
(e) the supply variations.

Regulation 26 states that if a distributor is not satisfied that a consumer's installation or other distributor's network is so constructed, installed, protected and used or arranged for use so as to prevent, so far as is reasonably practicable, danger or interference with any network, or with the supply to other consumers, he may issue a notice in writing, and if at the end of the period the works are not carried out he may disconnect or refuse to connect. Notices are not required if the disconnection can be justified on grounds of safety.

The detailed requirements for PME networks of the repealed Electricity Supply Regulations are not found in the ESQC Regulations. Regulation 9 provides some brief requirements. However, the guidance document from the DTI makes specific reference to the Energy Networks Association's Engineering Recommendation G12/3: 'Requirements for the application of protective multiple earthing to low voltage networks' dated 1995.

Guidance on planning limits for voltage fluctuations caused by industrial, commercial and domestic equipment is given in Engineering Recommendation P28 available from the Energy Networks Association. Compliance with these requirements would surely provide compliance with the requirements of Regulation 26 of the ESQC Regulations.

3.7 Supplies for safety services and standby systems
Regulation 313.2

Guidance on the installation of standby supplies can be obtained from Engineering Recommendation G.59/3, 'Recommendations for the connection of generating plant to the distribution systems of licensed distribution network operators', and Technical Report No. 113, 'Notes of guidance for the protection of private generating sets of up to 5 MW for operation in parallel with electricity board distribution networks'. These should be consulted and advice sought from the local electricity distributor on their particular requirements. This is particularly important if the standby supplies are to operate in parallel with the public supply. Supplies for safety services such as fire alarms and emergency lighting are required to comply with Chapter 56 of BS 7671. Decisions will need to be made on the earthing system to be adopted (see Regulation 560.5.3). BS 7671 requires emergency lighting installations to comply with BS 5266 and EN 1838 *Lighting applications Emergency lighting*, and fire detection and fire alarm systems to comply with BS 5839 *Fire detection and fire alarm systems for buildings*. The cables to be used in such installations are discussed in Chapter 5, 'Protection against thermal effects'.

3.8 Division of the installation
Section 314 (see also Regulation 132.3)

Regulation 314.1 requires that every installation be divided into circuits as necessary to:

(a) avoid hazards and minimize inconvenience in the event of a fault;
(b) facilitate safe inspection, testing and maintenance;
(c) take account of hazards that may arise from the failure of a single circuit such as a lighting circuit;
(d) reduce the possibility of unwanted tripping of RCDs due to excessive protective conductor currents produced by equipment in normal operation;
(e) mitigate the effects of electromagnetic disturbances; and
(f) prevent the indirect energizing of a circuit intended to be isolated.

With respect to the installation of RCDs, designers should ensure that only appropriate sections of an installation will be disconnected in the event of a fault or unwanted operation. A fault on an item of equipment or on one final circuit should not result in the disconnection of the complete installation.

The requirement to minimize inconvenience, as well as to avoid danger, requires the designer to segregate circuits adequately in terms of protection, isolation and control. Considerations will include the facility to:

(a) discriminate between circuits in the event of high fault currents, e.g. line-to-line faults, and low-current faults to earth, e.g. those requiring RCDs;

(b) isolate for repair and maintenance without switching off over much of the installation; and
(c) switch sufficiently discrete sections of load, e.g. lighting, for the convenience of the user.

3.9 External influences
Chapter 32

There is no Chapter 32 in BS 7671, as the work of the International Electrotechnical Commission continues to be insufficiently advanced for adoption. Regulation group 512.2 has general requirements and refers to Appendix 5, 'Classification of External Influences', where there is a list of external influences, which helps in the specification of equipment.

With reference to the presence of water and of foreign and solid bodies, codes AD and AE of Appendix 5, guidance on the selection of equipment is given by reference to the International Protection Code (IP code), as described in BS EN 60529:1992 (2013). The IP classification system is summarized in Figure 4.12.

3.10 Compatibility
Chapter 33

3.10.1 General
Regulation 331.1

BS 7671 requires the designer to carry out an assessment of the characteristics of equipment likely to have harmful effects upon other electrical equipment or other services, or likely to impair the supply.

Equipment subject to frequent load change (e.g. welders, compressor motors) may cause unacceptable interference in the supply to other equipment. For example, the light output from filament lamps may 'flicker'. The starting currents of large motors may trip small motors on no-voltage releases if correct starting sequences are not ensured.

The use of equipment with switched-mode power supplies fed from uninterruptible power supplies will need to be discussed with equipment manufacturers.

Guidance as to whether equipment is likely to interfere with the electricity supply sufficiently to cause disturbances to other electricity customers is given by the Energy Networks Association publication Engineering Recommendation P28. This publication deals with the flicker effects of rapidly changing loads, and must be considered if large motors, welding plant and similar equipment are to be installed. Interference can also be caused by power cable signal transmissions. BS 7484:1991, *Guide to electromagnetic environment for low frequency conducted disturbances and signalling in public power supply systems,* provides guidance on precautions necessary in these circumstances.

3.10.2 Electromagnetic compatibility
Section 332

There is considerable demand from designers for information on EMC, particularly on the interactions between power systems and data systems, and the effectiveness of 'clean earths'. This matter is considered in some detail in section 8.3.

3.11 Maintenance
Chapter 34

This brief chapter requires the designer to give consideration to maintenance of the installation. Facilities must be built into every installation to enable it to be maintained in a safe condition. The Electricity at Work Regulations 1989 require that electrical installations be maintained in a safe condition, and it is probable that the designer himself has a responsibility under these Regulations and the Construction (Design and Management) Regulations 2015. The more obvious areas where there is need for consultation with the client and architect include:

(a) replacement of lamps in luminaires; and
(b) accessibility of switchgear and other equipment for both repair and maintenance as well as operation (frequency of maintenance, inspection and test).

The designer is presumed to assume a frequency of inspection and testing, and a frequency of maintenance, and build this into the design, including selection of equipment. This information must be passed on to the client or user in the recommendation made under 'Next inspection' on the Electrical Installation Certificate of Appendix 6 of BS 7671. After discussion with the client and assuming degrees of abuse, frequency of inspection, testing and repair, the presumptions with respect to maintenance will be built into the design.

3.12 Construction

There is no chapter on construction in BS 7671. However, the Construction (Design and Management) Regulations 2015 require the designer to give thought to how installations might be constructed, how they might be maintained throughout their life, and how they might be disassembled and disposed of at the end of their working life. The more obvious examples of the requirements here would be the facility for installing large plant, such as transformers, switchgear and motors, and how they might be replaced reasonably safely. There is also a responsibility to consider the disposal of the construction products. Unfortunately, designers are often not aware in advance of problems that in later years are recognized as health and safety issues, e.g. asbestos and PCBs, both commonly used in electrical installations in the past. Concerns are being expressed that low-frequency electromagnetic radiation might be harmful, leading to the question whether or not the designer should take account of this.

Perhaps the least that can be done is that if two options have equal cost and design efficiency, and one is likely to provide lower radiation, then the low-exposure option should be taken (see also IET Public Affairs Board Report, 'The Possible Harmful Biological Effects of Low-level Electromagnetic Fields of Frequencies up to 300 GHz', IET Position Statement – May 2006).

3.13 Continuity of service
Chapter 36

Chapter 36 is also brief, but important as it requires the designer to consider reliability of supply under fault conditions, particularly with respect to life-support systems. It provides emphasis to Section 314 'Division of installation'.

Chapter 4
Protection against electric shock

4.1 The physiology of electric shock
Chapter 41

The prime requirement of any rules for electrical installations is to protect persons against electric shock and fire. In this chapter the mechanisms of electric shock and electric burns are discussed, and consideration given as to how BS 7671 deals with these hazards.

Electric shock and burns are not caused by voltage, but by the passage of electric current through the body. This current is dependent on a number of factors, including voltage, body impedance, contact impedance and frequency. The other major factor in electric shock and burns is time, for time affects the damage caused and the probability of dangerous effects. Ventricular fibrillation is considered to be the main cause of death by electric shock. There is also some evidence of death due to asphyxia or cardiac arrest. Pathophysiological effects such as muscular contractions, difficulty in breathing, rising blood pressure, disturbances of formation and conduction of impulses in the heart, including atrial fibrillation and transient cardiac arrest, may occur without ventricular fibrillation. Such effects are not lethal and are usually reversible, but current marks can occur. With currents of several amperes lasting more than seconds, deep-seated burns or other serious injuries (which can be internal) and even death are likely to occur.

4.1.1 Ventricular fibrillation

The normal steady pumping of the heart can be disturbed by electric current. If an electric current is applied to the heart during a vulnerable period (see Figure 4.1), the heart will fibrillate and cease its pumping action. Blood pressure then falls and blood is not circulated to the brain, resulting in death. The threshold of ventricular fibrillation depends on physiological parameters (anatomy of the body, state of the cardiac function etc.) as well as on electrical parameters such as duration of the current, the current path and the current magnitude. With sinusoidal alternating currents, there is considerable increase in the risk of fibrillation if the current flow continues for more than one cardiac cycle. For shock durations below 0.1 s, fibrillation may occur with currents in excess of 500 mA, and is likely to occur with currents of the order of amperes only if the shock falls within the vulnerable period – see Figures 4.1 and 4.2.

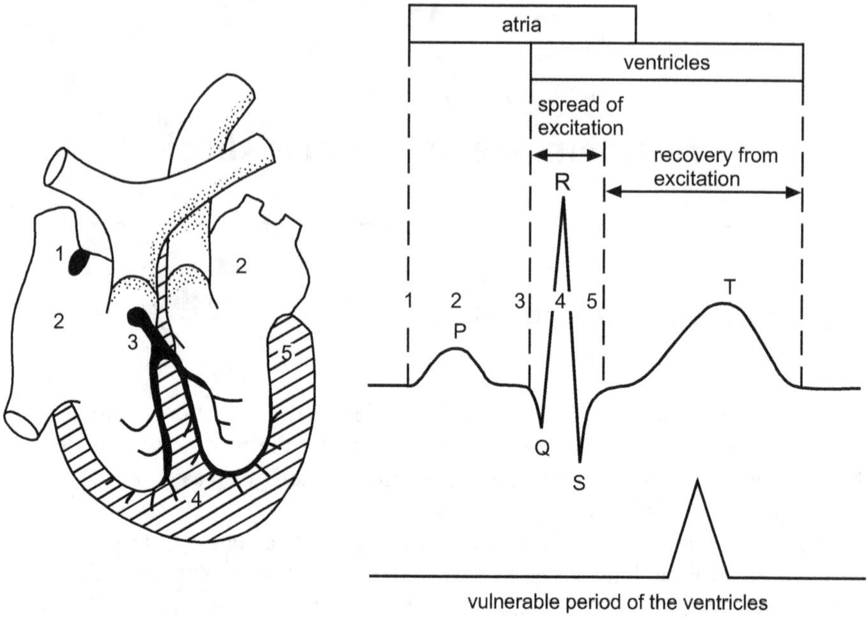

Figure 4.1 *The vulnerable period of the cardiac cycle (Figure 17 of DD IEC/TS 60479-1: 2005). The numbers designate the sequence of propagation of the excitation, and wave of muscular tension that produces the pumping action*

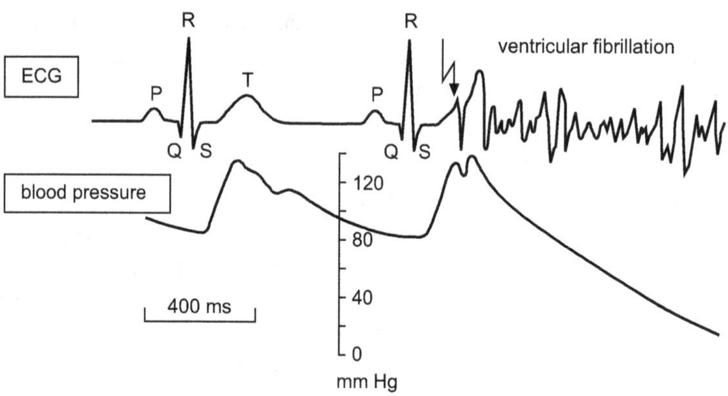

Figure 4.2 *Triggering of ventricular fibrillation in the vulnerable period: effects on electrocardiogram (ECG) and blood pressure [after Figure 18 of DD IEC/TS 60479-1: 2005]*

Figure 4.3 and Table 4.1 show the physiological effects for various alternating body currents flowing for a duration *t*. Note that, with increase of current and its duration, the probability of ventricular fibrillation increases.

It is important for all persons working with electricity to appreciate that, just because they have survived an electric shock on one or more previous occasions, they do not necessarily have a high resistance to electric shock. It may have been simply chance – and the next time...?

The threshold of perception and reaction to currents depends on many parameters such as the area of the body in contact with the live part, the conditions of the contact and the physiological characteristics of the person or animal concerned. A general value for threshold of perception of 0.5 mA independent of time is often assumed for alternating currents and 2 mA for direct currents. The threshold of let-go current is the current below which a person can remove himself from the source of electric shock. The threshold of let-go depends on the same sort of parameters as the threshold of perception. It is generally accepted that the threshold of let-go for alternating currents is around 10 mA. Unlike with a.c., there is no definable threshold of let-go for d.c. The making and breaking of the current leads to painful and cramplike contractions of the muscles. As discussed for a.c., the threshold of ventricular fibrillation with d.c. depends on physiological as well as electrical parameters.

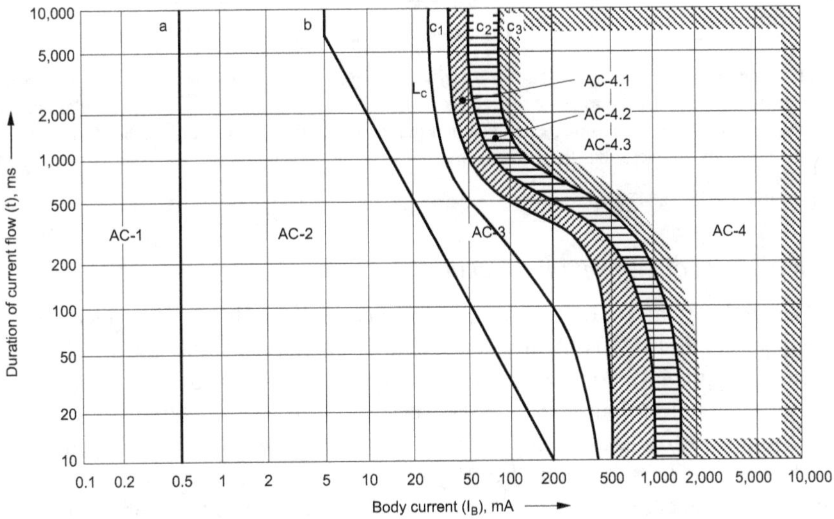

Figure 4.3 Conventional time/current zones of effects of alternating currents 15 Hz to 100 Hz on persons for a current path corresponding to left hand to feet (for explanations, see Table 4.1) [after Figure 20 of IEC 60479-1]

Table 4.1 Time/current zones for alternating currents 15 Hz to 100 Hz

Zone Designation	Zone limits	Physiological effects
AC-1	Up to 0.5 mA line a	Perception possible but usually no 'startled' reaction.
AC-2	0.5 mA up to line b	Perception and involuntary muscular contractions likely but usually no harmful electrical physiological effects.
AC-3	curve c_1	Strong involuntary muscular contractions. Difficulty in breathing. Reversible disturbances of heart function. Immobilization may occur. Effects increasing with current magnitude. Usually no organic damage to be expected.
AC-4	Above curve c_1	Pathophysiological effects may occur such as cardiac arrest, breathing arrest and burns or other cellular damage. Probability of ventricular fibrillation increasing with current magnitude and time.
	$c_1 - c_2$	AC-4.1: Probability of ventricular fibrillation increasing up to about 5 %.
	$c_2 - c_3$	AC-4.2: Probability of ventricular fibrillation up to about 50 %.
	Beyond curve c_3	AC-4.3: Probability of ventricular fibrillation above 50 %.

Note: For durations of current flow below 200 ms, ventricular fibrillation is initiated only within the vulnerable period if the relevant thresholds are surpassed. As regards ventricular fibrillation, this figure relates to the effects of current that flows in the path left hand to feet. For other current paths, the heart current factor has to be considered.

Source: Table 11, IEC 60479-1

4.1.2 Effects of current on the skin

The effect of current on the skin is generally considered to be dependent on the current density. Below 10 mA/mm^2, in general no alterations of the skin are observed (zone 0 of Figure 4.4). For longer durations, the skin below the point of contact may be of greyish white colour, with a coarse surface. Between 10 mA/mm^2 and 20 mA/mm^2, a reddening of the skin occurs, with a wavelike swelling of whitish colour along the edges of the electrode (zone 1 of Figure 4.4). Between 20 mA/mm^2 and 50 mA/mm^2, a brownish colour develops below the point of contact (zone 2 of Figure 4.4) . For longer durations of current, several tens of seconds, blisters are to be observed. Above 50 mA/mm^2, carbonization of the skin can occur.

4.1.3 Impedance of the human body

The impedance of the human body is dependent on the voltage applied, its frequency, the wetness of the contact, the current path and the surface area of contact. Figure 4.5

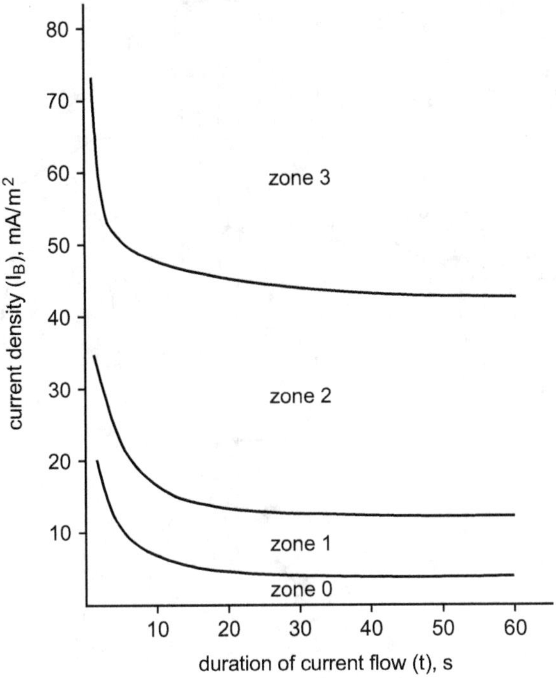

Figure 4.4 Dependence of the alterations of the human skin on current density and duration of current flow [Figure 14, IEC 60479-1]

shows the relative impedances of the various parts of the body, as a percentage of the hand-to-foot path. The impedance of the skin falls as the current is increased. Again, the value of the impedance depends on voltage, frequency, duration of current flow, surface area of contact, pressure of contact, moisture at the point of contact, temperature and, obviously, type of skin. For touch voltages up to 50 V a.c., the impedance of the skin will vary widely, even for an individual, depending on the area of contact, temperature, sweating, respiration rate etc. For touch voltages over approximately 50 V a.c., the skin impedance decreases considerably and becomes negligible when the skin breaks down. The impedance of the skin decreases when the frequency increases.

4.1.4 Total impedance of the human body

For touch voltages up to about 50 V, because of the considerable variations of the impedance of the skin, the impedance of the human body Z_t varies very widely. For higher voltage, the total impedance depends less on the impedance of the skin, and its value approaches that of the internal impedance of the body Z_I. The internal impedances can be considered as being mostly resistive and depend primarily on the current path, and to a much lesser extent on the area of contact. Figure 4.6 shows how the total body impedance for a current path hand-to-hand, for 50 Hz and large

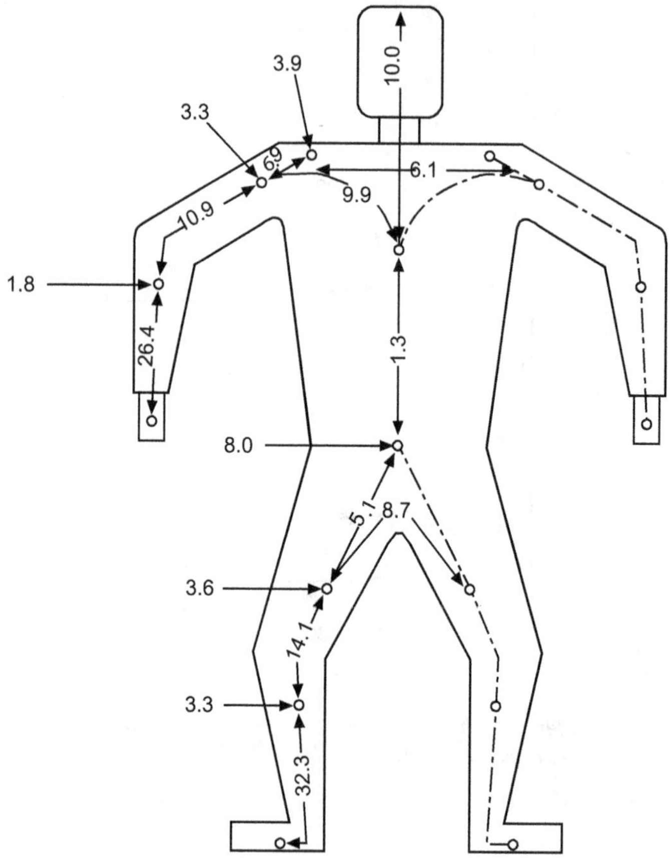

Figure 4.5 Percentage of internal impedances of the human body

The numbers indicate the percentage of the internal impedance of the human body for the part of the body concerned, in relation to the path hand-to-foot.

Note: In order to calculate the total body impedance Z_t for a given current path, the internal partial impedances Z_{ip} for all parts of the body of the current path have to be added as well as the impedances of the skin of the surface areas of contact. The numbers outside the body show internal portions of the impedance to be added to the total, when the current enters at that point.

surface areas of contact, varies with the touch voltage. Note in Table 4.2 how the value of total body impedance varies throughout the population. Note also that body impedance reduces as the touch voltage increases, the effect being most pronounced for small contact areas.

BS 7671 uses the concepts described above in setting protection measures against electric shock. Consideration is now given to these protective measures as given in Part 4 of BS 7671.

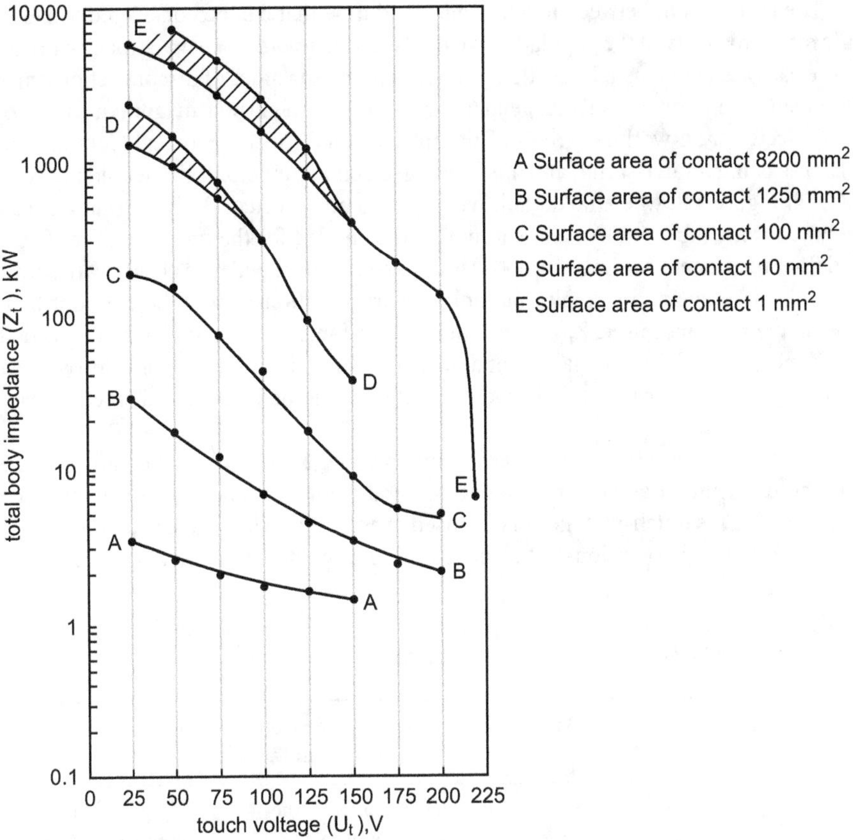

Figure 4.6 Dependence of the total impedance of the human body on the surface area of contact and the touch voltage (50 Hz)

4.1.5 Touch voltage

Prospective touch voltage is defined as the voltage between simultaneously accessible conductive parts when those conductive parts are not being touched by a person or an animal; and *touch voltage* is defined as the voltage between simultaneously accessible conductive parts when those conductive parts are being touched by a person or an animal.

Touch voltages in a healthy installation will be close to zero; under fault conditions they may be dangerous (see Figure 4.13). The protective measure automatic disconnection of supply that is the subject of Section 411 is intended to prevent a person being subjected to a dangerous touch voltage for a time sufficient to cause injury in the event of an insulation fault.

In order to meet this requirement, in the event of such a fault the circuit protective device must interrupt the resulting fault current sufficiently quickly to prevent the touch voltage persisting long enough to be dangerous.

The relationship between touch voltage and time that can be considered as being safe for the majority of the population is called the *touch-voltage curve.* A touch-voltage curve can be used in installation design to set maximum values of disconnection times for given touch voltages. To this end, a current duration/current magnitude curve L_c was selected by a working group of the International Electrotechnical Commission (IEC) as being a safety limit, and this is endorsed on Figure 4.3. It lies within the zone AC 3 of Figure 4.3, where no organic damage is to be expected. Body impedances can be taken from Table 4.2. An element needs to be added for the presence of footwear and floor resistance. A value of 1 000 Ω is assumed. This impedance clearly varies over a very wide range and is applicable only for dry situations where there is footwear. The body impedance used, $Z_{t5\%}$ from Column 2 of Table 4.2, is that which is exceeded by 95 per cent of the population, and, to allow for a double two-hands-to-two-feet current path, a factor of 0.5 is introduced. Effective impedance Z is then equal to 1 000 + $0.5Z_{t5\%}$ Ω.

A maximum duration of prospective touch voltage under normal conditions can now be determined as shown in Table 4.3 and a touch voltage curve drawn (see Figure 4.7). This touch-voltage curve is used in assessing what is a safe touch voltage under normal dry conditions. The chance of ventricular fibrillation occurring with

Table 4.2 Total body impedance Z_t for a current path hand-to-hand, a.c. 50/60 Hz, for large surface areas of contact in dry conditions

Touch voltage (V)	Values for the total body impedance Z_t (Ω) that are not exceeded for		
	5% of the population $Z_{t5\%}$	50% of the population $Z_{t50\%}$	95% of the population $Z_{t95\%}$
1	2	3	4
25	1 750	3 250	6 100
50	1 375	2 500	4 600
75	1 125	2 000	3 600
100	990	1 725	3 125
125	900	1 550	2 675
150	850	1 400	2 350
175	825	1 325	2 175
200	800	1 275	2 050
225	775	1 225	1 900
400	700	950	1 275
500	625	850	1 150
700	575	775	1 050
1 000	575	775	1 050
Asymptotic value = internal impedance	575	775	1 050

Note: Some measurements indicate that the total body impedance for the current path hand-to-foot is somewhat lower than for a current path hand-to-hand (10 to 30 per cent).

Table 4.3 Derivation of touch-voltage curve

Touch voltage (V)	Assumed body impedance plus contact impedance (Ω) (Note 1)	Body current (mA) (Note 2)	Maximum duration (ms) (Note 3)
1	2	3	4
50	1 700	30	no limit
100	1 500	67	400
225	1 400	162	180
400	1 350	296	60
500	1 315	380	40

Notes:

1. Given by 0.5 $Z_{t5\%}$ + 1000 Ω.
2. Column 3 = column 1/column 2.
3. From curve L_c of Figure 4.3.

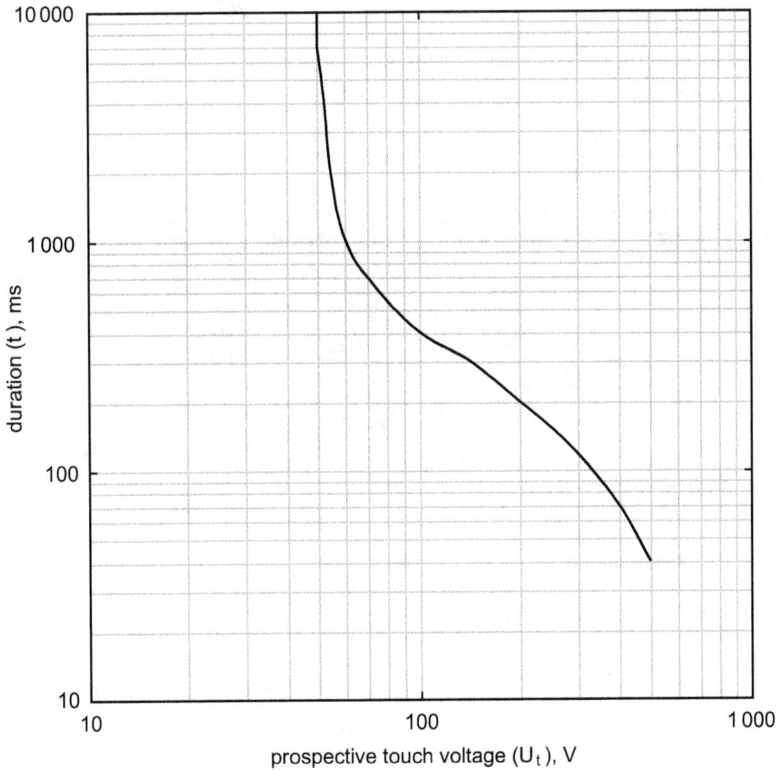

Figure 4.7 Maximum duration of prospective touch voltage U_t for normal dry situations (from IEC TR 61200-413:1996 (withdrawn))

these touch voltages is low, particularly if note is made that the body impedance assumed is exceeded by 95 per cent of the population.

4.2 Some definitions
Part 2

Before protection systems can be considered, the fundamental definitions have first to be understood. The expressions *exposed-conductive-part, extraneous-conductive-part* and *protective conductor* occur frequently. Understanding of the Regulations is somewhat hindered by exposed-conductive-parts and extraneous-conductive-parts having special meanings in terms of the Regulations, and not their common English meanings. Their special meaning is indicated by the hyphens linking the words.

4.2.1 Exposed-conductive-parts

Exposed-conductive-parts are not simply conductive parts that are exposed or accessible, such as a metal table. The definition of an exposed-conductive-part is a conductive part of equipment that can be touched and which is not normally live but which can become live when basic insulation fails. Further reference to definitions will indicate that equipment means electrical equipment and electrical equipment is an item that generates, converts, transmits, distributes or utilizes electrical energy. To summarize, an exposed-conductive-part must:

- be part of electrical equipment;
- be able to be touched;
- not be normally live; and
- be able to become live under fault conditions (failure of basic insulation).

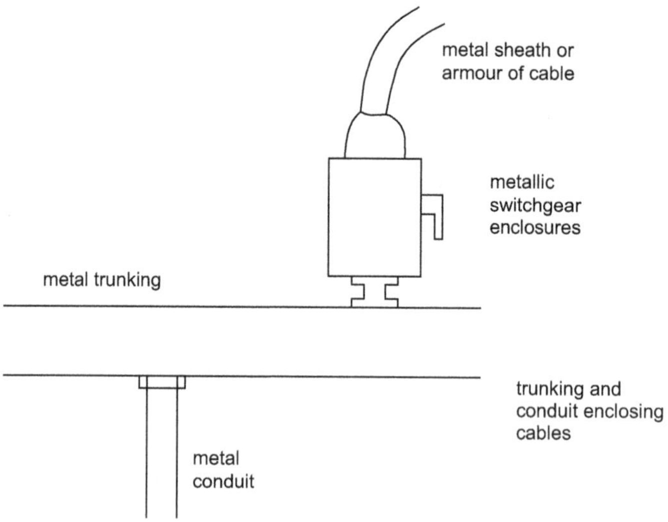

Figure 4.8 Examples of exposed-conductive-parts

Examples of exposed-conductive-parts, illustrated in Figure 4.8, are the armouring of cables, the enclosures of electrical equipment and metal trunking and conduits enclosing cables. Items that are *not* exposed-conductive-parts include metal radiators, metal baths and metal tables.

4.2.2 Extraneous-conductive-parts

In a similar manner to exposed-conductive-parts, reference needs to be made to the definition of an extraneous-conductive-part, for it is not simply any piece of extraneous conductive material. An extraneous-conductive-part is a conductive part liable to introduce a potential, generally Earth potential, and not forming part of the electrical installation. The essence of an extraneous-conductive-part is that it is likely to introduce Earth potential into a building. Examples include metal service pipes, gas, water, fuel oil, metal waste pipes and building metalwork in general contact with the ground. The essential requirement is that extraneous-conductive-parts be exposed, conductive and in general contact with the mass of Earth.

4.2.3 Protective conductors

A protective conductor is a conductor that is used for protection against electric shock and intended for connecting together any of the following:

- exposed-conductive-parts;
- extraneous-conductive-parts;

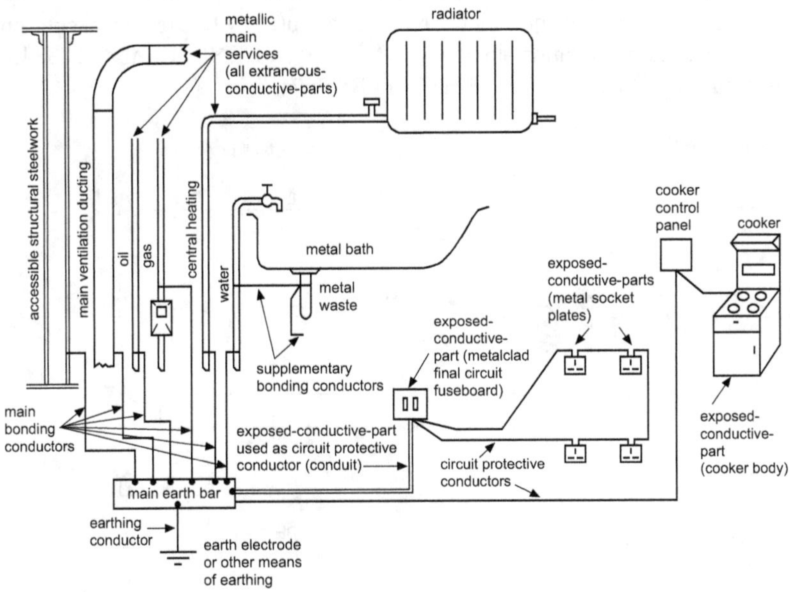

Figure 4.9 Typical arrangement of various protective conductors connecting exposed-conductive-parts and extraneous-conductive-parts and the means of earthing

- the main earthing terminal;
- the earth electrode; and
- the earth point of the source or an artificial neutral.

Protective conductors include:

- circuit protective conductors;
- combined neutral and protective conductors;
- earthing conductors;
- main protective bonding conductors;
- supplementary bonding conductors; and
- earth-free local protective bonding conductors.

Figure 4.9 shows different types of protective conductor.

Supplementary bonding conductors are protective conductors used to connect the various exposed-conductive-parts and extraneous-conductive-parts together, so that they are substantially at the same potential. Supplementary equipotential bonding is required where automatic disconnection cannot be achieved in the required time (Regulation 411.3.2.6), in certain conditions of external influence and in certain specified locations (see Part 7 of BS 7671).

4.2.4 Protective measures and provisions
Section 410

The terminology of IEC 61140, 'Protection against electric shock – Common aspects for installation and equipment', has been adopted by IEC technical committee TC 64 in electrical installation standard IEC 60364, by CENELEC for HD 384.4.41 and in its turn by UK committee JPEL/64 in BS 7671.

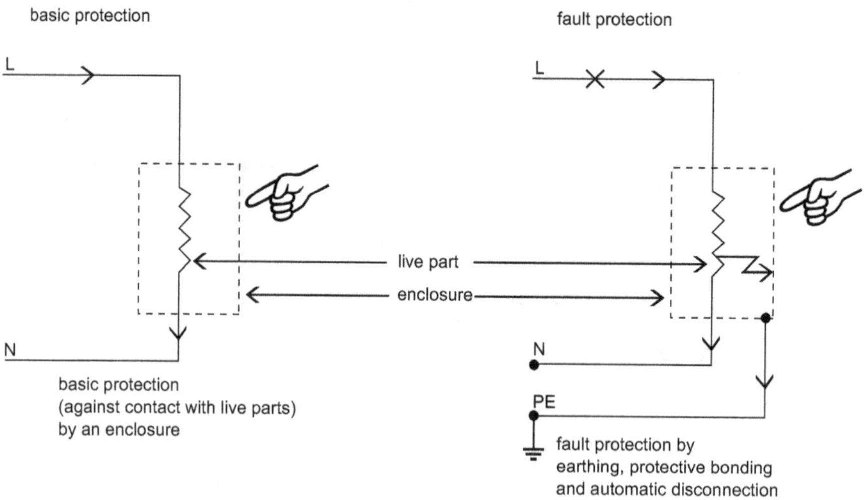

Figure 4.10 Protective measure: automatic disconnection of supply

IEC 61140 coordinates the protection measures of the system standards, e.g. BS 7671, with the equipment and appliance standards, e.g. BS EN 60335 series (household and similar appliances).

BS 7671 requires two lines of defence against electric shock: a provision for basic protection (e.g. basic insulation of live parts) and an independent provision for fault protection (e.g. automatic disconnection); the combination is a protective measure (see Figure 4.10). Alternatively, protection can be provided by an enhanced protective provision such as reinforced insulation (see Table 4.4).

Table 4.4 Protective measures

Protective measures	Protective provisions	
	Basic protective provision	Fault protective provision
Automatic disconnection of supply (411)	Insulation of live parts, barriers or enclosures	Protective earthing, automatic disconnection, protective bonding
Double insulation (412)	Basic insulation	Supplementary insulation
Reinforced insulation (412)	Reinforced insulation	
Electrical separation for one item of equipment (413)	Insulation of live parts	One item of equipment, simple separation from other circuits and earth
Extra-low voltage (SELV and PELV) (414)	Limitation of voltage, protective separation, basic insulation	
	For supervised installations	
Non-conducting location (418.1)	Insulation of live parts, barriers or enclosures	No protective conductor; insulating floor and walls, spacings/ obstacles between exposed-conductive parts and extraneous-conductive-parts
Earth-free local equipotential bonding (418.2)	Insulation of live parts, barriers or enclosures	Protective bonding, notices etc.
Electrical separation with more than one item of equipment (418.3)	Insulation of live parts	Simple separation from other circuits and earth, separated protective bonding, etc.

4.3 Provisions for basic protection
Sections 416 and 417

4.3.1 Introduction

Basic protection (against direct contact with live parts) can be provided by one or more of the following basic protective provisions:

(i) basic insulation of live parts (416);
(ii) barriers or enclosures (416);
(iii) obstacles (417); and
(iv) placing out of reach (417).

Basic insulation of live parts and barriers or enclosures are measures for general application with a fault protective provision. Obstacles and placing out of reach can be applied with or without fault protection; they must be controlled by skilled persons. See Figure 4.11.

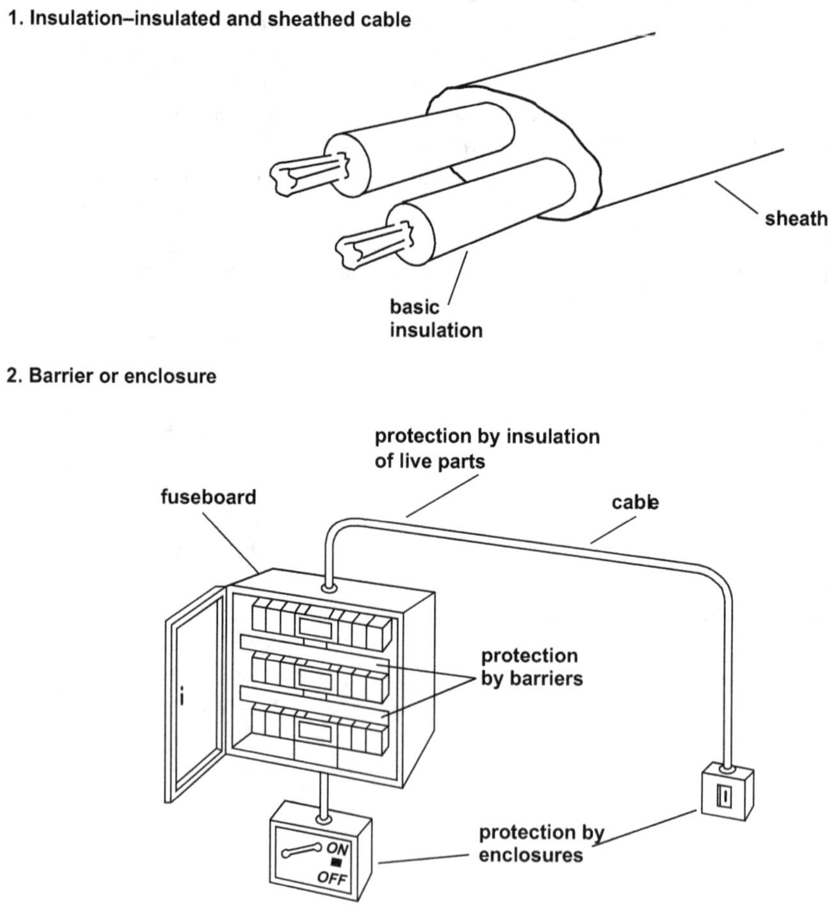

Figure 4.11 Certain provisions for basic protection (in Sections 416 and 417)

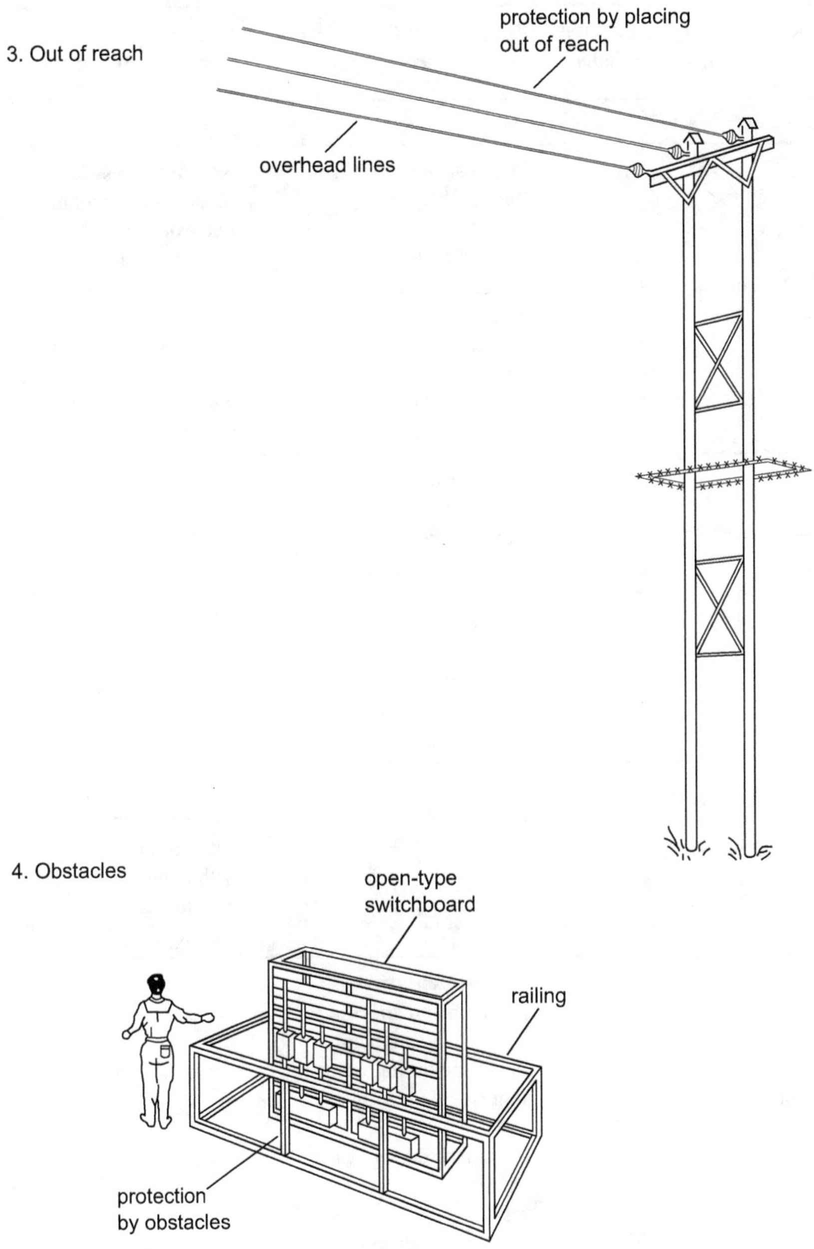

Figure 4.11 Continued

4.3.2 Basic protective provision: insulation of live parts Regulation 416.1

Persons may be protected against direct contact with live parts by insulation. Such insulation should be able to be removed only by destruction, and it must be capable of durably withstanding not only the electrical, but mechanical, thermal and chemical, stresses to which it may be subjected in service.

Element	Numerals of letters	Meaning for the protection of equipment	Meaning for the protection of persons
Code letters	IP	–	–
		Against ingress of solid foreign objects	**Against access to hazardous parts inside enclosures**
First characteristic numeral	0	non-protected	non-protected
	1	of ≥50 mm diameter	with back of hand
	2	of ≥12.5 mm diameter	with finger
	3	of ≥2.5 mm diameter	with tool >2.5 mm thick
	4	of ≥1.0 mm diameter	with wire >1.0 mm thick
	5	dust-protected	with wire >1.0 mm thick
	6	dust-tight	with wire >1.0 mm thick
		Against ingress of water	
Second characteristic numeral	0	non-protected	
	1	vertical dripping	
	2	dripping (tilted up to 15°)	
	3	spraying	
	4	splashing	
	5	jets	
	6	powerful jets	
	7	temporary immersion	
	8	continuous immersion	
			Against access to hazardous parts
Additional letter (optional)	A		with back of hand
	B		with finger
	C		with tool
	D		with wire
		Supplementary information specific to	
Supplementary letter (optional)	H	high-voltage apparatus	
	M	in motion during water ingress test	
	S	stationary during water ingress test	
	W	weather conditions	

Figure 4.12 International Protection code (IP code): Classification of degrees of protection provided by enclosures

4.3.3 Basic protective provision: barriers or enclosures
Regulation group 416.2

The term *enclosure* is readily understood as being a construction that prevents live parts being touched from any direction. *Barriers* are, because of their common English usage, sometimes confused with *obstacles.* A barrier in the sense of these Regulations is similar to an enclosure in that it is a part providing a defined degree of protection against contact with live parts, with the difference that the protection is provided only in the usual direction of access. A barrier may not have protection against contact from above, for example in a construction such as a very large switchboard where approach from above is not possible because of the height, perhaps because the live parts are inaccessible from that direction. In general, the requirement for protection by barriers or enclosures is that an opening larger than IPXXB or IP2X should not be allowed. This is increased to IPXXD or IP4X for horizontal top surfaces of barriers or enclosures which are readily accessible. The basic requirements of the International Protection Code are shown in Figure 4.12.

BS EN 60529:1992(2004) describes a system for classifying the degrees of protection provided by the enclosures of electrical equipment. The manufacturer should be consulted to determine the types of protection available and the parts of the equipment to which the stated degree of protection applies.

The basic requirement for IPXXB or IP2X is that the live part should be inaccessible to a finger and the basic requirement for IPXXD or IP4X is that it should be inaccessible to a wire exceeding 1 mm diameter. The higher degree of protection is required on top of an enclosure to prevent small items falling through. Regulation 416.2.3 requires any barriers or enclosures to be firmly secured in place and to have sufficient stability and durability to maintain the required degree of protection in the known conditions. In general, barriers or enclosures should be able to be removed or opened only by the use of a key or tool, or if the removal of the opening disconnects the supply, or there is a further intermediate barrier providing a similar degree of protection.

It has been custom and practice within the UK to use certain equipment that does not provide this degree of protection, in that tools are not required for the removal of the cover of, for example, ceiling roses, pull-cord switches, bayonet lampholders or Edison screw lampholders. A note to Regulation 416.2.4 exempts this type of equipment, provided that it complies with the appropriate British Standard.

4.4 Protective measure: automatic disconnection of supply
Section 411

4.4.1 Touch voltages in TN systems

The protective measure automatic disconnection of supply is the most commonly used protective measure. It requires:

(a) a basic protective provision: basic insulation, or barriers or enclosures (or, infrequently, obstacles and placing out of reach);

(b) a fault protective provision comprising:
  (i) protective earthing,
  (ii) protective equipotential bonding, and
  (iii) automatic disconnection in the event of a fault.

The fault protective provision (b) is the protective measure that used to be called protection against indirect contact.

4.4.1.1 Protective earthing

Protective earthing provides a circuit for fault currents to flow in the event of a fault to an exposed-conductive-part. Without this earthing the overcurrent device (fuse, circuit-breaker, RCD) will not operate in the event of a fault.

4.4.1.2 Protective equipotential bonding

Protective equipotential bonding can reduce the touch voltages within the building.

4.4.1.3 Automatic disconnection

The fault current (I_f) flowing through the earth fault loop (see Figure 4.13), results in the operation of the protective device, thereby disconnecting the supply.

Danger can arise from the touch voltage generated by the fault current. The circuit must be so designed that the touch voltage persists only for the time allowed by Regulation 411.3.2.

In Figure 4.13, where there is protective equipotential bonding, the touch voltage is given by:

$$\text{Prospective touch voltage } U_t = I_f Z_2$$

If there were no protective equipotential bonding, the prospective touch voltage would be given by:

$$U_t = I_f (Z_2 + Z_{PEN})$$

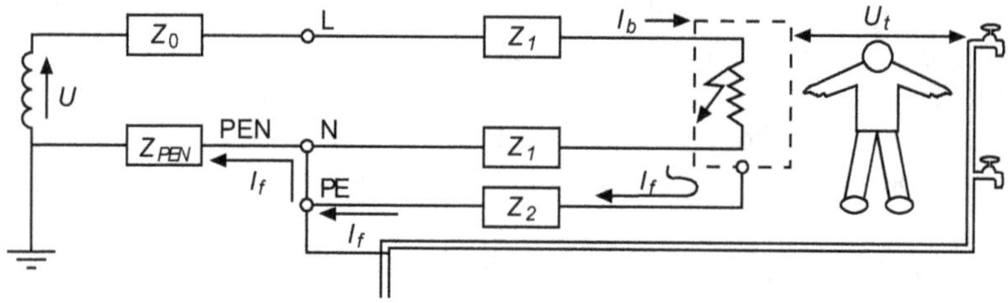

where:
Z_0 = line impedance of the supply
Z_{PEN} = impedance of the PEN conductor of the supply
Z_1 = line impedance of the installation (neutral impedance similar)
Z_2 = impedance of the protective conductor of the installation
U = supply system voltage
I_b = load current
I_f = fault current
U_t = prospective touch voltage

Figure 4.13 Touch voltages in TN systems

From Figure 4.13, where there is a fault between the line conductor and an exposed-conductive-part, the resulting earth fault current I_f is given by:

$$I_f = \frac{U}{Z_0 + Z_1 + Z_2 + Z_{PEN}}$$

The prospective touch voltage appearing between the exposed-conductive-parts and earth is given by:

For installations with main protective bonding

$$\text{prospective touch voltage } U_t = I_f \times Z_2 = \frac{U}{Z_0 + Z_1 + Z_2 + Z_{PEN}} \times Z_2$$

For final circuits, Z_1 and Z_2 are resistive and are usually represented by R_1 and R_2, and $Z_0 + Z_{PEN}$ by Z_e; then:

$$U_t = \frac{UR_2}{Z_e + R_1 + R_2}$$

where Z_e is that part of the earth fault loop impedance external to the installation

$$\text{or} \quad U_t = \frac{UR_2}{Z_s}$$

where $Z_s = Z_e + R_1 + R_2$, the earth fault loop impedance.

For TN-C-S installations without main bonding

$$U_t = \frac{(Z_{PEN} + R_2)U}{Z_e + R_1 + R_2}$$

For TN-C-S systems the multiple bonding required by the Electricity Safety, Quality and Continuity Regulations means Z_{PEN} is low and generally negligible compared with R_2 for final circuits; and main protective bonding will not reduce touch voltages to any significant degree for faults on the installation.

Note: For faults on the distribution system, e.g. an open-circuit PEN conductor, main protective bonding reduces touch voltages between the metalwork of the installation and true Earth.

For TN-S systems with Z_{PE} perhaps of the order of 0.5 ohm (Z_e of 0.8 ohm), main protective bonding will significantly reduce touch voltages for faults on the installation:

$$U_t = \frac{(Z_{PE} + R_2)}{Z_e + R_1 + R_2}$$

where Z_{PE} is the impedance of the protective conductor of the supply.

In UK installations, use is often made of reduced cross-sectional area protective conductors. This increases touch voltages. The higher the ratio of protective conductor resistance to line conductor resistance (per metre), the greater the touch voltage.

The external impedance Z_e also affects touch voltage: the higher the external impedance the lower the touch voltage.

4.4.1.4 Examples of touch voltage calculations

Consider a PME supply $Z_e = 0.35\ \Omega$ and a circuit of 20 m wired in 2.5 mm^2 cable with 1.5 mm^2 protective conductor ($R_1 = 7.41$ mΩ/m, $R_2 = 12.1$ mΩ). $U = 240\ V$.

Then using $U_t = \frac{UR_2}{Z_e + R_1 + R_2}$

$$U_t = \frac{(20 \times 12.1)/1000 \times U}{0.35 + (20(7.41 + 12.1))/1000}$$

$$U_t = 0.33\text{U}$$

$$U_t = 79\text{V}$$

From the touch voltage curve Figure 4.7, the disconnection time would need to be of the order of 0.5 second.

If Z_e is so low as to be ignored

then $U_t = \frac{(20 \times 12.1)/1000 \times U}{0 + (20(7.41 + 12.1))/1000}$

$$U_t = 0.62\text{U}$$

$$U_t = 149\text{V}$$

From the touch voltage curve Figure 4.7, the disconnection time would need to be of the order of 0.3 second.

4.4.2 Touch voltages in TT systems

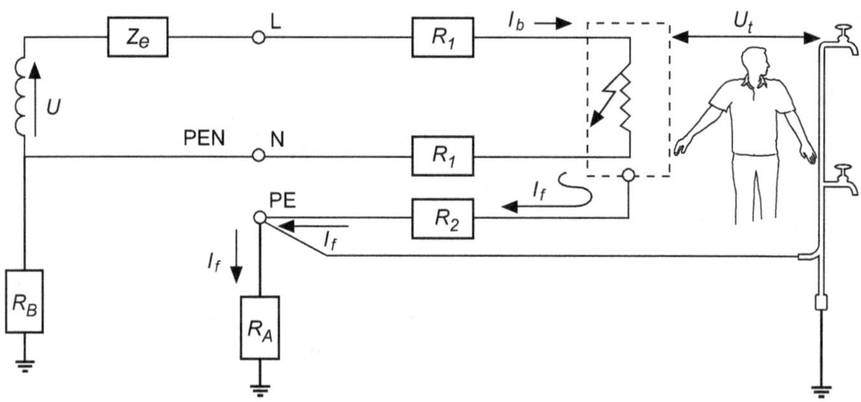

touch voltage $U_t = I_f R_2$

where:
Z_e = impedance of the supply
R_1 = line resistance of the installation (neutral resistance similar)
R_2 = resistance of the protective conductor of the installation
R_B = resistance of the supply transformer earth electrode
R_A = resistance of the installation earth electrode
I_f = fault current
U_t = prospective touch voltage

Figure 4.14 Touch voltages in a TT system

From Figure 4.14:

For installations with main protective bonding:	For installations without main protective bonding:
$U_t = R_2 \times I_f$	$U_t = (R_2 + R_A) \times I_f$
$U_t = \frac{R_2 \times U}{Z_e + R_1 + R_2 + R_B + R_A}$	$U_t = \frac{(R_2 + R_A) \times U}{Z_e + R_1 + R_2 + R_B + R_A}$
As Z_e, R_1 and R_2 are small compared with R_A and R_B the equation becomes	As Z_e, R_1 and R_2 are small compared with R_A and R_B the equation becomes
$U_t = \frac{R_2 \times U}{R_B + R_A}$	$U_t = \frac{R_A \times U}{R_B + R_A}$

Consider an installation with $R_A = 100\ \Omega$ and $R_B = 20\ \Omega$ and for a 20 m circuit with 1.5 mm^2 protective conductor $R_2 = 20 \times 12.10$ mΩ.

Touch voltage	
With main bonding	Without main bonding
$U_t = \frac{R_2 \times U}{R_B + R_A}$	$U_t = \frac{R_A \times U}{R_B + R_A}$
$U_t = \frac{0.242 \times 240}{20 + 100}$	$U_t = \frac{100 \times 240}{20 + 100}$
$U_t = 0.46$ V	$U_t = 200$ V

This indicates the particular importance of main bonding in TT installations and the particular hazard for TT installations for circuits outdoors or exposed-conductive-parts accessible from outdoors in the event of a fault, for example, for caravans.

4.4.3 *Automatic disconnection times*
Regulation 411.3.2

To limit the hazards of touch voltages in the event of a fault for installations where protection against electric shock is being provided by automatic disconnection, maximum disconnection times are set in Regulation 411.3.2.

From the estimates above we determine disconnection times as follows for a.c. installations. For TT installations the assumption is that there will be no effective main bonding as for persons outside the building, then U_t approx = 0.8 U_0. For TN with low Z_e then U_t approx = 0.5 U_0. See Table 4.5.

Table 4.5 Derivation of a.c. disconnection times (seconds)

System		120 V		230 V		400 V		1000 V	
	Touch voltage	Touch voltage	Disc time	Touch voltage	Disc time	Touch voltage	Disc time	Touch voltage	Disc time
TN	0.5 U_0	60	0.9	115	0.4	200	0.2	500	Off curve
TT	0.8 U_0	96	0.4	184	0.2	320	0.1	800	Off curve

These disconnection times are similar to those published in BS 7671 as below:

411.3.2.2 The maximum disconnection time stated in Table 41.1 shall be applied to final circuits not exceeding 32 A.

Table 41.1 Maximum disconnection times

System	$50\text{ V} < U_0 \leq 120\text{ V}$ (seconds)		$120\text{ V} < U_0 \leq 230\text{ V}$ (seconds)		$230\text{ V} < U_0 \leq 400\text{ V}$ (seconds)		$U_0 > 230\text{ V}$ (seconds)	
	a.c.	d.c.	a.c.	d.c.	a.c.	d.c.	a.c.	d.c.
TN	0.8	NOTE 1	0.4	5	0.2	0.4	0.1	0.1
TT	0.3	NOTE 2	0.2	0.4	0.07	0.2	0.04	0.1

Where, in a TT system, disconnection is achieved by an overcurrent protective device and protective equipotential bonding is connected to all the extraneous-conductive-parts within the installation in accordance with Regulation 411.3.1.2, the maximum disconnection times applicable to a TN system may be used. U_0 is the nominal a.c. rms or d.c. line voltage to Earth.

Where compliance with this regulation is provided by an RCD, the disconnection times in accordance with Table 41.1 relate to prospective residual fault currents significantly higher than the rated residual operating current of the RCD.

Note: Disconnection is not required for protection against electric shock but may be required for other reasons, such as protection against thermal effects.

411.3.2.3 In a TN system, a disconnection time not exceeding 5 s is permitted for a distribution circuit and for a circuit not covered by Regulation 411.3.2.2.

411.3.2.4 In a TT system, a disconnection time not exceeding 1 s is permitted for a distribution circuit and for a circuit not covered by Regulation 411.3.2.2.

In TN systems, BS 7671 requires for standard 230 V final circuits up to a rating of 32 A a 0.4 second disconnection time. This is relaxed to 5 seconds for final circuits exceeding 32 A and for distribution circuits.

There may be circumstances where 0.4 second disconnection is not appropriate or achievable for circuits rated up to 32 A. In these circumstances supplementary bonding will enable the requirements of BS 7671 to be met (see subsection 4.4.4).

4.4.4 Supplementary equipotential bonding
Regulations 411.3.2.6, 415.2

Regulation 411.3.2.6 requires that, where automatic disconnection in accordance with Table 41.1 cannot be achieved, supplementary bonding shall be provided in accordance with Regulation 415.2. This reduces the touch voltage generated by the fault current (see Figure 4.14).

> **415.2.2** Where doubt exists regarding the effectiveness of supplementary equipotential bonding, it shall be confirmed that the resistance R between simultaneously accessible exposed-conductive-parts and extraneous-conductive-parts fulfils the following condition:
>
> $R \leq 50\ \text{V}/I_a$ in a.c. systems
>
> $R \leq 120\ \text{V}/I_a$ in d.c. systems
>
> where I_a is the operating current in amperes of the protective device –
>
> for RCDs, $I_{\Delta n}$.
> for overcurrent devices, the current causing automatic operation in 5 s.

It is to be noted that I_a is the current to cause operation of the device in five seconds or less, not the rating I_n. This is considered in more detail in subsection 11.6.2, where maximum lengths of supplementary bonding conductor are tabulated. These lengths may well be less than the circuit length.

It may be difficult to achieve the required disconnection times for type C and D circuit-breakers, and this facilitates their use.

The effect of compliance with this requirement is that, if the fault current is less than that necessary to operate the protective device, even if all the fault current flows along the supplementary bonding conductor, the touch voltage will not exceed 50 volts.

4.4.5 Additional protection
Regulation 411.3.3

4.4.5.1 Scope of the additional RCD protection

A substantial change has been made in Amendment No. 3 requiring additional protection by means of an RCD for:

(a) Socket-outlets with a rated current not exceeding 20 A;, and
(b) Mobile equipment with a current rating not exceeding 32 A for use outdoors.

All socket-outlets, whether in domestic, commercial or industrial premises, are therefore required to be protected by an RCD having a rated residual operating current ($I_{\Delta n}$) not exceeding 30 mA. There are two permitted exceptions for (a), and these are discussed in 4.4.5.3.

4.4.5.2 Meaning of additional

Regulation 411.3.3 requires additional protection by means of a 30 mA RCD. That is, additional to the requirements of Regulation 411.3.2.1 for automatic disconnection in the case of a fault. It is to be noted that this additional protection is not required for reduced low voltage systems or for FELV systems. It is also not required for protective measures not using automatic disconnection in the case of a fault as the means of fault protection.

There has been debate, both nationally and internationally, as to whether 'additional protection' means protection additional to that given by the device providing automatic disconnection of supply in accordance with Table 41.1 as in figure 4.15, or whether the requirements both of Regulations 411.3.2 (automatic disconnection in case of a fault) and 411.3.3 (additional protection) can be met by one 30 mA RCD alone. At an IEC TC 64 meeting, the Dutch delegates proposed that two devices were required. However, the international committee decided that one 30 mA RCD would meet the requirement.

This has far reaching implications. For example, only one RCD is required for TT installations, and an installation with type C or D circuit-breakers that do not provide automatic disconnection, if protected by a 30 mA RCD, will meet the shock protection requirements.

4.4.5.3 Commercial and industrial installations

The changes made to Regulation 411.3.3 in Amendment No. 3 resolve the debate as to the meaning of 'under the supervision of skilled or instructed persons'. The 30 mA RCD may only be omitted (from other than a specifically labelled or otherwise suitably identified socket-outlet...) where a documented risk assessment determines that the RCD is not necessary.

4.4.6 Automatic disconnection, TN systems
Regulation 411.4

To achieve the required disconnection times, the loop impedance of each final circuit Z_s must be sufficiently low to ensure that the fault current I_f exceeds the current to cause operation of the device within the required time of 0.4 or 5 s. These currents are given the designation I_a:

$$I_f \geq I_a$$

where I_a is the current to cause operation of the protective device within the maximum disconnection time given in Table 41.1 of BS 7671 or Regulation 411.3.2.3.

In Paragraph 2.5 of Appendix 4 of BS 7671 now includes reference to CENELEC Technical Report PD CLC/TR 50480. This recommends calculating the maximum earth fault loop impedance (Z_s) to allow disconnection of supply within the maximum disconnection time using the following equation:

$$Z_s \leq \frac{C_{min}\, U_o}{I_a}$$

where C_{min} is a voltage factor to allow for the worst-case condition, for example, under full load conditions.

CENELEC Technical Report CLC/TR 50480: Determination of cross-sectional area of conductors and selection of protective devices, recommends a C_{min} value of 0.95; see the table below. This is not a particularly conservative value.

CLC/TR 50480 Table 7 Voltage factor C

Nominal voltage	Voltage factor C	
	C_{max}	C_{min}
100 V to 1000 V	1.1	0.95

C_{max} is to be used in calculations of maximum earth fault and maximum short-circuit currents.

The values of I_a for the most common devices can be found in Appendix 3 of BS 7671, and so Z_s can be calculated for all the devices in the appendix.

Example calculation of maximum earth fault loop impedances for type B circuit-breakers:

Device type	Device rating I_n A	I_a for 0.4 s disconnection	Maximum $Z_s = \frac{C_{min}\, U_o}{I_a}$ Ω note 1
B	6	30	7.28
B	10	50	4.37
B	16	80	2.73
B	I_n	$5I_n$	$0.95*230/5I_n$

Note: These are the values which appear in Table 41.3 (a) of BS 7671, due to Amendment No. 3.

4.4.7 *Automatic disconnection, TT systems*
Regulation 411.5

In TT systems disconnection in the event of a fault can be provided by overcurrent devices or RCDs, as for TN systems. However, the impedance R_A of the connection with Earth at the consumer's premises and to a lesser extent R_B, the impedance of the connection with Earth of the supply transformer neutral (see Figure 4.15), will normally result in a loop impedance ($Z_s = Z_D + R_1 + R_2 + R_A + R_B$) so high that overcurrent devices cannot provide disconnection within 0.2 or 1 s.

Regulation 411.5.3 requires that where protection is provided by an RCD the following conditions shall be met:

(i) $R_A \times I_{\Delta n} \leq 50$ V; and also
(ii) disconnection in the time required by Table 41.1, i.e. 0.2 s (or 1s for final circuits exceeding 32 A rating and for distribution circuits)

where:

R_A = sum of the resistances of the earth electrode (R_A) and the protective conductor (R_2) (see Figure 4.15)
$I_{\Delta n}$ = is the rated residual operating current of the RCD.

The choice of the symbol R_A in BS 7671 is somewhat confusing, but as R_2 (say 0.1 Ω) is small compared with R_A (say 20 Ω) $R_A + R_2 \approx R_A$.

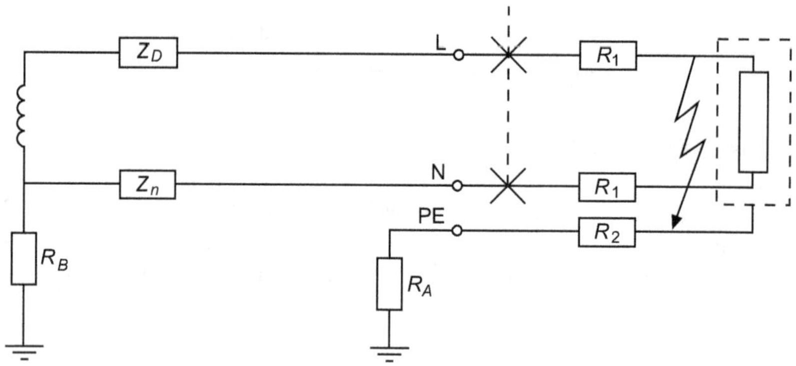

Figure 4.15 TT system showing earthing impedances R_A and R_B

Compliance with equation (a) above will ensure that, in the event of the fault current being insufficient to operate the device, the touch voltage will not exceed 50 V.

The use of equation (a), $R_A I_{\Delta n} \leq 50$ V will suggest maximum earth electrode resistances as given in Table 4.6, the values being calculated as follows:

$$30\text{ mA devices } R_A \leq \frac{50}{30 \times 10^{-3}} = 1.66\text{k } \Omega$$

$$100\text{ mA devices } R_A \leq \frac{50}{100 \times 10^{-3}} = 500\ \Omega$$

$$300\text{ mA devices } R_A \leq \frac{50}{300 \times 10^{-3}} = 167\ \Omega$$

Table 4.6 Calculation of maximum earth fault loop impedance for RCDs

Rated residual operating current (mA)	R_A to meet $R_A \leq \frac{50}{I_{\Delta n}}$
30	1667[1]
100	500[2]
300	167
500	100

Note:

1. From Table 1 of BS EN 61008-1.
2. Electrode resistances over 200 Ω are likely to be unstable as small changes in the soil, drying out or movement, cause bigger changes in resistance than occurs if there is good electrical contact with Earth, hence Note 2 to Table 41.5 of BS 7671.

4.4.8 Neutral-to-earth faults in RCD-protected TT installations

In TT installations RCDs *may* be tripped by the spill of the load current to earth if there is a neutral-to-earth fault. However, this is not certain, and it is more likely to occur if R_A is low and the load current fairly high.

In these installations RCDs protecting circuits with a N-E first fault will generally provide protection for second fault to earth, if R_A is kept low as recommended; see Figure 4.16. However, protection against direct contact with live parts will not necessarily be provided if there is a neutral-to-earth fault (see Figure 4.17). If a person suffers a direct contact shock in such an installation, a 30 mA RCD is unlikely to offer protection.

Assume a low body impedance of, say, 1500 Ω. The body current I_f is then approximately 230/1500 A = 153 mA. If there were no neutral-earth fault, this would be more than sufficient to trip a 30 mA RCD. The fault splits the current between

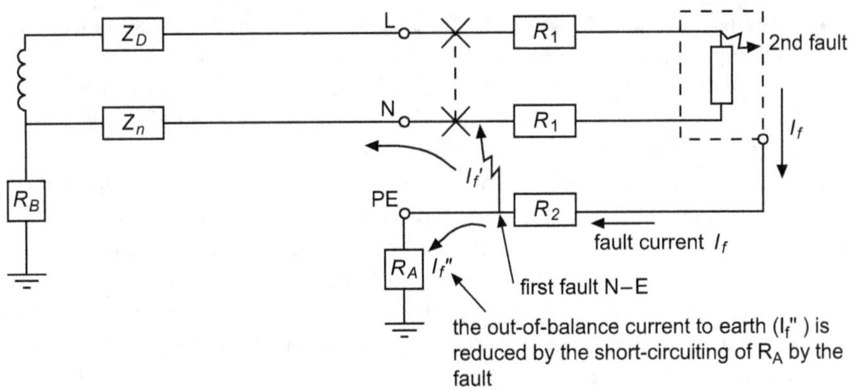

Figure 4.16 TT Installation with an RCD and neutral-to-earth first fault and line-to-earth second fault

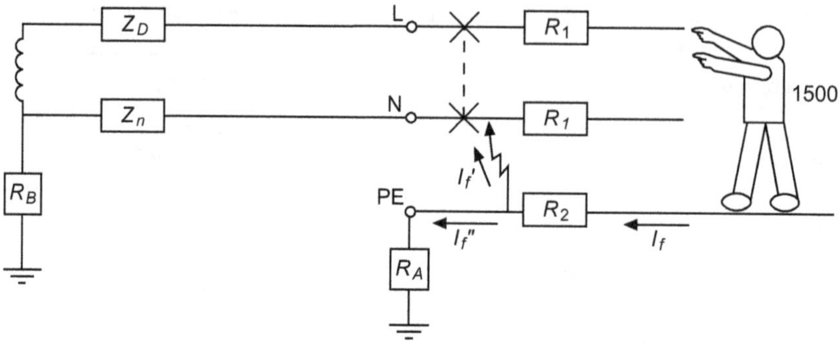

Figure 4.17 TT Installation with an RCD and N-E first fault with direct contact shock risk

the installation earth R_A and the neutral conductor. The current taking the path through R_A, I_f'' is given by:

$$I''_f = I_f \times \frac{Z_n}{Z_n + R_A + R_B}$$

Unless R_A is as low as 4 × Z_n, the RCD will not operate. Z_n is unlikely to exceed 0.35 Ω, so R_A would need to be less than 1.4 Ω. (For this reason it is most important to locate and rectify neutral to earth faults.)

4.4.9 *Automatic disconnection, IT systems* Regulation 411.6

Designers and installers may require knowledge of IT systems. IT systems are used in some continental European countries for electricity distribution, in locations such as medical centres and mines, where the supply must be maintained even in the event of a fault, and where connection with Earth is difficult, for example, mobile generators.

IT systems are either unearthed or are earthed only at the source and through a sufficiently high impedance (see Figure 4.18).

The Regulations require monitoring to indicate a fault to Earth or to exposed-conductive-parts (Regulation 411.6.3.1). This allows the fault to be rectified at the first opportunity while maintaining supplies.

The value of the earthing impedance Z is selected to limit the first fault current to a safe value, but sufficient to readily initiate the monitoring device, normally of the order of 250 mA. Where the installation is not connected to Earth, the first fault current arises because of the capacitance of the conductors to Earth (Figure 4.18).

It is strongly recommended that neutrals should not be distributed in IT systems, as this will tend to negate the benefits of the system (Regulation 411.6.1). A first fault of neutral to Earth is not easily detected, and the second fault will then represent a shock risk. Operation of the overcurrent device will be dependent on the first fault being maintained, which is an unsatisfactory situation.

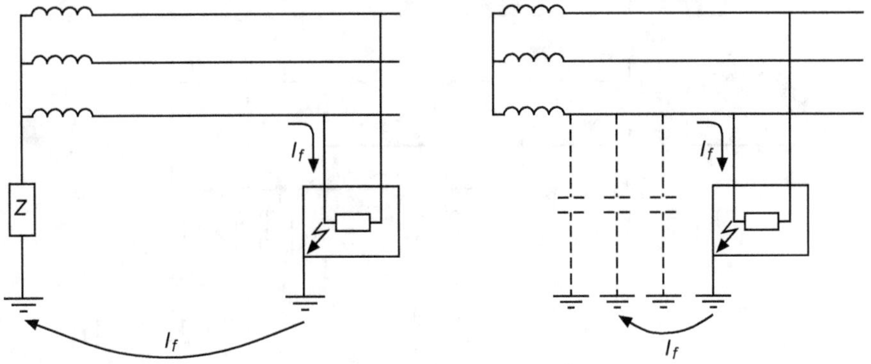

IT systems with impedance earth (impedance Z limits earth fault currents).

IT system unearthed (first fault current flows as a result of conductor capacitance).

Figure 4.18 IT systems

The requirement to be fulfilled where a neutral is not distributed (three-phase three-wire distribution) (Regulation 411.6.4) is:

$$Z_s \le \frac{C_{\min} U}{2I_a}$$

or, where the neutral conductor or midpoint conductor, respectively, is distributed (three-phase four-wire distribution), is:

$$Z_s^1 \le \frac{C_{\min} U_0}{2I_a}$$

where:

Z_s is the impedance in ohms of the earth fault loop comprising the line conductor and the protective conductor of the circuit;

Z'_s is the impedance in ohms of the earth fault loop comprising the neutral conductor and the protective conductor of the circuit; and

I_a is the current in amperes causing operation of the protective device within the time specified in Table 41.1 when applicable (see Regulation 411.3.2.2), or within 5 s for all other circuits when this time is allowed (see Regulation 411.3.2.3) as for TN systems.

These equations may be derived as described below.

4.4.9.1 IT system, where a neutral is not distributed (consider Figure 4.19)

The first fault will not result in the operation of any overcurrent device, as the supply is either not earthed or has an impedance earth. There is no shock risk, as no (or very little) current can flow to earth.

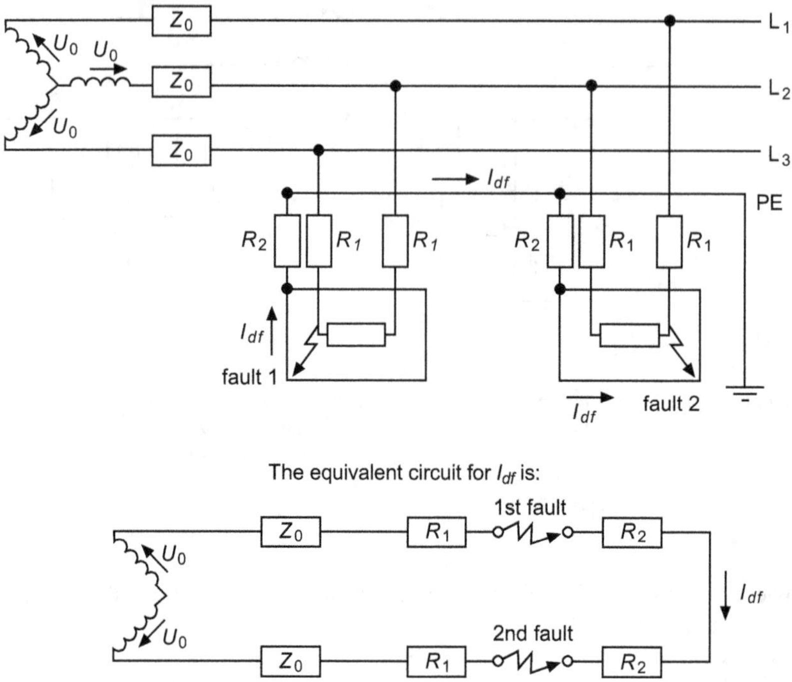

Figure 4.19 IT system, neutral not distributed

The first fault considered is line to earth (there can be no line-to-neutral or neutral-to-earth fault, as there is no neutral conductor); and the second fault is another line to earth.

A double fault current I_{df} will flow (see Figure 4.19).

To simplify the equivalent circuit, the loops comprising the load resistance have been omitted, as they have no influence on I_{df}. Line and protective conductor impedances are assumed to be the same in each installation.

$$I_{df} = \frac{C_{\min}U}{2Z_0 + 2R_1 + 2R_2}$$

If I_{df} is to result in disconnection within a specified time, it must be equal to or greater than I_a:

$$I_{df} \geq I_a$$

then:

$$\frac{C_{\min}U}{2Z_0 + 2R_1 + 2R_2} \geq I_a$$

and if $Z_s = Z_0 + R_1 + R_2$

then:

$$\frac{C_{\min}U}{2s} \geq I_a$$

or, transposing the terms:

$$Z_s \leq \frac{C_{\min} U}{2I_a}$$

where $Z_s = Z_0 + R_1 + R_2$, the earth fault loop impedance as usually defined. Z_0 is small compared with $R_1 + R_2$; hence Z_s may be defined as the impedance of an earth fault loop comprising only the line conductor and protective conductor, as in Regulation 411.6.4.

4.4.9.2 IT system, where a neutral is distributed (consider Figure 4.20)

The first fault assumed is neutral to earth, as this is difficult to detect. The second fault assumed is line to earth.

The equivalent circuit is:

$$I_{df} = \frac{C_{\min} \times \text{line to neutral } U_0}{Z_0 + Z_n + R_1 + R_n + 2R_2}$$

As R_n is never less than R_1, although it may be greater than R_1, and as $(Z_0 + Z_n)$ is small compared with the circuit resistance $(R_1 + R_2)$,

$$I_{df} \text{ approximates to } \frac{C_{\min} U_0}{2(R_n + R_2)}$$

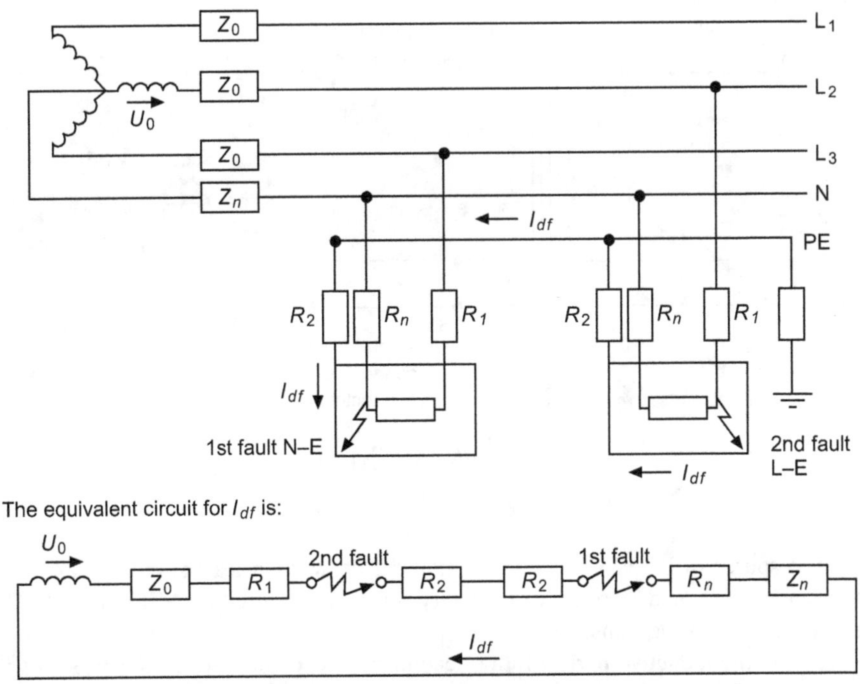

Figure 4.20 IT system with a distributed neutral

If the overcurrent device is to operate in a specified time, $I_{df} \geq I_a$

$$\text{hence: } \frac{C_{\min} U_0}{2(R_n + R_2)} \geq I_a$$

$$\text{now: } Z_s^1 = R_n + R_2$$

$$\text{hence: } Z_s^1 \leq \frac{C_{\min} U_0}{2I_a}$$

where Z'_s is the impedance of the earth fault loop comprising the neutral conductor and earth conductor of the circuit; see Regulation 411.6.4.

4.4.10 Functional extra-low voltage
411.7

Functional extra-low voltage (FELV) is a protective measure employing basic protection with fault protection by extra-low voltage and automatic disconnection of supply (see Figure 4.21).

Basic protection is provided for the secondary circuit by insulation and barriers to meet the requirements for the primary voltage, and protection against faults is provided either by the extra-low voltage of the secondary or, for a fault between primary and secondary, automatic disconnection on the primary side.

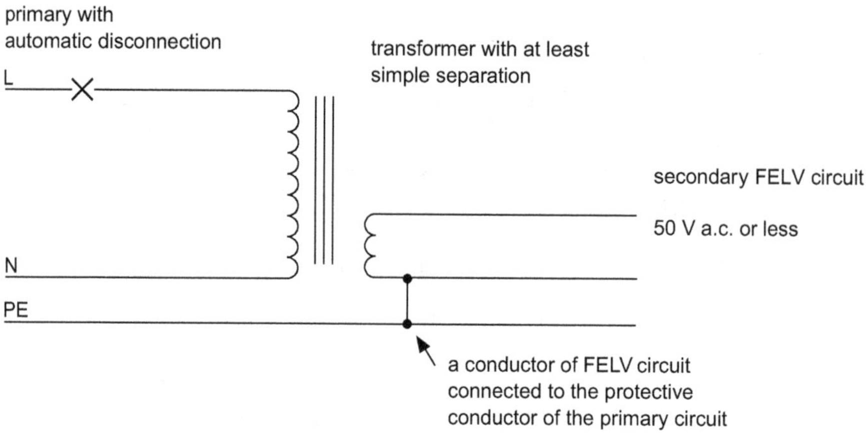

Figure 4.21 Functional extra-low voltage (FELV)

The source of the FELV is either a transformer with at least simple separation between primary and secondary or a safety source as required for SELV and PELV, e.g. a safety isolating transformer.

Exposed-conductive-parts of the secondary are connected to the protective conductor of the primary and a conductor of the FELV system may be connected to the protective conductor (see Figure 4.21).

4.4.10.1 Reduced low voltage systems

Regulation group 411.8

Reduced low voltage systems are more commonly known as 110 V centre-tap earthed systems, and they are widely used for supplying mobile equipment and hand-held tools in work situations where there is a high risk of damage to equipment, such as on building sites and in engineering workplaces. The widespread use of the system in the UK is due in no small part to its excellent safety record. The high level of safety arises from the voltage to earth being 55 V (63.5 V for three-phase supplies) and the line-to-line voltage being 110 V. See Figure 4.22. It also has the most significant advantage of being supplemented by automatic disconnection of supply in the event of a fault. A protective conductor is run back to earth and to a centre point of the transformer secondary, resulting in automatic disconnection of the supply at the first fault.

The system has its origins in the 1949 Annual Report of HM Chief Inspector of Factories, which recommended the system for use on building and construction sites and other applications involving large-scale use of portable electric tools. The system was described in British Standard Code of Practice, CP 1017:1969, *Distribution of electricity on construction and building sites* (now BS 7375:2010). Equipment was specified in BS 4363:1969 *Specification for distribution assemblies for electricity supplies for construction and building sites* (now BS EN 60439-4): 2004 *Particular requirements for assemblies for construction sites (ACS).*

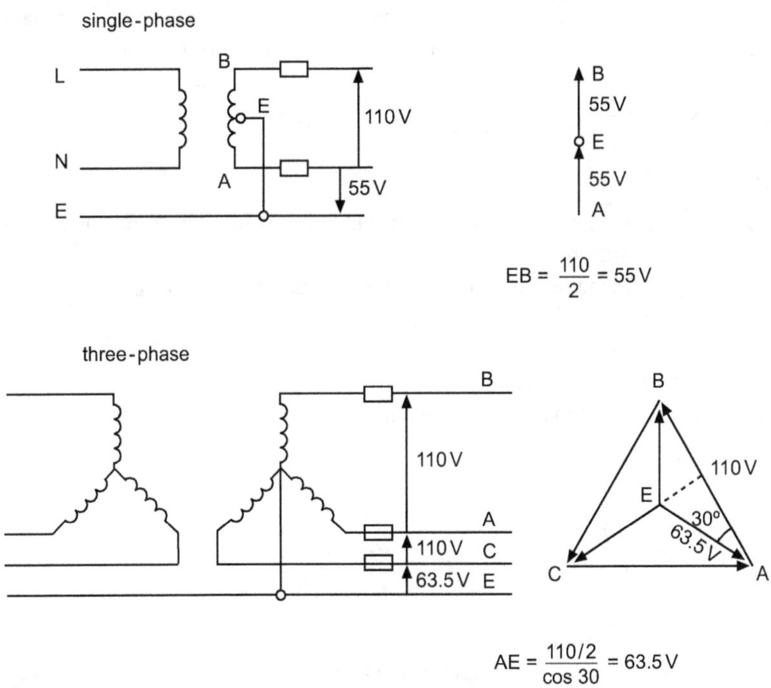

Figure 4.22 Reduced low voltage supplies

The United Kingdom's entry in 1973 into what was then known as the Common Market has resulted in the acceptance by the Health and Safety Executive that the continental practice on building sites of using 30 mA RCDs to provide additional protection to users of portable tools has also to be permitted. BS 7671 has been amended to reflect this. However, there are believed to be no recorded electrical fatalities associated with the reduced low voltage system; it has been widely adopted; it is quite practical; and it should continue to be used.

The system is designed to provide 110 V power supplies to tools and lamps. Where the supply is from a single-phase transformer, this results in a voltage of 55 V to earth, and where the supply is from a three-phase transformer, this results in a voltage to earth of 63.5 V (see Figure 4.22 and the accompanying vector diagrams).

The system is specified in Regulation 411.8, as it is an application of the protective measure automatic disconnection of supply, but having the additional advantage of the reduced voltage levels described above.

Basic protection is provided by insulation or barriers. Fault protection is provided by automatic disconnection (an overcurrent device in each line conductor with a requirement for disconnection in 5 seconds or by RCD).

Table 41.6 of BS 7671 provides maximum earth fault loop impedances to achieve a 5 second disconnection at the reduced voltages. These are the loop impedance values given in Table 41.3 for circuit-breakers and Table 41.4 for fuses multiplied by 0.239, i.e. 55/230 for single-phase, and by 0.276, i.e. 63.5/230 for three-phase installations.

If it is wished to ascertain the loop impedance by calculation, the impedance at the end of a circuit fed from the secondary of a step-down transformer is given by:

$$Z_{\text{sec}} = Z_P \times \left(\frac{V_s}{V_p}\right)^2 + \frac{Z\,\%\,\text{tran}}{100}\,\frac{(V_s)^2}{VA} + (R_1 + R_2)_s$$

where:

Z_p = loop impedance of the primary circuit including that of the source of supply Z_e
$Z\%$ tran = percentage impedance of the step-down transformer
V_A = rating of the step-down transformer
V_s = secondary voltage
V_p = primary voltage
$(R_1 + R_2)_s$ = the secondary circuit line and protective conductor resistances.

This may be simplified to:

$$Z_{\text{sec}} = (Z_e + 2R_1)_p \times \left(\frac{V_s}{V_p}\right)^2 + (R_1 + R_2)_s$$

where:

Z_e = external loop impedance
R_1 = primary circuit line conductor resistance.

4.5 Protective measure: double or reinforced insulation
Section 412

4.5.1 General

BS 7671 requires that where this protective measure is the sole protective measure in the installation or circuit, the installation shall be under effective supervision so that double or reinforced insulation equipment is not replaced by equipment requiring another protective measure, for example, earthing and automatic disconnection.

Basic protection is provided by basic insulation and fault protection by another layer of insulation. The two layers may be combined in one layer of reinforced insulation.

This measure requires that all equipment shall be either:

(i) Class II equipment; or
(ii) declared in the relevant product standard to be equivalent to Class II.

Class II equipment is best introduced by first considering Class I. Class I equipment has two lines of defence:

(i) basic insulation, which may be solid or a maintained air gap; and
(ii) a metal case connected to the main earth of the installation by a protective conductor.

See Figure 4.23.

The general concept of Class II construction is again that there are two protective measures:

(i) basic insulation; and
(ii) supplementary insulation.

See Figure 4.24.

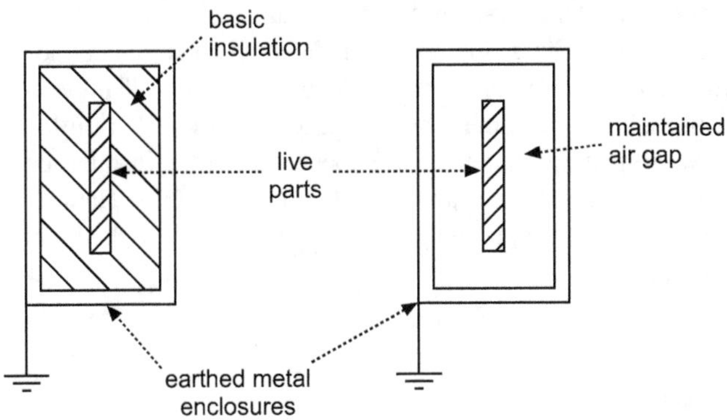

Figure 4.23 Class I equipment construction

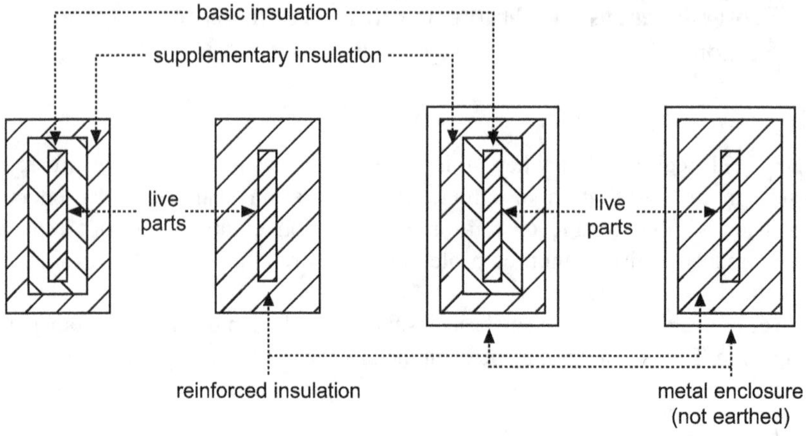

Figure 4.24 Some Class II equipment constructions

The basic and supplementary insulation may be combined as one in an enhanced layer of insulation called reinforced insulation (again, see Figure 4.24).

There is no dependence on a connection with earth for fault protection.

As a general concept, Class II insulation would be all-insulated. However, the construction does allow such equipment to have metal enclosures, but there would be two layers of insulation or equivalent between the metal and the live parts and the metal enclosure would have no provision for earthing it, as no attempt should be made to do so.

When an installation is to consist entirely of Class II equipment, a protective conductor is not required, but such an installation is required to be under effective supervision in normal use, so that no changes can be made, such as installing Class I equipment, which would impair the safety of the installation. Supervision in this sense does not mean 24-hour-a-day attendance, but supervision that would require the authority of a competent person before any changes were made to the installation. An installation that uses Class II as its sole means of protection cannot have socket-outlets (Regulation 412.1.3), as there would be no means of preventing a person attempting to use Class I equipment by plugging it into a socket-outlet. The use of a socket-outlet would prevent the effective supervision of the installation.

It is expected that the requirements will be met by the installation of equipment that is type-tested and marked to the relevant standard and marked with the Class II equipment symbol:

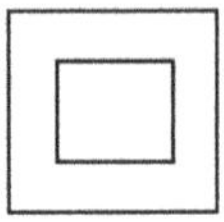

Regulation 412.2.1.2 recognises the addition of supplementary insulation to equipment having basic insulation only, applied during erection, while Regulation

412.2.1.3 permits reinforced insulation to be applied in the erection of electrical equipment but only where constructional features prevent the application of double insulation.

4.5.2 Class II equipment in installations using the protective measure automatic disconnection
Regulation 412.2.3.2

Where a circuit other than one complying with all the requirements for the protective measure double or reinforced insulation, including effective supervision, supplies Class I or Class II equipment, a protective conductor shall be run to and terminated at each point of the wiring, and at each accessory. This is to allow Class II equipment to be replaced with Class I equipment at a later date.

4.5.3 Class II appliances

Class II household appliances constructed to BS EN 60335 will not have a protective conductor connection. However, other types of equipment may require a functional will require an earth if they are to function. The basic concept of Class II equipment is that it does not require a connection with earth for its safety.

4.6 Protective measure: electrical separation to one item of equipment
Section 413

Section 413 gives the requirements for the protective measure of electrical separation for the supply to one item of equipment (see Figures 4.25 and 4.26).

Electrical separation is a protective measure in which:

- basic protection is provided by basic insulation or barriers; and
- fault protection is provided by simple separation of the separated circuit from the primary circuit, other circuits and Earth.

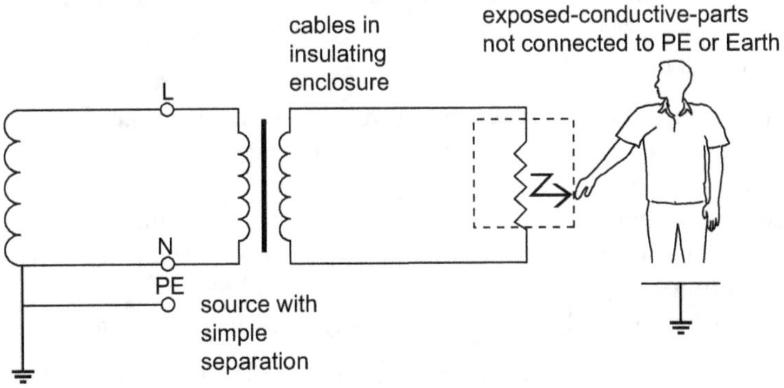

Figure 4.25 Protective measure: electrical separation for one item of equipment– source: a transformer

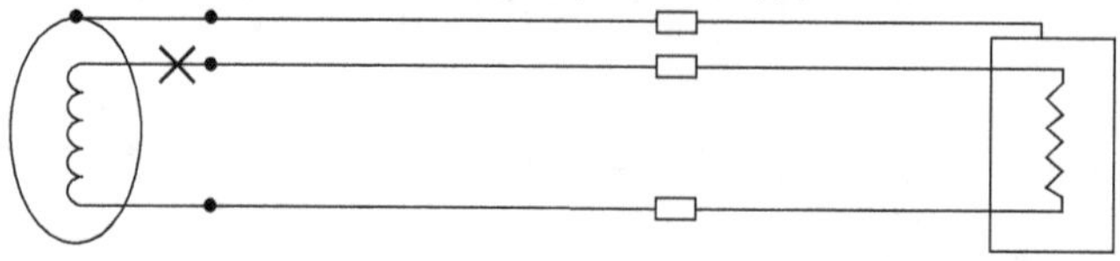

Figure 4.26 Protective measure: electrical separation for one item of equipment–source: a generator

Where there is more than one item of equipment the requirements of Regulation 418.3 apply, including supervision by skilled or instructed persons and a prominent warning notice (see section 4.10.3).

In the event of failure of the basic insulation, protection is provided by the simple separation from the primary circuit, so fault current cannot flow to Earth and a shock cannot be received. However, the fault is not detected and in the event of a second fault a shock risk arises. This is unlikely with one item of equipment.

As no connection to Earth is provided for equipment, equipment requiring a functional earth cannot be used with this protective measure.

The protective measure electrical separation described in the 17th Edition differs from that in the 16th, as a safety isolating transformer is not required (but may be used). A transformer with simple separation (basic insulation between windings) is required.

4.7 Protective measure: extra-low voltage (SELV or PELV)
Section 414

4.7.1 General

The protective measure extra-low voltage may use SELV (separated extra-low voltage) or PELV (protective extra-low voltage) systems, see Figures 4.27 and 4.28. The measure requires:

(i) limitation of voltage (not to exceed 50 volts a.c. or 120 volts d.c.; lower voltages may be required in particular locations or situations);
(ii) a safety source (e.g. a safety isolating transformer);
(iii) protective separation of the SELV or PELV system circuits (e.g. 300/500 V cables with mechanical protection of a non-metallic sheath or non-metallic conduit/trunking); and
(iv) for SELV systems only, basic insulation between the SELV system and Earth.

SELV systems are allowed in locations of particular shock risk, such as bathrooms and swimming pools, because they operate at extra-low voltage, are separated from the primary circuits by safety isolating transformers, and there is no connection with the protective conductor of the primary circuits (so that touch voltages arising from the faults elsewhere in the installation cannot be imported into the location).

SELV and PELV are also notable for allowing access to live parts in certain locations, if especially low voltage limits are met (see section 4.7.2).

4.7.2 SELV and PELV access to live conductors (basic protection) Regulation 414.4.5

If the nominal voltage of the system exceeds 25 V a.c. rms or 60 V ripple-free d.c., or if the equipment is immersed, insulation or barriers shall be provided.

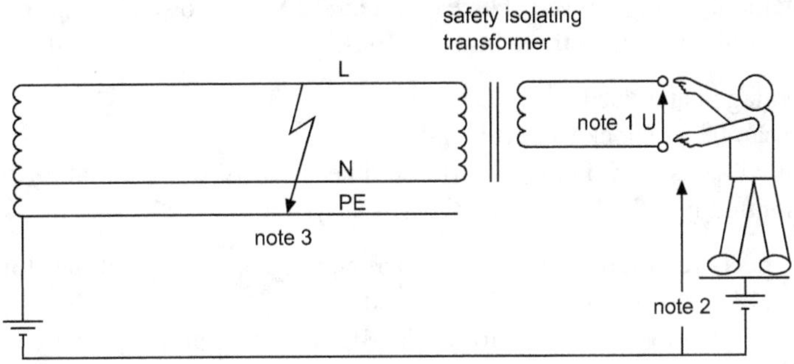

Notes:

1. Secondary voltage if not exceeding 25 V a.c. or 60 V d.c. is too low to be hazardous in normal dry situations.
2. A shock to earth cannot occur for a single fault as a circuit to earth cannot be made, since the secondary has no connection with Earth.
3. Touch voltages in the event of a fault are not transmitted from the primary, as there is no protective conductor.

Figure 4.27 Protective measure SELV

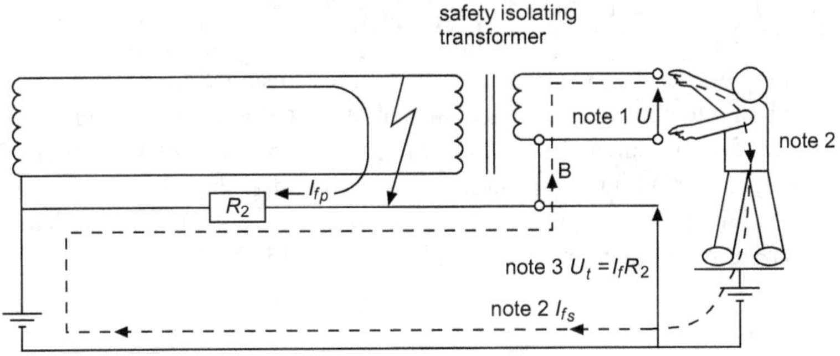

Notes:

1. Secondary voltage if not exceeding 25 V a.c. or 60 V d.c. is too low to be hazardous normal in dry situations.
2. There can be shocks to earth as the connection B provides an earth loop.
3. Fault voltages to earth U_t on the primary are introduced into the secondary circuits via the protective conductor ($U_t = I_f R_2$).

Figure 4.28 Protective measure PELV

If the nominal voltage of the system does not exceed 25 V a.c. rms or 60 V ripple-free d.c. in normal dry situations, the conductors may be accessible to the touch.

In other situations, if the nominal voltage of the system does not exceed 12 V a.c. rms or 30 V ripple-free d.c., again the conductors may be accessible to the touch.

4.7.3 SELV

SELV systems are often allowed in locations of increased shock risk, where other systems are not allowed. For example, SELV is allowed in bathrooms and swimming pools. There are three basic reasons:

(a) the voltage is reduced;
(b) there can be no shock to Earth; and
(c) touch voltages from faults to earth on other parts of the installation cannot be transmitted into the location by the SELV system.

PELV meets requirement (a), but not (b) and (c), so it is not allowed in certain locations of high risk where SELV is allowed.

There is no protective conductor in the SELV system and exposed-conductive-parts of the system are not allowed to be connected to:

- earth;
- an exposed-conductive-part of another circuit; or
- a protective conductor of another circuit.

As a result, faults to earth on the primary system cannot result in shock risks in the area supplied by the SELV system.

4.8 Additional protection in accordance with Section 415

Section 415, Regulation 411.3.3

Section 415 describes two protective provisions (installation of residual current devices and supplementary equipotential bonding) that are specified elsewhere as part of protective measures or specified for special locations in Part 7 of BS 7671. For example, for the additional protective provision, installation of residual current devices is required by Regulation 411.3.3 (see section 4.4.5) and supplementary bonding is required in Section 702 (Regulation 702.415.2).

4.8.1 RCDs

Section 415 details the provisions. For example, it specifies that RCDs used for additional protection should have a rated operating current ($I_{\Delta n}$) not exceeding 30 mA and an operating time not exceeding 40 ms at a residual current of $5I_{\Delta n}$.

RCDs are recognized as providing protection against contact with live parts in the event of failure of basic insulation. This was called *direct contact* in the 16th Edition. Where the prospective touch voltage is 230 V the disconnection time (see Figure 4.7) must be of the order of 200 ms for probable avoidance of harmful physiological effects such as ventricular fibrillation. Unfortunately, the body impedance is such that, even in wet conditions, insufficient current will flow either line to neutral or line to earth to result in the operation of an overcurrent device. In a normal dry situation, body impedance will be of the order of 1500 Ω and as a consequence about 150 mA will flow at 230 V. If the body is wet, its impedance will be about 300 Ω and the current about 750 mA. This will not operate even a 5 A fuse, as it is much lower than any load currents, and so the shock will be maintained until the person removes him/herself, if able, from the live part. It can be seen that a sensitive residual current circuit-breaker, such as a 30 mA RCD to BS EN 61008 or BS EN 61009, could provide protection against contact with live parts in that it will operate for currents exceeding 30 mA and will have an operating time not exceeding 40 ms at a residual current of 150 mA. However, it will not provide any protection against line-to-neutral shocks, since shock current for line-to-neutral contact is indistinguishable from load current. It is to be noted that BS 7671 allows the use of RCDs only as an additional measure of protection against direct contact with live parts. The sure operation of an electromechanical device cannot be guaranteed even when life depends on it.

Regulation 415.1.2 prohibits the use of RCDs as the sole means of protection and their use does not obviate the need for one of the protective measures (basic protection plus fault protection).

4.8.2 Supplementary equipotential bonding

Supplementary bonding is required by Regulation 411.3.2.6 when automatic disconnection times cannot be achieved by the use of overcurrent devices and the installation of an RCD is not suitable.

The detailed requirements for the supplementary bonding are found in this section.

The requirement of Regulation 415.2.2 for a.c. systems is that the resistance of the bonding conductor R between simultaneously accessible exposed-conductive-parts and extraneous-conductive-parts must comply with:

$$R \leq \frac{50}{I_a}$$

where I_a is the current in amperes (A) causing operation of the protective device – for RCDs, $I_{\Delta n}$.

for overcurrent devices, the current causing automatic operation in 5 s.

The touch voltage U_t , if there is no supplementary bonding, is $I_f R_2$.

If the supplementary bond R is applied, the touch voltage will be reduced. In the worst case of R_2 being very high and the resistance R of the supplementary bonding being the maximum allowed of $50/I_a$ and the disconnection time of the overcurrent

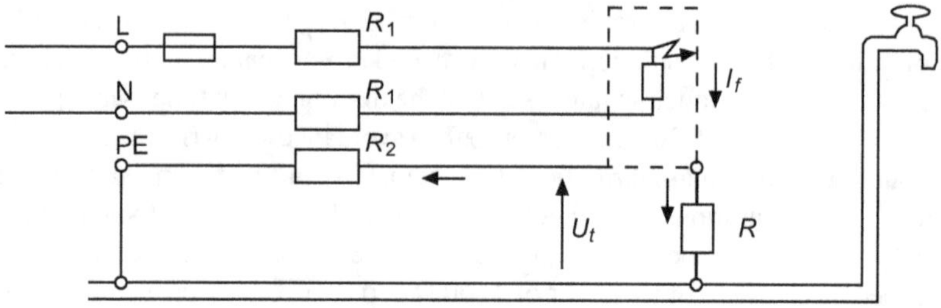

Figure 4.29 Effect of supplementary bonding

device being the maximum of 5 s, then the touch voltage will be limited to 50 V at that disconnection time and current. The touch voltage will be higher for higher fault currents and the disconnection time shorter, and general compliance with the touch voltage curve will be obtained.

4.9 Protective measures controlled or supervised by skilled persons
Regulation 410.3.5, Section 417

The protective measures of obstacles and placing out of reach are to be applied only in installations where access is limited to skilled persons or instructed persons under the supervision of skilled persons. The measures provide basic protection only (Regulations 417.1 and 410.3.5 refer).

4.9.1 Protective measure: obstacles
Regulation group 417.2

Obstacles do not necessarily provide secure protection against contact with live parts, as can barriers or enclosures, as they protect only against unintentional approach to live parts and may be able to be circumvented. When using protection by obstacles, the designer has to bear in mind the requirements of the Electricity at Work Regulations and be confident that it is reasonable to use this method of protection in the particular location.

4.9.2 Protective measure: placing out of reach
Regulation group 417.3

Bare or insulated overhead lines must be installed as required by the Electricity Safety, Quality and Continuity Regulations 2002 (as amended) Schedule 2 (see Table 4.7). Where overhead lines are used, the minimum height is required to be 3.5 m in positions inaccessible to vehicular traffic and 5.8 m in positions accessible to vehicles. At road crossings, the height is required to be increased to 5.8 m.

Protection by placing out of reach (unless minimum heights of live parts meet the requirements for overhead lines) is allowed only in locations where access is restricted to skilled persons or instructed persons under the supervision of skilled persons. The installations are also required to be controlled or supervised by skilled persons. In such locations, a bare live part shall not be within arm's reach. Arm's reach is defined in Part 2 and Fig. 417 of BS 7671. The requirement for live parts to be 'out of reach', as defined, is increased in places where bulky or long conducting objects are normally handled (Regulation 417.3.3).

It is quite clear that placing out of reach as a means of protection against direct contact with live parts, particularly taken in the context of the Electricity at Work Regulations 1989, should be applied only when it is reasonable that such a measure of protection could be used. The designer must ask whether, should an accident occur, he or she will be able to demonstrate that it was a reasonable protective measure to have taken, even allowing for the fact that only skilled or instructed persons would be allowed in to the areas where it is used.

Table 4.7 Maximum lengths of span and minimum heights above ground for overhead wiring between buildings etc.

		Minimum height of span above ground (m)		
Type of system	Maximum length of span (m)	At road crossings	In positions accessible to vehicular traffic, other than crossings	In positions inaccessible to vehicular traffic
1	2	3	4	5
Cables sheathed with thermoplastic (PVC) or having an oil-resisting and flame-retardant or HOFR sheath, without intermediate support	3	5.8	5.8	3.5
Cables sheathed with thermoplastic (PVC) or having an oil-resisting and flame-retardant or HOFR sheath, in heavy gauge steel conduit of diameter not less than 20 mm and not jointed in its span.	3	5.8	5.8	3
Thermoplastic (PVC) covered overhead lines on insulators without intermediate support.	30	5.8	5.8	3.5
Bare overhead lines on insulators without intermediate support.	30	5.8	5.8	5.2

4.10 Protective measures under control or supervision of skilled or instructed persons
Regulation 410.3.6, Section 418

The protective measures:

- non-conducting location;
- earth-free local equipotential bonding; and
- electrical separation for the supply to more than one item of current-using equipment

are to be applied only where the installation is under the supervision of skilled or instructed persons so that unauthorized changes cannot be made.

4.10.1 Protection by non-conducting location
Regulation group 418.1

Regulation 418.1 states that protection by non-conducting location is not recognized for general application, and Regulation 410.3.6 requires that it shall be applied only in special situations, which are under effective supervision.

Non-conducting locations are to provide for safety when parts may be at different potentials through failure of basic insulation (note to Regulation 418.1); this might include the use of Class 0 equipment.

Class 0 equipment has only basic insulation (Figure 4.30) and there is no secondary line of protection, as there is for Class I, where metal cases are earthed, and Class II, where there is a second layer of insulation (or equivalent).

Protection is afforded by ensuring that, within the location that the equipment is installed, there are no exposed- conductive-parts or extraneous-conductive-parts simultaneously within reach of a person. See Figure 4.31. Walls and floors have to be insulated so that, in the event of the breakdown of basic insulation, there is no shock risk. However, this form of protection suffers from a number of serious disadvantages. While, in the event of failure of basic insulation, persons will not suffer a shock on the first fault, there is no clearance of the fault by the operation of an overcurrent device. Persons touching exposed-conductive-parts that are live as a result of a fault will suffer no shock, as the fault current will be very low, and consequently, of course, overcurrent and even earth leakage current protective devices will not operate. The location must be under effective supervision, as the introduction at a later date of Class I equipment or metal pipes (extraneous-conductive-parts) introduces a shock risk. Protection, in the event of a second fault, is provided by requiring all exposed-conductive-parts to be out of reach of other exposed-conductive-parts.

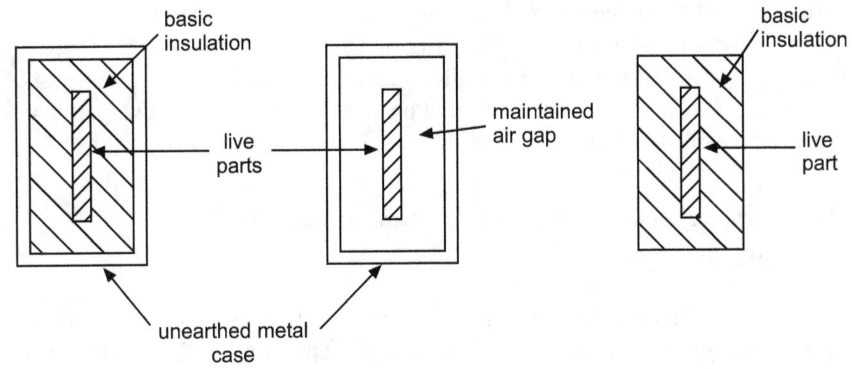

Figure 4.30 Class 0 equipment

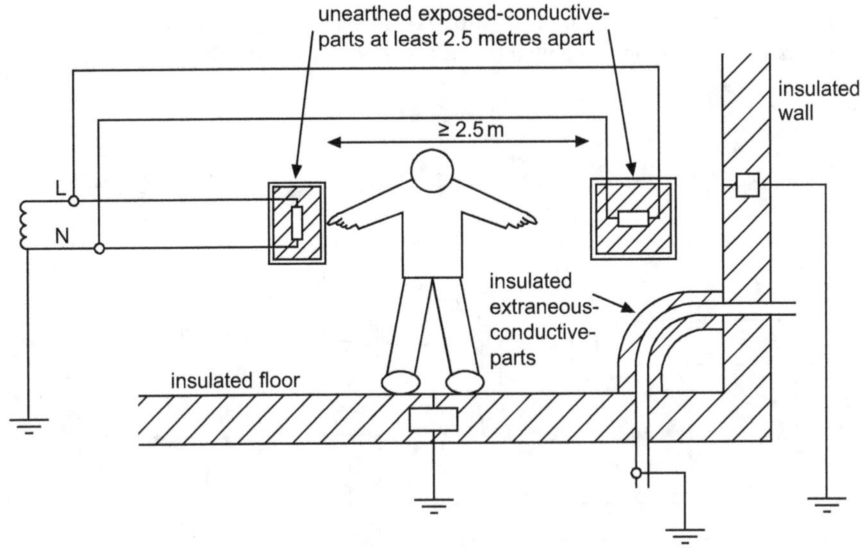

Figure 4.31 Protection by a non-conducting location

Regulation 418.1.4 specifies how the general requirements can be met by ensuring that the distance from one exposed-conductive-part to another exposed-conductive-part and from an exposed-conductive-part to an extraneous-conductive-part is not less than 2.5 m. For parts out of the zone of arm's reach, the separation shall be 1.25 m. The interposition of effective obstacles between exposed-conductive-parts and extraneous-conductive-parts is allowed provided that they extend the distance to be surmounted to the values stated above. These obstacles shall not be connected to earth or to exposed-conductive-parts, and they shall as far as possible be of insulating material. Any extraneous-conductive-parts in the location shall be insulated (Figure 4.31). This insulation would allow compliance with the requirements of Regulation 418.1.7 to prevent any potential on an extraneous-conductive-part being transmitted outside the location.

4.10.2 Protection by earth-free local equipotential bonding Regulation group 418.2

The very first sentence in this regulation requires that the method be used only in special circumstances. The measure prevents the appearance of dangerous touch voltages by local protective bonding. The local protective bonding is not connected with Earth, either through exposed-conductive-parts or through extraneous-conductive-parts (Regulation 418.2.3). The effect of earth-free local equipotential bonding is to make a Faraday cage. In the event of a fault to exposed-conductive-parts (there can be no extraneous-conductive-parts), the whole cage becomes live and there is no means of obtaining a potential difference, nor hence a shock within the cage.

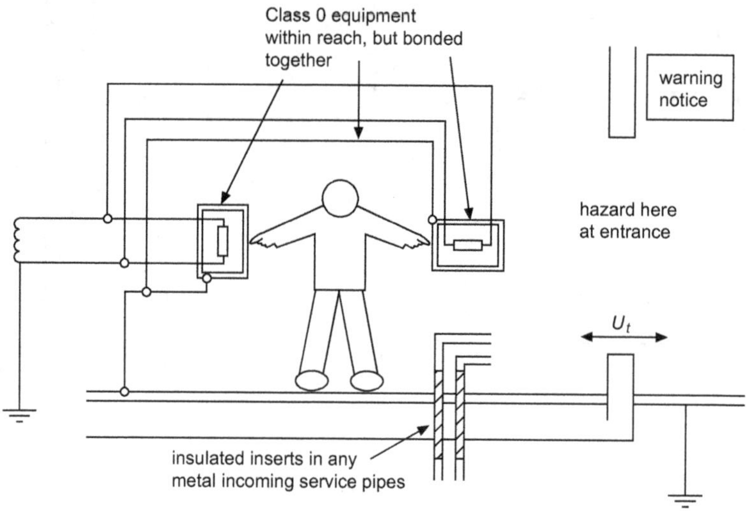

Figure 4.32 Protection by earth-free local equipotential bonding

In a similar way to a non-conducting location, this method of protection has inherent dangers. In the event of a fault making exposed-conductive-parts live, there is no mechanism for disconnecting the supply, as there is in earthed protective bonding and automatic disconnection of supply; likewise with a second fault. Regulation 418.2.4 points out a particular hazard in such a situation, which is the point at which an earth-free local protective bonding installation meets with a standard installation using, say, protective equipotential bonding and automatic disconnection of supply. In the event of a fault, the earth-free local equipotential bonding will be at a potential relative to Earth, and there is no mechanism within the system to disconnect the fault.

Because of the potential hazard when entering an earth-free zone, where this measure is applied, a warning notice complying with Regulation 514.13.2 shall be fixed in a prominent position adjacent to every point of access to the location.

4.10.3 Protective measure: electrical separation to more than one item of equipment
Regulation group 418.3

The protective measure electrical separation when applied to more than one item of equipment requires compliance with Section 413 for electrical separation for one item of equipment plus requirements of Regulation 418.3 including (Figure 4.33):

(i) supervision of the installation by skilled or instructed persons;
(ii) warning notices in a prominent position adjacent to every point of access to the location;

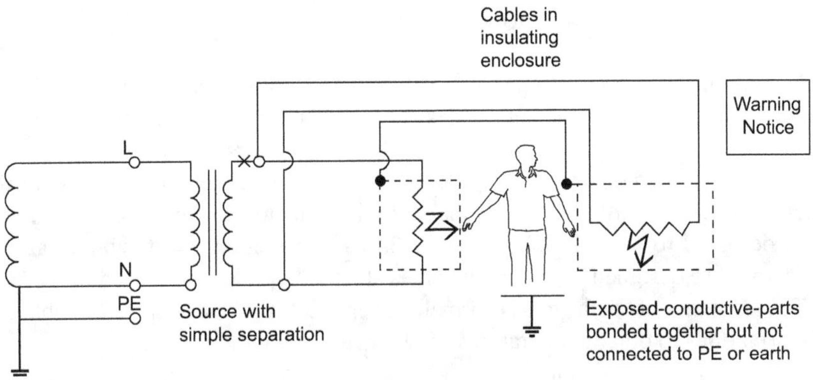

Figure 4.33 Electrical separation via transformer to more than one item of equipment

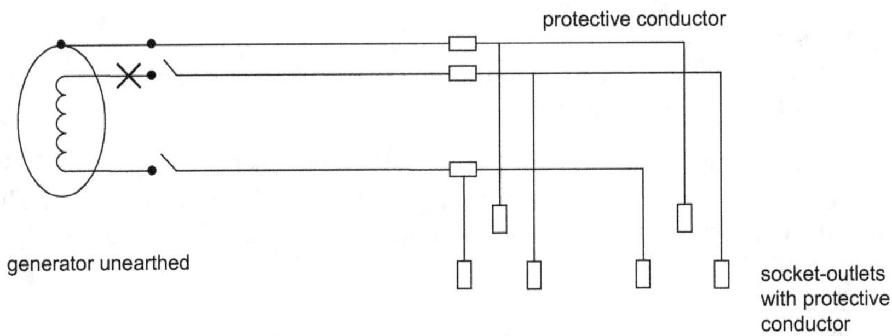

Figure 4.34 Electrical separation to more than one item of equipment supplied by generator

(iii) the connecting together of exposed-conductive-parts of the items of equipment; and

(iv) automatic disconnection per Regulation 411.3.2 in the event of two faults from conductors of different polarity.

These exposed-conductive-parts are not to be connected to a protective conductor, nor to exposed-conductive-parts of other circuits nor to any extraneous-conductive-parts.

The particular risk with this protective measure is that first faults are not detected and multiple faults can result in shock risks.

This method of protection is commonly encountered with portable generators (see Figure 4.34). These are usually used to provide supplies to individual items of equipment. There is no means of earthing any point of the generator, and protection is by electrical separation. It is recommended that protective conductors be installed for these systems from the unearthed metal case of the generator to exposed-conductive-parts of equipment. The generator output needs overcurrent protection in order to meet the requirements of Regulation 418.3.7.

4.11 Conductor resistance

Notes to Tables 41.2 to 41.4 and 41.6

Note 2 to these tables requires the designer to take account of the effect of conductor temperature. The note states that the circuit loop impedance given in the table should not be exceeded when the line conductors are at their appropriate maximum permitted operating temperature, as given in Table 52.1. To allow for the selection of thermosetting cables (capable of a conductor operating temperature of 90°C) designed to operate at 70 °C, note 3 has been added. Most switchgear and accessories are designed for use with cables at a temperature not exceeding 70 °C.

The resistance of conductors is normally given at 20 °C. For example, see Table 4.8. Corrections to other temperatures can be made using the formula

$$R_t = \{1 + \alpha(t - 20)\}R_{20}$$

where:

R_t = resistance at temperature t.

R_{20} = resistance at 20°C.

α = temperature coefficient of resistance (values of α for common materials at 20°C are given in Table 4.9).

When circuits are protected against overload by semi-enclosed fuses to BS 3036, a derating factor of 0.725 is required to be applied to the tabulated cable ratings. This reduces the conductor operating temperature to 51°C instead of 70°C (see Chapter 16 for the calculation).

4.12 Conductor resistance and testing

When carrying out loop impedance measurements as in Regulation 612.9, the loop impedances given in Tables 41.2, 41.3 etc. cannot be used directly. As noted below the tables, they are the loop impedances at normal operating temperature, and corrections to ambient temperature need to be made.

Correction factors for correcting from conductor normal operating temperature to 20°C ambient are given in Table 4.10.

Consider a circuit with a loop impedance Z_{41} at maximum conductor operating temperature. The maximum loop impedance Z_t allowed if testing 70 °C PVC cables at, say, 5 °C ambient is:

$$Z_t \leq (Z_{41} - Z_e) \times \frac{0.94}{1.20} + Z_e$$

where:

Z_{41} = loop impedance given by Tables 41.1, 41.2 etc.
Z_e = external (supply) loop impedance.
Z_t = test impedance.
1.20 is factor from Table 4.10.

It is presumed that under full load conditions the supply loop impedance will be unchanged.

If Z_e is not known it can be assumed to be zero, and $Z_t \leq Z_s \times 0.94/1.20$.

Table 4.8 Resistance of copper and aluminium conductors at 20 °C

Cross-sectional area (mm²)		Resistance/m or ($R_1 + R_2$)/m (mΩ/m)	
Line conductor	Protective conductor	Copper	Aluminium
1	–	18.1	
1	**1**	**36.2**	
1.5	–	12.1	
1.5	1	30.2	
1.5	**1.5**	**24.2**	
2.5	–	7.41	
2.5	1	25.51	
2.5	**1.5**	**19.51**	
2.5	2.5	14.82	
4	–	4.61	
4	**1.5**	**16.71**	
4	2.5	12.02	
4	4	9.22	
6	–	3.08	

Continues

Table 4.8 Continued

Cross-sectional area (mm^2)		Resistance/m or $(R_1 + R_2)$/m (mΩ/m)	
Line conductor	Protective conductor	Copper	Aluminium
6	**2.5**	**10.49**	
6	4	7.69	
6	6	6.16	
10	–	1.83	
10	**4**	**6.44**	
10	6	4.91	
10	10	3.66	
16	–	1.15	1.91
16	**6**	**4.23**	–
16	10	2.98	–
16	16	2.30	3.82
25	–	0.727	1.20
25	10	2.557	–
25	16	1.877	–
25	25	1.454	2.40
35	–	0.524	0.87
35	16	1.674	2.78
35	25	1.251	2.07
35	35	1.048	1.74
50	–	0.387	0.64
50	25	1.114	1.84
50	35	0.911	1.51
50	50	0.774	1.28

Table 4.9 Coefficients of resistance for conductors

Material	Coefficient of resistance α at 20 °C
Annealed copper	0.00393*
Hard-drawn copper	0.00381
Aluminium	0.00403*
Lead	0.00400
Steel	0.0045

* An average value of 0.004 is often used for both copper and aluminium.

Table 4.10 Corrections to conductor resistances for various cable types for copper and aluminium conductors

Insulation type	Conductor maximum operating temperature (°C) (Note 1)	Correction factor from 20 °C to conductor operating temperature (Notes 2 and 3)
60 °C thermoplastic or thermosetting	60	1.16
85 °C rubber	85	1.26
70 °C thermoplastic (PVC)	70	1.20
90 °C thermosetting	90	1.28
Mineral thermoplastic covered or bare to touch	70*	1.20
Mineral bare but not exposed to touch	105*	1.34

Notes:

1. From Table 52.1 or the headings to current rating tables in Appendix 4 of BS 7671.
2. Correction factor = {(col 2 − 20) × 0.004 + 1}.
3. The temperature coefficient of resistance used, 0.004 is an average figure for copper and aluminium from BS 6360.
4. See also Table C17.

* Sheath temperature.

Table 4.11 Ambient temperature multipliers

Ambient temperature (°C)	Correction to 20 °C
5	0.94
10	0.96
15	0.98
20	1.00
25	1.02
30	1.04

Chapter 5
Protection against thermal effects

Ref HD 384.4.42S1 A2 1994, 384.4.482 S1 1997
(Adviser: Colin Reed I Eng MIIE)

5.1 Introduction
Chapter 42

Chapter 42 is concerned with protecting people, livestock and property against the harmful effects of heat and thermal radiation developed by electrical equipment, generally in the normal course of its use. It includes:

(a) requirements for protection against fire caused by electrical equipment (421.1) that are applicable as a minimum to all locations; and
(b) requirements for particular precautions to be taken where there are particular risks associated with fire:
  (i) for escape routes in high-rise buildings and locations open to the public (422.2);
  (ii) in locations where flammable materials are processed or stored (422.3);
  (iii) in locations constructed of combustible materials (422.4);
  (iv) in locations whose shape or dimensions propagate fire (422.5);
  (v) in locations with assets of particular value (422.6).

Chapter 42 does not deal with the thermal effects of overcurrent. This is found in Chapter 43 of BS 7671 and is considered in Chapter 6 of this Commentary.

Section 20 of the London Building Acts (Amendment) Act 1939 imposes particular requirements for fire alarms, transformers, substations and equipment in car parks in buildings in the London area. Other local authorities may have similar requirements.

5.2 Fires in occupied buildings

Tables 5.1 and 5.2 (see table on p. 91) are taken from the Fire Statistics, United Kingdom, published by the Department for Communities and Local Government. The data indicate the importance of fire prevention and provide pointers to situations requiring particular care.

Fires associated with smokers' materials and cooking are notable. The fires associated with electric cookers are probably fat fires. The oil in frying pans can be set alight by the radiant ring or plate of an electric cooker if the oil is overheated. Cooker hobs should never be left unattended when in use, particularly when frying.

Table 5.1 Sources of ignition for accidental fires in dwellings[1], with casualties for the year 2013/14

	Fires	Fatal casualties		Non-fatal casualties	
	Total	Total	Per 1,000 fires	Total	Per 1,000 fires
Total accidental	35,024	219	6	6872	196
Smokers' materials	2360	80	34	673	285
Cigarette lighters	217	6	23	93	429
Matches	232	6	26	57	246
Cooking appliances	18,092	30	2	3642	201
Space heating appliances	1256	20	16	313	249
Central and water heating appliances	394	0	–	35	89
Blowlamps, welding and cutting equipment	237	0	–	22	93
Electrical distribution	4032	18	4	419	104
Other electrical appliances	4277	13	3	682	159
Candles	951	11	12	385	405
Other	2207	15	7	3764	165
Unspecified	769	21	27	187	243

[1] *Includes caravans, houseboats, mobile homes and other non-permanent structures used solely as a permanent dwelling (see explanatory notes).*

Source: 'Fire Statistics Great Britain April 2013 to March 2014', Department for Communities and Local Government

The designer may well be able to influence fire statistics that at first sight appear to be outside his control. Sufficient and well-located sockets may reduce fires associated with flexes and electric heaters. The number of fires associated with lighting, particularly in retail premises, might perhaps be reduced by improved design. General compliance with the Regulations and selection of correctly rated equipment is of obvious importance.

Fires associated with the fixed installation are recorded. They are not as high as they used to be: 'The greatest element of safety is therefore the employment of skilled and experienced electricians to supervise the work' – 1st Edition of the Wiring Regulations, as published in June 1882.

The benefit of fitting smoke detectors and alarms in dwellings is shown by Table 5.3 (see tables on p. 92).

5.3 Protection against fire caused by electrical equipment
Section 421, Regulation 559.5.1.204

5.3.1 Introduction

The regulations in Chapter 42 are concerned with fixed equipment. Control of the harmful effects of fixed equipment is the responsibility of the designer.

Table 5.2 Fatal casualties in fires by location, Great Britain, 1981/82 - 2013/14p

Year	Total	Dwellings	Other Residential	Non-Residential	Other buildings	Total outdoor	Other buildings	Other
1981/82	937	745	–	–	79	113	66	47
1982/83	892	708	–	–	81	103	52	51
1983/84	851	679	–	–	75	97	52	45
1984/85	921	716	–	–	75	130	68	62
1985/86	967	684	–	–	119	164	63	101
1986/87	859	681	–	–	62	116	72	44
1987/88	834	638	–	–	92	104	76	28
1988/89	855	664	–	–	67	124	79	45
1989/90	841	622	–	–	73	146	109	37
1990/91	853	642	–	–	54	157	105	52
1991/92	767	594	–	–	57	116	72	44
1992/93	693	523	–	–	58	112	73	39
1993/94	686	518	–	–	55	113	92	21
1994/95	604	467	–	–	36	101	68	33
1995/96	712	548	–	–	36	128	74	54
1996/97	700	548	–	–	38	114	76	38
1997/98	696	550	–	–	28	118	77	41
1998/99	603	464	–	–	30	109	63	46
1999/00	621	446	–	–	41	134	86	48
2000/01	554	426	–	–	34	94	72	22
2001/02	583	461	–	–	42	80	61	19
2002/03	522	398	–	–	21	103	64	39
2003/04	576	416	–	–	51	109	67	42
2004/05	483	371	–	–	27	85	58	27
2005/06	469	349	–	–	37	83	59	24
2006/07	430	311	–	–	30	89	61	28
2007/08	458	353	–	–	35	70	41	29
2008/09	404	324	–	–	17	63	38	25
2009/10	416	319	–	–	26	71	49	22
2010/11	404	318	12	9	21	65	44	21
2011/12	397	298	4	25	29	70	37	33
2012/13r	342	257	7	11	18	67	39	28
2013/14p	322	258	5	12	17	47	33	14

Notes:

'–' Not available

r = revised

p = provisional

Definitions and Explanatory Notes for these tables may be found in the pdf document, pages numbered 52 to 55

https://www.gov.uk/government/organisations/department-for-communities-and-local-government/series/fire-statistics-great-britain

The source of the data is DCLG Incident Recording System

https://www.gov.uk/government/publications/incident-recording-system-for-fire-and-rescue-authorities

Table 5.3 Number of fires and fire casualties by presence of fire alarms in dwellings[1] by presence and operation of smoke alarms for the United Kingdom

Year	Present, operated & raised the alarm	Present, operated, but did not raise the alarm	Present, but did not operate	Absent	Unspecified	Total
Fires						
2002	17660	2705	7535	37126	7	65033
2005	19995	3144	7183	27421	10	57753
2007/8	18906	3220	6832	21444	5	47466
2008/9	17715	3558	10176	15998	19	47466
2010/11	16382	4433	7780	16407		45002
2011/12	16226	4551	7881	14936		43594
2012/13	16199	4136	7716	12803		41034
Fatal casualties						
2002	60	45	84	240	1	430
2005	63	41	69	203	–	376
2007/8	67	43	102	159		371
2008/9	70	55	103	112		340
2010/11	83	43	76	16		318
2011/12	73	52	62	111		298
2012/13	68	61	47	81		257
Non-fatal casualties						
2007/8	4177	799	1882	3989	–	10847
2008/9	3994	903	903	2845	6	9822
2009/10	–	–	–	–	–	10316
2010/11	3787	1417	1147	2599		9094
2011/12	3901	3901	1198	2468		9000
2012/13	3785	3785	1062	2078		7758

Notes:

1. Includes caravans, houseboats and other non-building structures used solely as a permanent dwelling.
2. Data not available due to incomplete records from one Fire and Rescue Authority in 2009/10.

While no reference is made in Section 421 to portable equipment such as table lamps or portable electric heaters, the installation designer must give some consideration to these matters. If space or water heating is not being provided by fixed equipment, either electrical or other fuel, the designer must allow for electric heaters being plugged into the sockets in the current rating of circuits and the type of accessories supplied. Examples that spring to mind are dwellinghouses without heating, where it can be assumed that radiant heating may be used, and consequently sufficient circuits of sufficient capacity must be provided. Similarly, the facility for heating hot-water cylinders in an airing cupboard may be appropriate. It is generally recognized that a 13 A plug and socket-outlet is not suitable for supplying the 3 kW load of an immersion heater, in the high ambient of an airing cupboard; and a double-pole switched flex outlet should be provided for an immersion heater.

Regulation 421.1 requires any relevant installation instructions of the equipment manufacturer to be taken into account. The designer/installer has a duty to read and comply with these instructions when installing fixed equipment. Regulation 421.1.4 concerns the fixing of equipment causing a focusing or concentration of heat, which must be at a sufficient distance from walls, ceiling, furnishings etc. to prevent dangerous temperatures. The British Standards for equipment, e.g. BS EN 60335-1 for household and similar appliances and BS EN 60950 for information-technology equipment, have temperature rise limitations in specified test conditions. If equipment is installed in more onerous situations, e.g. by being enclosed or installed in a high ambient temperature, overheating may arise for which the protective devices within the equipment have not been designed.

Where fixed electrical equipment is installed, having in normal operation a sufficient surface temperature to cause a risk of fire etc., it may be installed provided that the precautions of Regulation 421.1.2 are complied with. The 15th Edition of the IEE Wiring Regulations provided minimum clearances for heat sources exceeding 90 °C, from other materials. These are repeated as Figure 5.1.

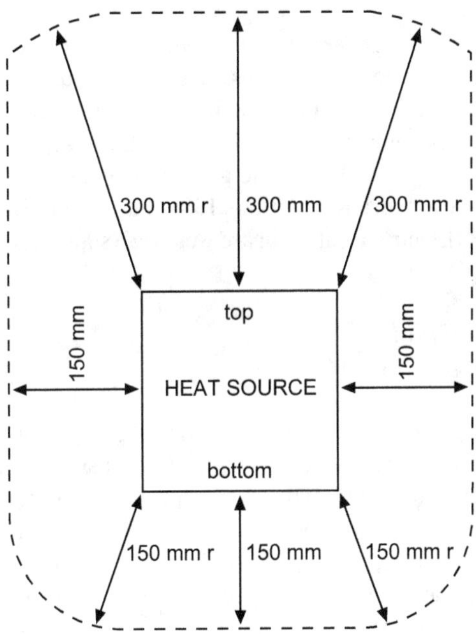

Figure 5.1 Minimum clearances for heat sources

Attention must be given to recessed or semi-recessed luminaires and luminaire controlgear mounted in ceiling voids, to ensure that heat is properly dissipated and that thermal insulation materials or any other material that may be installed cannot restrict the cooling of the equipment. This means that recessed luminaires protruding into loft and similar spaces should be suitably protected against being covered. Regulation 559.5.1.204 requires bayonet lampholders B15 and B22 to comply with BS EN 61184, *Bayonet lampholders,* and to have the temperature rating T2 of that standard (T2 marking, suitable for temperatures on the lamp cap up to and including 210 °C). Where lampholders are other than B15 or B22, their rated operating temperature must be suitable for the particular application.

The heat output of high-wattage luminaires such as those with tungsten halogen lamps must be considered. Heat will be focused by the reflector, so the luminaire must meet the focusing requirements of Regulation 421.1.4. Protection is required against conducted, convected and radiated heat (see Regulation 421.1.2 regarding the risk of fire and harmful effects to adjacent materials, and Regulation 423.1 with respect to protection against burns).

Table 55.3 of BS 7671 gives explanations of symbols used in luminaires, including those symbols indicating suitability or otherwise for mounting on flammable or combustible surfaces.

5.3.2 Arcing or the emission of high-temperature particles Regulation 421.1.3

Regulation 421.1.3 has general requirements for protection against arcing or the emission of high-temperature particles by electrical equipment. Generally, this is guarded against by the installation of equipment constructed to the appropriate British Standard. Permanently installed electric welding sets might be considered to need to comply with BS 7671. Electric arc welding is covered by Regulations 7 and 8 of the Electricity at Work Regulations 1989. Further guidance is given in HSE publication HSG118 'Electrical safety in arc welding' (supersedes PM 64 'Electrical safety in arc welding').

5.3.3 Flammable liquids Regulation 421.1.5

Regulation 421.1.5 requires that, where electrical equipment contains flammable liquid in significant quantity (which is generally considered to be in excess of 25 litres), adequate precautions shall be taken to prevent the spread of burning liquid, flame and the products of combustion. Clearly, a designer will need to take advice if he is faced with this situation, and all the requirements that may be necessary cannot be described here. Fire authorities have specific requirements for such equipment as oil-filled transformers and fuel tanks.

5.3.4 Switchgear assemblies Regulation 421.1.201

The IET were asked by the London Fire Brigade to consider precautions to reduce the incidence of fires originating in consumer units. The first response was to consider introducing a 960°C glow-wire flammability test as defined in BS EN 60695-2-11. However, on reflection of the evidence in a BRE report (commissioned by the Department for Communities and Local Government) and a report prepared by Electrium, it was considered that this was insufficient and the following wording was agreed:

Regulation 421.1.201 Within domestic (household) premises, consumer units and similar switchgear assemblies shall comply with BS EN 61439-3 and shall:

(i) *have their enclosure manufactured from non-combustible material, or*
(ii) *be enclosed in a cabinet or enclosure constructed of non-combustible material and complying with Regulation 132.12.*

NOTE 1: Ferrous metal e.g. steel is deemed to be an example of a non-combustible material.

Whilst this basically requires metalclad consumer units, it leaves the door open for new materials with similar fire-resistant properties cs metal.

5.3.5 Terminations
Regulations 421.7, 526.3, 526.5

Regulation 526.5 requires every termination of live conductors or joints between them to be contained within an enclosure, in accordance with Regulation 526.5. This requirement essentially means such terminations and joints should be made within accessories or equipment complying with an appropriate British Standard, or in enclosures that could be partially formed or completed with building material which is non-combustible when tested to BS 476-4. This requirement to enclose terminations includes ELV terminations, as, although they may not represent a shock risk, such terminations need to be enclosed to reduce the risk of fire.

Poor and loose terminations are the cause of many fires of electrical origin. Regulation 526.3 requires every connection and joint to be accessible for inspection and testing and maintenance, with the exception of joints that are considered to be of a permanent nature – a joint designed to be buried, a compound-filled joint, connection between cold tails and heating elements, and joints made by welding, soldering, brazing or a compression tool. BS 7671 also recognizes that connections can be made within a maintenance free accessory, provided certain conditions are complied with; see 9.11.1. For similar reasons, all joints are required to be enclosed to limit the fire risk should there be overheating.

5.4 Precautions where particular risks of fire exist
Section 422

5.4.1 General

In locations where a particular risk of fire exists there are general requirements:

(i) Electrical equipment shall be restricted to that necessary for the location;

(ii) Equipment shall be selected and erected so that normal temperature rise shall not cause a fire;
(iii) Temperature cut-outs shall have manual reset.

For particular situations, Section 422 specifies particular performance requirements for cables in the locations (see Table 5.4 for a summary).

Table 5.4 Cable fire performance requirements of Section 422

Performance requirement	Regulation	Location	Suitable cables (see note 1)
BS EN 50266 (i)	422.2.1	Escape routes	BS 6724 (iv) BS 7846 (v) BS 7835 (vi)
	422.3.4	Flammable stored/ processed materials and where cables bunched, or long vertical runs	
	527	Particular risks	
BS EN 61034-2 (ii) (was BS EN 50268-2, was BS 7622-2)	422.2.1	Escape routes	BS 6724 (iv) BS 7846 (v) BS 7835 (vi)
	711.521	Exhibitions and shows where no fire alarm system	
BS EN 60332-1-2 (iii)	422.3.4	Flammable stored/ processed materials	Most cables meet this requirement other than polyethylene sheathed cables and some rubber cables
	422.4	Combustible constructional materials	

Note: Only British Standards that require the cables to meet the stated performance have been listed. Manufacturers may have tested cables to meet a higher performance than required by a BS and therefore may be suitable for use.

(i) BS EN 60332-3-24 *Tests on electric and optical fibre cables under fire conditions.*
(ii) BS EN 61034-2:2005 *Measurement of smoke density of cables burning under defined conditions. Test procedure and requirements* – replaced BS EN 50268-2.
(iii) BS EN 60332-1-2:2004 *Tests on electric and optical fibre cables under fire conditions. Test for vertical flame propagation for a single insulated wire or cable. Procedure for 1 kW premixed flame.*
(iv) BS 6724 *Thermosetting insulated, armoured cables for voltages of 600/1000 V and 1900/3300 V, having low emission of smoke and corrosive gases when affected by fire.*
(v) BS 7846 *Thermosetting insulated, armoured, fire-resistant cables of rated voltage 600/1000 V, having low emission of smoke and corrosive gases when affected by fire.*
(vi) BS 7835 *Armoured cables with thermosetting insulation for rated voltages from 3.8/6.6 kV to 19/33 kV having low emission of smoke and corrosive gases when affected by fire.*

BS EN 60332-3 series contains fire-propagation tests where a number of cables are bunched together in a vertical formation on a ladder rack (similar to the old CEGB power station ladder test), and is published in six parts as shown below.

Standard	Cable volume category	Non-metallic cable volume (litres/metre length)
BS EN 60332-3-10	Apparatus	–
BS EN 60332-3-21	A F/R (front/rear)	7
BS EN 60332-3-22	A	7
BS EN 60332-3-23	B	3.5
BS EN 60332-3-24	C	1.5
BS EN 60332-3-25	D small cables	0.5

The cable standards BS 6724, BS 7846 and BS 7835 require that cable manufactured to these standards meet category C to BS EN 60332-3-24. As is the case for all cable standards, the manufacturer of a product can test to more categories than are required by its British Standard and then declare compliance to these additional categories. For example, BS 7629 does not require compliance with any category of test to BS EN 60332-3 series; however, most manufacturers produce and test their cables to meet category C.

The historic specially produced reduced flame propagation armoured PVC (thermoplastic) cables are no longer commonly manufactured. They have been replaced with armoured low smoke cables (BS 6724, BS 7846 and BS 7835) which must meet reduced the propagation requirement of BS EN 60332-3-24.

BS EN 61034-2 was formerly BS EN 50268-2 and BS EN 7622-2. The test is a 3-metre cube smoke test, which all low smoke cables have to meet.

5.4.2 Requirements for escape routes in high-rise buildings and locations open to the public etc.
Regulation group 422.2

To provide for safety of persons evacuating a building in an emergency, wiring systems installed in escape routes categorized as BD2, BD3 or BD4 must be non-flame propagating. For categories see Table 5.5.

Cables are required to comply with BS EN 60332-3 series and BS EN 61034-2, see Table 5.4.

5.4.2.1 Conduits and trunking
Regulation 422.2.1

Where conduit and trunking systems are used in locations where particular risks of fire exist, they must be to the appropriate British Standard and be classified as fire-resistant by these standards. For conduits the standard referred to is BS EN 61386-1. For trunking the standard referred to is BS EN 50085. In amendment No.1 reference to BS EN 61537 for cable tray and ladder rack and BS EN 61534 for Powertrack systems was added.

Table 5.5 Categories BD2, BD3 and BD4

Code	Class	Characteristics	Applications and examples
BD2	Difficult	Low density occupation, difficult evacuation	High-rise buildings
BD3	Crowded	High density occupation, easy evacuation	Locations open to the public (theatres, cinemas, department stores, etc.)
BD4	Difficult and crowded	High density occupation, difficult evacuation	High-rise buildings open to the public (hotels, hospitals, etc.)

5.5 Locations with risks of fire due to the nature of processed or stored materials
Regulation group 422.3

Same of the requirements of Regulation group 422.3 are described here.

Cables are required as a minimum to comply with the test requirements of BS EN 60332-1-2 *Tests on electric and optical fibre cables under fire conditions.*

This is a single cable, vertical flame propagation test (bunsen burner). Most cables meet this requirement other than polyethylene sheathed cables and some rubber cables.

The requirements of Regulation group 422.3 may be summarized as follows:

(a) Allowance must be made for the accumulation of material, dust, chippings, etc., in the selection and rating of equipment;
(b) Enclosures must be at least IP4X or IP5X if in the presence of dust; or IP6X in the presence of electrically conductive dust;
(c) Cables, if not embedded, must be to BS EN 60332-1-2;
(d) Where the risk of flame propagation is high, e.g. for long vertical runs, cables shall meet the requirements of BS 60332-3 series (see section 5.4);
(e) Except for mineral insulated cable systems and busbar trunking or powertrack systems, wiring is to be protected by an RCD having a rated residual current not exceeding 300 mA;
(f) An RCD having a rated residual operating current not exceeding 30 mA is are required where a resistive fault may cause a fire; the example given is for overhead heating with film elements;
(g) All isolators are required to switch the neutral;
(h) Minimum distances are specified for spotlights, etc., from combustible materials. This clearly requires some anticipation of circumstances by the designer, particularly with respect to a build-up of materials and likely installations following fitting out. Persons carrying out annual inspections need to be diligent in these respects.

5.6 Locations with combustible constructional materials
Regulation group 422.4

The requirements apply for construction code CA2, buildings mainly of combustible materials, for example wooden buildings.

The basic requirements here are as follows:

(a) Cables and cords are required to comply with BS EN 60332-1-2. This is the vertical flame propagation test, and most cables meet this requirement other than polyethylene sheathed cables and some rubber cables;
(b) Where conduit and trunking systems are used they must be to the British Standard and be classified as fire-resistant by these standards;
(c) Electrical equipment installed into or on a combustible wall shall be totally enclosed and not make use of the wall materials for part of the enclosure. Allowance has to be made for temperature rise within the equipment arising from enclosure by the building materials;
(d) Luminaires are to be fixed at an adequate distance from combustible materials. It is noted that luminaires to BS EN 60598-1 suitable for flammable surfaces are marked:

(e) In prefabricated hollow walls containing a pre-installed wiring system including accessories, all boxes and enclosures shall have a degree of protection of at least IP3X where the wall is liable to be drilled during erection.

5.7 Wiring systems in escape routes
521.11

Because of concerns raised directly to the IET by the London Fire Brigade and the Electricity Safety First a new regulation was agreed to be introduced into Amendment No. 3. Firefighters had died in circumstances where part of the problem was that cables fixed to the ceilings of escape routes had become detached and the firemen had become entangled. Cables have commonly been fixed to the ceilings of corridors particularly following rewires. As the notes to the regulation explain, the requirement precludes the sole use on non-metallic cable clips, cable ties, and conduit or trunking systems. The designer must select a system that will prevent cables and equipment obstructing escape routes even when the fire is well established. Metal conduit and trunking are likely to be preferred in such locations.

5.8 Fire propagating structures
Regulation 422.5.1

In buildings where the shape and dimensions facilitate the spread of fire, precautions are required to ensure that the electrical installation, e.g. via ducts, trunking, penetrations of the structure, cannot propagate fire.

5.9 Locations of significance
Regulation 422.6

The designer is required 'to consider' the use of cables with fire-resisting characteristics in buildings of national importance, with assets, facilities, items, equipment etc. of significant value. This perhaps is a requirement to bring to the client's attention the recommendation for fire-resisting cables. Mineral-insulated cables are given as an example.

5.10 Protection against burns
Section 423

Table 42.1 of BS 7671 lists maximum temperatures for accessible parts of equipment. The temperatures tabulated are maximum values and it must be remembered that contact with any surface temperatures at or above 70 °C may cause dangerous reflex actions. Regulation 423.1 requires equipment with temperatures exceeding those given in Table 42.1 to be guarded.

BS EN ISO 13732-1:2006 *Ergonomics of the thermal environment. Methods for the assessment of human responses to contact with surfaces. Hot surfaces* provides temperature thresholds for burns when human skin is in contact with hot surfaces and methods for the assessment of the risks of burning, and is intended to be used for determining acceptable maximum temperatures for controls and working surfaces of heated domestic equipment.

Surface temperatures that lead to burns (burn thresholds) depend on the surface material and the duration of contact (see Figure 5.2).

A person's ability to cope with high surface temperatures is dependent on age. Very young, aged or infirm persons need more protection against surface temperatures than other persons. It may be necessary to guard heaters in nursery areas, and some appliances, such as heaters and cookers, have surface temperatures considered necessary for proper functioning by the appropriate British Standard, but which may cause burns. The designer will have to take these considerations into account when he is responsible for the selection and installation of equipment with heated surfaces. Similarly, the aged or infirm will not be able to respond quickly to hot surfaces, and the very young, very old or infirm may not be able to escape from a hot surface should they fall against it. Surface temperatures that are perfectly safe for someone who is capable of free movement, could be hazardous to the elderly or infirm.

NHS Health Information Sheet: Managing the risks from hot water and surface temperatures in health and social care.

This explains the risks associated with hot water and hot surfaces in health and social care premises and sets out guidance to help control them. It covers the risks vulnerable people may be exposed to when bathing or showering, or where there are hot surfaces such as pipes or radiators.

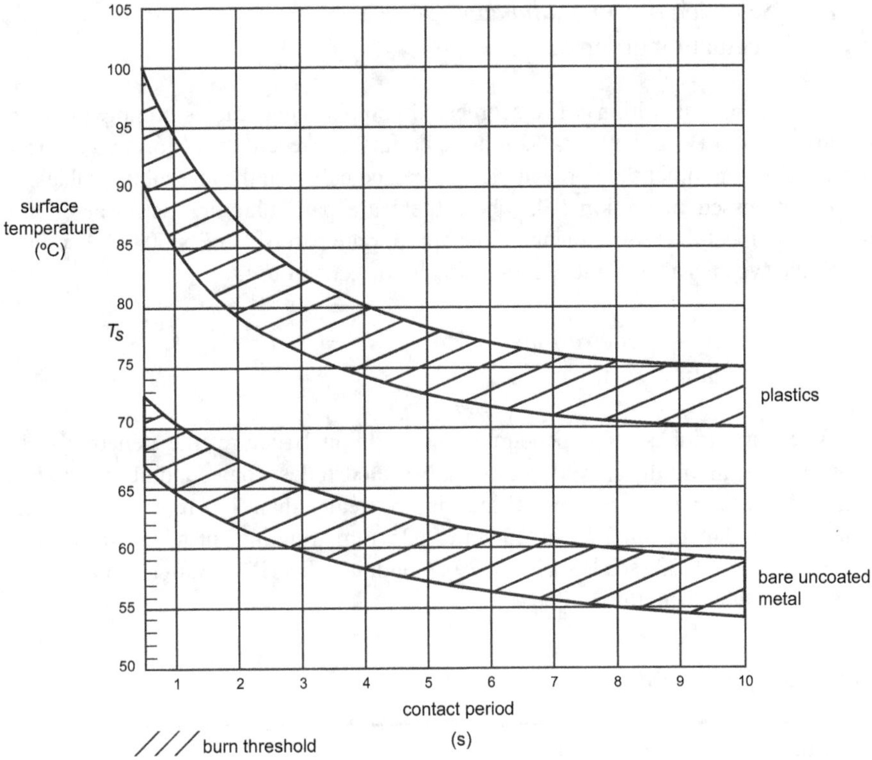

Figure 5.2 Burn threshold spread when the skin is in contact with a hot smooth surface [Source: ISO 13732-1]

5.11 Minimization of the spread of fire
Section 527

5.11.1 Introduction

There is a general requirement to minimize the spread of fire by the selection of appropriate materials and by appropriate erection procedures, including sealing of penetrations. When selecting a cable with respect to fire hazard, there are two important factors to be considered:

(a) the contribution the cables themselves may make to the spread of the fire, evolution of smoke and fume emission; and
(b) the need for the integrity of any safety circuit to be maintained under fire conditions (see Regulation 560.8.1).

The electrical installation is required neither to reduce the building structural performance (strength) nor to reduce the fire withstand of the building.

5.11.2 Selection of cables and cable systems Regulation group 527.1

For general use within a fire-segregated compartment, cables complying with requirements of BS EN 60332-1-2 require no further precautions. Most cables meet this requirement other than polyethylene sheathed cables and some rubber cables.

As discussed in section 5.4, where there are particular fire risks cables are required to meet the requirements of the appropriate part of BS EN 60332-3 series.

Cable systems as in Table 5.6 are suitable for general use.

5.11.3 Sealing of penetrations Regulation group 527.2

Where a wiring system penetrates the building structure, the penetration is required to be internally as well as externally sealed, unless for a conduit or trunking system the internal cross-sectional area is no greater than 710 mm^2, i.e. 32 mm diameter conduit or smaller, or 25 mm × 25 mm trunking, provided it and the terminations are IP33 (see BS EN 60529). Conduit to BS EN 61386-1 and trunking to BS EN 50085 meet this IP code.

Table 5.6 Cable systems

Standard	Description
BS EN 61386	Conduit systems for cable management
BS EN 50086	Specification for conduit systems for cable management
BS EN 60439-2	Low-voltage switchgear and controlgear assemblies, particular requirements for busbar trunking systems (bus ways)
BS EN 61543	Powertrack systems
BS EN 61537	Cable tray systems and cable ladder systems for cable management
BS EN 50085	Cable trunking and ducting systems

Chapter 6
Protection against overcurrent

(Adviser: Tony Haggis I.Eng MIET, Tony Huggis Consulting Ltd)
Chapter 43, Sections 533 and 536

6.1 General

Chapter 13, Regulations 131.4 and 131.5

It is a fundamental requirement of the Regulations (Regulations 131.4, 131.5) that persons and livestock shall be protected against injury, and property against damage, as a result of overcurrents, and that the installation shall be protected against damage arising from faults. Protective equipment must interrupt the supply without danger and facilitate ready restoration.

Overcurrents may be:

(i) overload currents; or
(ii) fault currents, including short-circuit.

6.1.1 Overload currents

It is important to note that overloads do not arise as a result of a fault. Many final circuits are not liable to overload, for example where the electrical load is fixed or where there is a load-controlling device. A final circuit supplying a shower of 7 kW rating cannot overload to 8 kW. A circuit supplying a motor fitted with overload protection is presumed not to overload. In both these circumstances, the designer may omit the overload protection to the final circuit, but not fault protection. In the example of a motor, the overcurrent device at the origin of the circuit will be providing protection only against fault current, as overload protection is provided by the motor starter. An overcurrent device rated to provide overload protection would be tripped or fused by the motor starting currents.

6.1.2 Fault currents

Fault currents arise as a result of a fault in either wiring or equipment. Fault currents are likely to be ten, a hundred or even a thousand times the rated current. Overload currents are likely to be of the order of one and a half to twice the rated current, although devices can be installed to provide closer overload protection.

Section 432 makes it clear in Regulation 432.1 that a device may provide protection against both overload current and fault current and this is the normal situation. There are other occasions when a device will provide protection against overload current only (Regulation 432.2), e.g. a motor overload device. There are also occasions, relatively common, where a device will provide for protection against fault current only (see Regulation 432.3), e.g. fuses backing up circuit-breaker boards.

6.1.3 Ageing of cables

Estimating the life of a cable can be only approximate because of the obvious difficulties in accumulating data. There is a general understanding that PVC cables with a continuous conductor operating temperature of 70°C have a life of twenty years. There is also a rough guide that, for each 8°C increase in core conductor continuous operating temperature above 70°C, the life of the cable will be halved. A PVC cable running with an overload such that its core conductor temperature is 78 °C will last for ten years.

IEC 943, now withdrawn, gave an equation for ageing of cables, which is reproduced here because of the insight it gives to the effects of overtemperature on cables, but see the advice of thc British Cables Association given below. The general equation given for ageing is:

$$\log_e t = \frac{A}{T} + A_1$$

where:

t = time in hours
T = absolute temperature K (273 + °C)
A = a constant 15,028 for PVC, 14,500 for EPR and PRC
A_1 = a constant −31.6 for PVC, −27.19 for EPR and PRC.

Table 6.1 provides further guidance.

Life termination is assumed to be on the appearance of cracks on samples of cables wound on their own diameter.

Cable loadings are rarely constant; estimates can be made of the combined effects of different loadings by the use of the formula below:

$$\frac{1}{L} = \frac{1}{24}\left\{\frac{a}{L_1} + \frac{b}{L_2} + \frac{c}{L_3}\right\}$$

where:

L_1, L_2 and L_3 = lives at specific temperature
a, b, c, etc. = hours in day at these temperatures.

Table 6.1 Life until deterioration against conductor core temperature (from withdrawn IEC 943:1989)

	Life until deterioration[1]			
	PVC cable[5]		EPR and PRC cable[5]	
Temperature (°C)[2]	Permanent rating[3]	Normal rating[4]	Permanent rating[3]	Normal rating[4]
70				
75			69 yrs	
80		69 yrs	39 yrs	69 yrs
85	23 yrs	37 yrs	23 yrs	40 yrs
90	12 yrs	20 yrs	13 yrs	24 yrs
95	7 yrs	11 yrs	8 yrs	15 yrs
100	4 yrs	6 yrs	5 yrs	9 yrs
105	2 yrs	43 mths	3 yrs	69 mths
110	14 mths	25 mths	23 mths	43 mths
115	8 mths	15 mths	14 mths	27 mths
120	5 mths	9 mths	9 mths	18 mths
125	3 mths	5 mths	6 mths	12 mths
130	2 mths		4 mths	
135				
140				
Temperature indices:				
Duration 5,000 h	101°C		133°C	
Duration 20,000 h	89°C		118°C	

Notes:

1. The values indicated are only orders of magnitude due to the different types of materials and the great dispersion of the complex ageing phenomena of these materials.
2. The temperature referred to is that of the cable conductor resulting from the ambient temperature and its own temperature rise.
3. Permanent rating – load/temperature maintained 24 hours a day.
4. Normal rating – load/temperature maintained 8 hours a day.
5. PVC – polyvinyl chloride; EPR – ethylene/propylene rubber; PRC – chemically reticulated polyethylene.

(From IEC 943, 1989)

6.1.3.1 British Cables Association advice on ageing of cables

The advice of the British Cables Association is most important and is as follows.

> Estimating the life of a cable can only be approximate. There is no definitive or simple calculation method that can be used to determine the life expectancy of a fixed wiring cable.

Many factors determine the life of a cable. For example:

- mechanical damage
- presence of water
- chemical contamination
- solar or infrared radiation
- number of overloads
- number of short-circuits
- effects of harmonics
- temperature at terminations
- temperature of the cable.

Provided an appropriate good quality cable has been selected, taking into account operating temperatures, installation conditions, equipment it is being attached to, and in accordance with appropriate regulations, it should meet or exceed its design life.

The design life of good quality fixed-wiring cables is in excess of twenty years, when appropriately selected and installed. This design life has been assessed on a maximum loading – that is, the cable running at the maximum conductor temperature for 24 hours a day, 365 days a year.

If an installation is not fully loaded all the time, the expected life of the cable would be greater than the design life of the cable. There are many instances of good quality fixed-wiring cables operating for in excess of forty years; this is mainly due to the cables not being abused and being lightly or periodically loaded.

6.2 Protection against overload current

Sections 433, 533, Appendix 4

6.2.1 General

The relevant symbols when considering overcurrent are:

I_b = the design current of the circuit e.g. the current intended to be carried in normal service

I_z = current-carrying capacity of a cable for continuous service under the particular installation conditions concerned (determined from Appendix 4 of BS 7671)

I_n = rated current (nominal) or current setting of the overcurrent protective device

I_1 = non-fusing or non-tripping current

I_2 = current causing effective operation of the device in conventional time

I_t = value of current tabulated in Appendix 4 of BS 7671 for the type of cable and installation method concerned, for a single circuit in an ambient temperature of 30 °C (20 °C for buried cables)

The requirements of Regulation 433.1 provide for coordination between the conductor cross-sectional area and the protective device. They are fundamental to protection against overload, and are repeated below.

(i) the rated current or current setting of the protective device I_n must be greater than or equal to the design current I_b of the circuit

$$I_n \geq I_b, \text{ and}$$

(ii) the as-installed current-carrying capacity I_z of the conductors of the circuit must be greater than or equal to the rated current or current setting of the overcurrent device I_n

$$I_z \geq I_n, \text{ and}$$

(iii) the current I_2 causing effective operation of the protective device does not exceed 1.45 times the lowest of the current-carrying capacity I_z of any of the conductors of the circuit

$$1.45 I_z \geq I_2.$$

These three conditions make for two fundamental rules:

(i) the nominal current rule $I_b \leq I_n \leq I_z$; and
(ii) the tripping current rule $I_2 \leq 1.45 I_z$.

These requirements are shown diagrammatically in Figure 6.1.

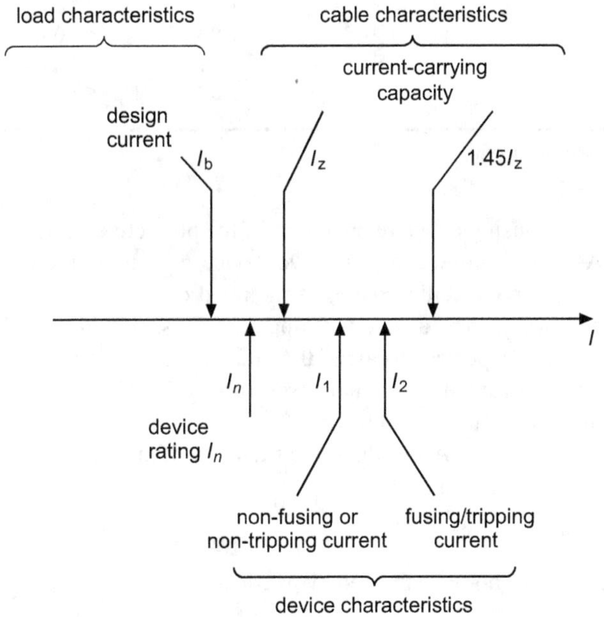

Figure 6.1 Coordination of the characteristics of conductors and protective devices for protection against overload

Table 6.2 Fusing (I_2) and non-fusing currents (I_1)

Device type	Rated current I_n (A)	Non-fusing* current I_1 or I_{nf} (A)	Fusing* current I_2 or I_f (A)	Conventional fusing* time (hours)
BS 88 fuse	<16	1.25 I_n for 1 hour	1.6 I_n	1 (Note 1)
(EN 60269-1:2007)	$16 < I_n \leq 63$	1.25 I_n for 1 hour	1.6 I_n	1
(including 88-2 and 88-3)	$63 < I_n \leq 160$	1.25 I_n for 2 hours	1.6 I_n	2
	$160 < I_n$ 400	1.25 I_n for 3 hours	1.6 I_n	3
	$400 < I_n$	1.25 I_n for 4 hours	1.6 I_n	4
BS 1361 type 1	$5 < I_n \leq 45$		1.5 I_n	4
type 2	$60 < I_n \leq 100$		1.5 I_n	4
BS 1362	$I_n \leq 13$	1.6 I_n	1.9 I_n	
1, 2, 3, 4 mcbs to BS 3871	≤10	1.0 I_n for 2 hours	1.5 I_n	1
	>10	1.0 I_n for 2 hours	1.35 I_n	1
B, C, D CBs to BS EN 60898	≤63	1.13 I_n	1.45 I_n	1
	≥63	1.13 I_n	1.45 I_n	2
BS 3036	5	1.8 I_n	≤2.0 I_n	0.75
	15	1.8 I_n	≤2.0 I_n	1.00
	30	1.8 I_n	≤2.0 I_n	1.25
	60	1.8 I_n	≤2.0 I_n	1.5
	100	1.8 I_n	≤2.0 I_n	2.0
	150	1.8 I_n	≤2.0 I_n	2.0
	200	1.8 I_n	≤2.0 I_n	2.5

* 'fusing' for fuses, 'tripping' for CBs.
Note: Under consideration.

The current I_2 causing effective operation of the protective device requires some explanation. As is well known, a protective device will allow the rated current to flow continuously. At a specified factor of the rated current, e.g. 1.6 I_n or 16 A for a 10 A fuse, the overcurrent device will operate in a specified time. For example, a 10 A BS 88 fuse is expected to carry 10 A indefinitely while an overload of 16 A is required to be interrupted in one hour (see Table 6.2). This time specified in the standard is called the conventional time.

The ratings of cables in Appendix 4 are such that, if an overcurrent device has a rating I_n less than or equal to I_z, that is:

if $I_n \leq I_z$ and $I_2 \leq 1.45\ I_z$

then the cable will not be damaged by overload, where:

$I_z = I_t C_g C_a C_s C_d C_i C_c$

I_z = current-carrying capacity of a cable for continuous service under the particular installation conditions concerned

I_t = tabulated current-carrying capacity of a cable
C_a = rating factor for ambient temperature, both air and ground (Tables 4B1 and 4B2 respectively)
C_d = rating factor for depth of burial (Table 4B4)
C_s = rating factor for soil resistivity (Table 4B3)
C_g = rating factor for grouping (Tables 4C1 to 4C5)
C_i = rating factor for thermal insulation
C_c = rating factor for cables buried in the ground (i.e. 0.9)
C_f = rating factor for semi-enclosed fuse to BS 3036 (i.e. 0.725)

See chapter 16 of this Commentary for further discussion.

Regulation 433.1.201 in effect states that, where the fuse is a general-purpose type gG fuse to BS 88-2, a fuse to BS 88-3, a circuit-breaker to BS EN 60898, a circuit-breaker to BS EN 60947-2 or a residual-current circuit-breaker with overcurrent protection (RCBO) to BS EN 61009-1, I_2 is deemed to be less than 1.45 I_z for these devices. There is one notable exception to this list of devices, and that is semi-enclosed (rewirable) fuses to BS 3036. The current to cause effective operation of a rewirable fuse is $2I_n$, and not $1.45I_n$, as for other devices. Therefore, for overload protection by a rewirable fuse $I_n \leq (1.45/2)I_z$ or $I_n \leq 0.725I_z$.

The ratio $(1.45/I_2)$ is given the designation rating factor C_f (see para 5.1 of Appendix 4).

The type of protection provided by a rewirable fuse is called *coarse protection* and that provided by devices where $I_2 \leq 1.45I_n$, *close protection.*

The values of I_2 and the conventional fusing times for the commonly encountered overcurrent devices are listed in Table 6.2. Designers may wish to make reference to this table if they are concerned about small overloads. I_2 is also called the fusing current I_f and is distinguished from I_1, sometimes called I_{nf}, the current that will not cause operation of the device within conventional time. Table 6.2 shows that I_l, I_2 and the conventional time depend on the device standard and rating. The value of I_2 is not always exactly 1.45 I_n and is sometimes in excess of this. That is why Regulation 433.1.201 is needed to state that, for the purposes of these Regulations, all the commonly used overcurrent devices are deemed to comply with $I_2 \leq 1.45\ I_n$. The note to Table 6.2 is also relevant. Regulation 433.1 requires that every circuit shall be designed so that small overloads of long duration are unlikely to occur. For small overloads say, between 1 and 1.4 times the rating of the device, the device will not operate within the conventional time.

Ageing and deterioration of insulation and connections increase rapidly as operating temperatures exceed the rated values. The basic limits set in the Regulations assume that any cable overloads and hence excess temperatures will be infrequent, and generally due to unforeseen circumstances. No practical limit can be placed on the permissible number of overload events in a given time. To avoid deterioration, it is essential that a designer should ensure that the designed rating of a circuit is high enough to include all foreseeable peak loads of a protracted nature. *It is never acceptable to use a fuse or circuit-breaker as a load limiting device.*

6.2.2 *Ring final circuits*
Regulation 433.1.203

The overload requirements of Regulation 433.1.1 are relaxed for ring circuits by Regulation 433.1.203. Ring circuit conductors that would otherwise need to be rated at or above 30 A or 32 A are allowed to have an installed rating of not less than 20 A, provided that, under the intended conditions of use, the load current in any part of the ring is unlikely to exceed for long periods the current-carrying capacity I_z of the cable as installed. Assuming the cable as installed has a rating of 20 amperes, the designer needs to be able to demonstrate that this current will not be exceeded for long periods. 'Long periods' is not defined but it needs to be considered in terms of the ageing of cable (see 6.1.3) and perhaps the conventional fusing time of overcurrent devices (see Table 6.2). In this context a long time would be of the order of, say, an hour.

In a domestic premises, it is preferable if water heaters and permanently connected heating appliances that are part of a comprehensive space-heating installation are not supplied from the ring circuit supplying kitchen appliances, and the distribution of the kitchen load needs consideration.

The load profiles of typical kitchen appliances are given in Figure 6.3 and it is presumed that they are all switched on at the same time.

In the example, the current in the first half-hour is 31.5 A and in the second half hour 10 A and the third 20 A. The rating of the cable would be exceeded in the first half-hour if the midpoint of the kitchen load was less than one-third of the total ring loop. However, even if all the load was at the end of the ring close to the consumer unit the 30 A overcurrent device would not operate. This is an unsatisfactory arrangement.

The load in each leg of the ring can be estimated as the current will divide in the ratio:

$$I_a = bI_L/(a + b) \text{ and } I_b = I_L - I_a$$

where:

I_L is the load of the socket
I_a is the load in leg a
I_b is the load in leg b
a and b are the cable lengths to the distribution board.

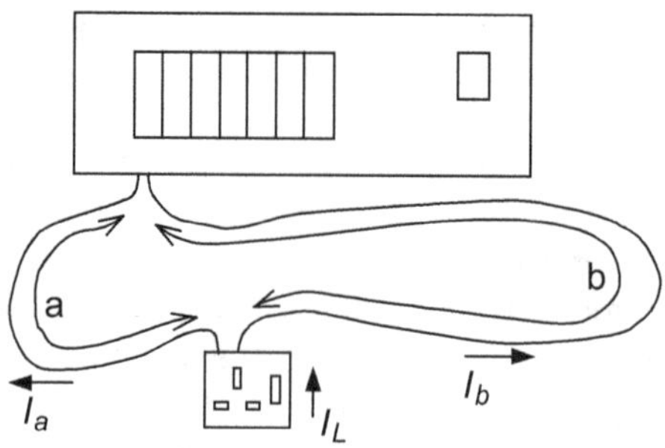

Figure 6.2 Current split in a ring

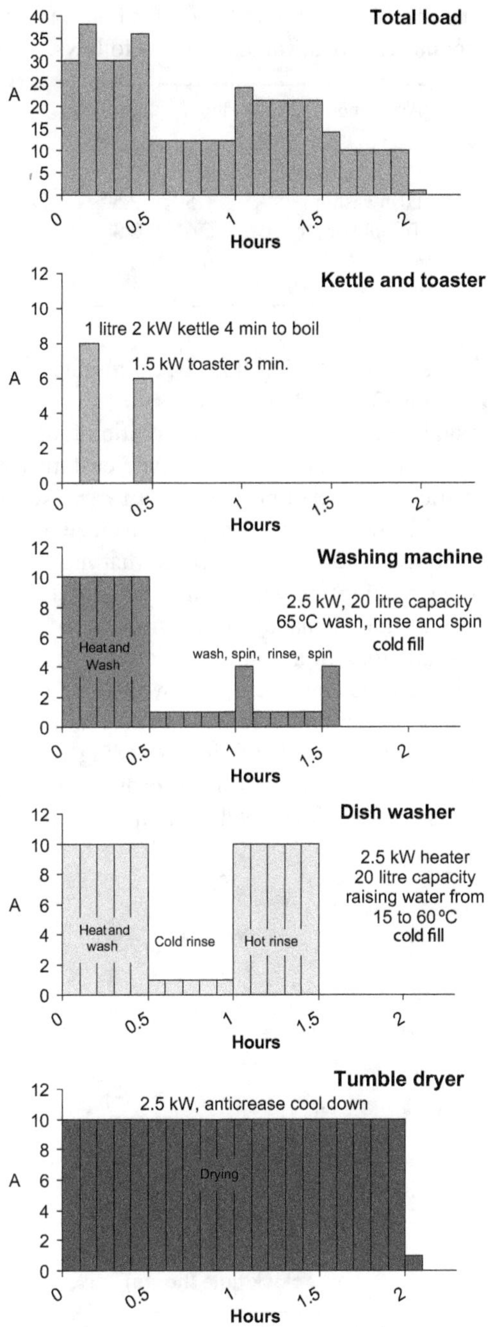

Figure 6.3 Kitchen appliance loads

Consider a washing machine ⅕ around the ring, a dishwasher a ¼ around and a tumble dryer ⅓ way round. The load would divide as follows:

Appliance	leg *a*	leg *b*
Washing machine	8	2
Dishwasher	7.5	2.5
Tumble dryer	6.6	3.3
Total	22.1	7.8

As this is a small overload persisting for, say, 30 minutes, this would meet the requirements of Regulation 433.1.204.

A 30 A or 32 A radial circuit is an alternative solution for kitchen sockets.

The half-hourly customer demands of a domestic customer by household size published by the Electricity Association (now the Energy Networks Association) are shown in Figure 6.4. The demands shown are an average winter weekday, which means that from time to time they would be somewhat higher than this. They are total demands, including cooking and water heating, if any. They do, however, give an insight to the likely half-hour demands of ring circuits.

With respect to installations in an office, as long as known fixed loads such as water heating and space heating are wired with their own circuits, designs based on the load per desk or per unit area will not present problems. In industrial installations, specific consideration of the likely loading must be made. Socket-outlet circuits are sometimes used in industrial locations to supply known loads and when this is the case calculations need to be made apportioning the load as described above.

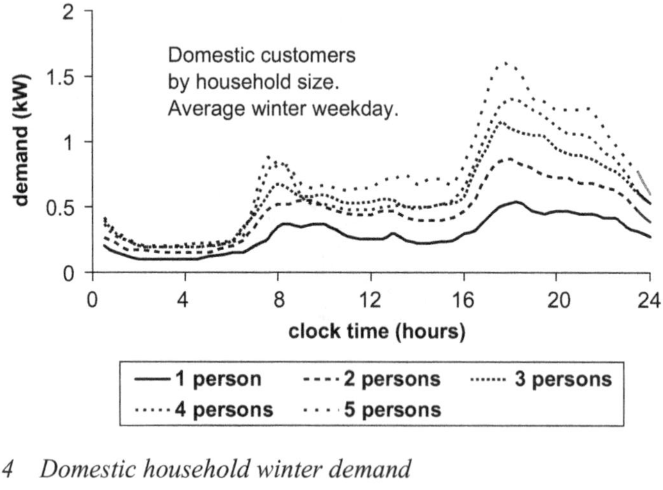

Figure 6.4 Domestic household winter demand

6.2.3 *Variable loads*
Regulation 533.2.1

For cyclic loads, Regulation 533.2.1 states that the values of I_n (the rated current of the overcurrent device) and I_2 (a current giving effective operation of the overload protective device) shall be chosen on the basis of values of I_b and I_z for the thermally equivalent constant load. This regulation is saying that cable and overcurrent device ratings can be selected on the basis of the sum of the squares of the current, over the cyclic period. This can be applied only when the cycle is regular, and if the maximum currents are not sustained for long periods. If the maximum currents are sustained, cables and protective devices must be sized according to the maximum current expected.

The determination of the thermal equivalent loads takes time and, as mentioned above, care must be taken in assessing whether such an approach can be allowed. Described below is a routine that will indicate whether it is worthwhile determining the thermally equivalent load, or if ratings must be based on the maximum load.

1. Determine the maximum current I_{max}, making sure that all load contributions are included, whatever their magnitude and duration (see Figure 6.5). Initially let $I_b = I_{max}$.
2. Select a conductor size to fulfil the condition $I_b \leq I_n \leq I_z$.
3. If the duration of the maximum current I_{max} is greater than the duration given in Table 6.3 ('Test for variable load') for the conductor size and type of cable selected, no advantage is gained by calculating the thermally equivalent current, and Step 2 will give the correct cable size. This applies to single pulses of load or to a repetitive load.
4. If the duration of I_{max} is less than the duration in Table 6.3, it may be possible to select a smaller conductor size. This will depend on the magnitude of the lower current and its duration.
5. Calculate the ratio X = (duration of I_{max})/(duration from Table 6.3) for the cable selected by 2 above ($I_b \leq I_n \leq I_z$). From Figure 6.6, determine the value Y.

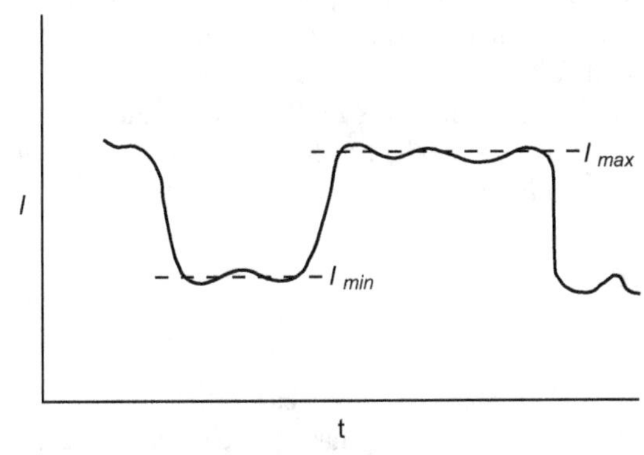

Figure 6.5 A variable load

Table 6.3 Test for variable load (duration of load current for equivalent constant load)

Conductor size (mm^2)	Duration of load current (minutes)			
	1-core cable	2-core cable	3-core cable	4-core cable
1.0	–	12	11	12
1.5	–	13	12	15
2.5	–	15	14	16
4	–	16	16	20
6	–	18	19	24
10	–	22	21	26
16	–	26	27	33
25	–	35	39	49
35	–	41	46	58
50	36	37	47	62
70	42	46	56	74
95	50	55	67	89
120	58	62	77	102
150	67	73	90	120
185	78	84	103	136
240	92	102	124	165
300	106	118	143	192
400	126	143	170	227
500	157	–	–	–
630	172	–	–	–

Note: These durations have been prepared for PVC-insulated copper conductor cables. They will be valid for thermosetting insulation, but not aluminium conductors.

6. If the duration of the lower current I_{min} is less than Y multiplied by (duration of I_{max}), no worthwhile reduction in conductor size will be possible. Step 2 will continue to give the correct size of conductor.
7. If the duration of I_{min} is greater than Y multiplied by (duration of I_{max}), calculation of the thermally equivalent current (which will be less than I_{max}) may lead to lower values of I_b and I_n, and possibly a smaller cable size.

6.2.4 *Peak currents*
Regulation 533.2.1

Regulation 533.2.1 states that, 'in certain cases, to avoid unintentional operation, the peak current values of the loads may have to be taken into consideration'. There

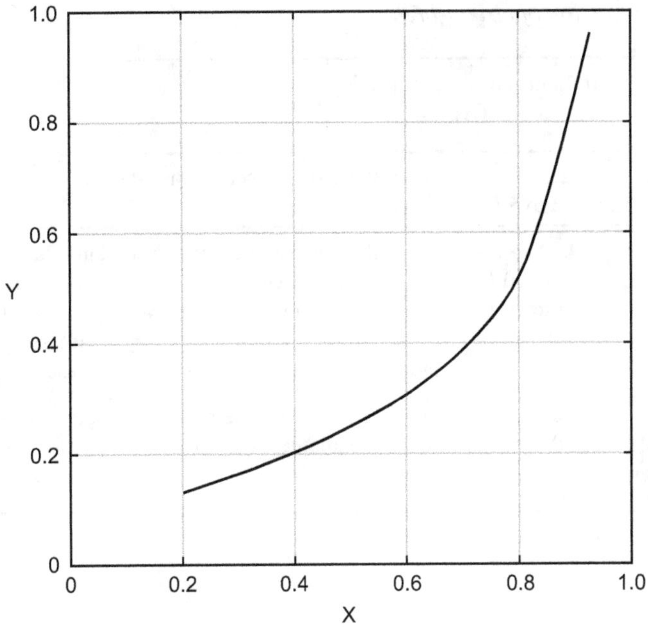

Figure 6.6 Test for variable load

are circumstances where short-time peaks occur in the current taken by equipment. The starting current of a direct-on-line motor and the switch-on surge of an inductive circuit (e.g. fluorescent luminaires) are examples. In most motor circuit applications, gM-type fuse links are used because otherwise the need to withstand motor starting currents would dictate a higher rating fuse link than would need to be selected on the basis of the motor full-load current.

In the special situation of motor protection, gM characteristic fuses give economies in fusegear size and cost. gM motor circuit fuse links have a dual rating – a maximum continuous rating based on the load of the motor on which the fuse-carrying base may be designed and an operational characteristic necessary to avoid tripping on starting. For example, a 32M50 fuse and its base have a continuous rating of 32 A and an operational characteristic of a 50 A fuse link. This means that the fuse will behave as a 50 A fuse, but it will fit into holders and enclosures rated at 32 A.

Regulation 435.2 requires that the energy let-through of the fuse or circuit-breaker shall not exceed the withstand capability of the motor starter. This coordination is described in BS EN 60947-4-1, and is discussed in Paragraph 6.9 below. Circuit-breakers must be selected in accordance with the switch-on characteristics of their load to avoid nuisance tripping. Table 6.4 provides guidance on this aspect of the selection of miniature circuit-breakers. Section 10.5 discusses discrimination between devices including those of different types, e.g. fuse to circuit-breaker.

Table 6.4 Selection of circuit-breakers

Circuit-breaker type	Instantaneous trip current (A)	Application
1 B	2.7 to 4 I_n 3 to 5 I_n	Domestic and commercial installations having little or no switching surge.
2 C 3	4.0 to 7.0 I_n 5 to 10 I_n 7 to 10 I_n	General use in commercial/industrial installations where the use of fluorescent lighting, small motors etc. can produce switching surges that would operate a Type 1 or B circuit-breaker. Type C or 3 may be necessary in highly inductive circuits such as banks of fluorescent lighting.
D 4	10 to 20 I_n 10 to 50 I_n	Suitable for transformers, X-ray machines, industrial welding equipment etc. where high inrush currents may occur.

I_n = rated current rating of the device.

6.2.5 *Position of devices for overload protection* Regulations 433.2.1 and 433.2.2

A device for protection against overload is required to be placed on the supply side of any change in conductor cross-sectional area, method of installation, type of cable or conductor, or environmental conditions, which results in a reduced current-carrying capacity. This requirement is inapplicable if the device at the origin of the circuit is of sufficiently low rating to provide protection in that part of the circuit with the lowest current rating.

The device protecting a conductor against only overload may be placed anywhere along the run of the conductor, provided that there are no branch circuits or outlets on the supply side of the device. However, overload devices may not be installed in building locations where there is a fire risk, or a special requirement or recommendations apply. If the device providing protection against overload is also providing protection against fault current, it must be installed at the origin of the circuit.

6.2.6 *Omission of devices for protection against overload* Regulations 433.3.1 and 433.3.3

Devices for overload must not be used where the unexpected disconnection of the circuit could cause danger. Regulation 433.3.3 provides examples of such situations, such as supply circuits to lifting magnets. The designer has to weigh the danger associated with omitting the overload protection against the hazards that could arise from providing it. A common situation where overload protection is also not required is where, because of the characteristics of the load or because of controlling mechanisms, the conductors will not be subjected to an overload current. (Regulation 433.3.1 refers.)

The difference between an overload and a fault current has to be considered again. A shower or electric heating element cannot overload (although it may fault); however, a motor can electrically overload as a result of increasing mechanical load unless it is fitted with overload protection in the starter.

6.2.7 Omission of overload protective devices in IT systems
Regulation 433.3.2.1

The omission of overload protective devices in an IT system is permitted only where conductors are protected by residual current devices, or where the protective measure double or reinforced insulation is adopted (Section 412). The overload protection in an IT system also provides protection against electric shock in the event of a second fault, and cannot be omitted.

6.3 Protection against fault current
Section 434 and Regulation 533.3

6.3.1 General

The Regulations consider two types of fault: (i) short-circuits and (ii) earth faults. Short-circuit currents flow as a result of a fault between live conductors, e.g. line to line or line to neutral. Earth fault currents flow when there is a fault from a live conductor to earth, e.g. line to earth or line to PEN conductor. The predicted fault current in a circuit is called the *prospective fault current.* The prospective fault current is dependent not only on the type of the fault, but also on the position in the circuit considered. Fault currents will be higher at the origin of an installation than at the extremities of final circuits where the fault currents are reduced by conductor impedances (see Figure 6.7).

6.3.2 Fault current protective devices
Regulations 131.5, 432.1

The fault current protective device characteristics must ensure that the fault current is interrupted before damage occurs to the installation. It must therefore be located at the origin of the installation or circuit it is protecting, and must be able to interrupt the highest expected fault current. In the selection of switchgear the maximum fault current at the origin of the circuit or installation is the limiting factor, e.g. the three-phase fault at position 'a' in Figure 6.7. The most onerous faults as far as cables are concerned are those at the extremity of a circuit, position 'b' in Figure 6.7, where the fault currents are lower and the disconnection times longer. Cables may be damaged by the energy let-through by the overcurrent device before the current is interrupted. The heating effect of current in a cable is proportional to I^2R, and the energy let-through by an overcurrent device is proportional to I^2t. Because of the characteristics of overcurrent devices, the energy let-through is less for large currents than for small, as will be seen from Table 6.5 showing the I^2t energy let-through (or joule integral) for a 100 A BS 88 fuse.

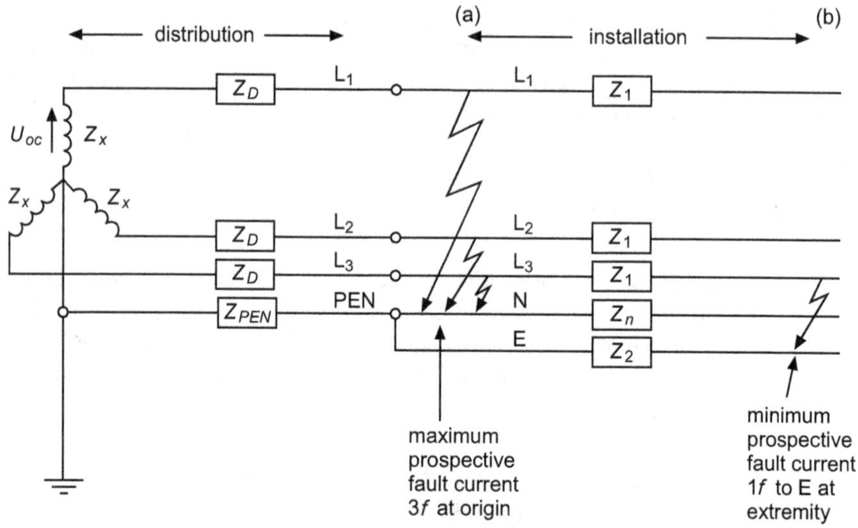

where:

U_{∞} = open circuit phase voltage
Z_x = impedance of the transformer or supply
Z_D = line impedance of the distribution cable
Z_{PEN} = impedance of the PEN conductor (the impedance to be added to $Z_D + Z_x$ to give the supply loop impedance)
Z_1 = line impedance of the circuit line conductor
Z_n = neutral impedance of the circuit neutral conductor (the impedance to be added to Z_1 to give the line neutral loop impedance of the circuit)
Z_2 = impedance of the protective conductor of the circuit (to be added to Z_1 to give neutral earth loop impedance of the circuit)

Figure 6.7 Maximum and minimum prospective fault currents (TN-C-S system)

Table 6.5 Joule integral for a BS 88 100 A fuse (Figure 3.3B of Appendix 3 of BS 7671)

Fault current I(A)	Disconnection time t(s)	Energy let-through I^2t* (A^2s)
1 400	0.1	196 000
980	0.4	384 160
550	5.0	1 512 500

* The calculation here is made using the instantaneous value of current at the disconnection time t. This overestimates the energy let-through.

6.3.3 *Determination of prospective fault current* Regulation 434.1

Regulation 434.1 requires the prospective fault current under both short-circuit and earth fault conditions to be determined at every relevant point of the complete installation. This means that, at every point where switchgear is installed, the maximum fault current must be determined to ensure that the switchgear is adequately rated to interrupt the fault currents. Conversely, to ensure that cables are properly protected, we need to calculate the lowest fault current on each cable run, e.g. at the extremity. We have commented in paragraph 6.3.2 that protective devices let most energy through at low fault currents (relative to the overcurrent device rating), and as a consequence it is at low fault currents and long disconnection times that cables, particularly reduced-section protective conductors, are most likely to suffer damage.

Regulation 434.1 states that the determination of fault levels may be carried out by calculation, by enquiry or by measurement. The wording on an Electrical Installation Certificate reflects this statement.

6.3.3.1 CLC TR 50480 "Determination of cross-sectional area of conductors and selection of protective devices"

Paragraph 2.5 of Appendix 4 to BS 7671, introduced by Amendment 3, makes reference to technical report CLC TR 50480. The main purpose of Technical report is to standardise the approach to the calculation of the various parameters of an electrical system including fault currents. The intention is that users of software programmes will have some understanding as to how the calculations have been carried out. The report introduces a "c" factor for estimating maximum and minimum voltages under fault conditions.

Section 6 of the technical report states:

Characteristics of the supply source

Voltage

The reference parameter is the line to neutral nominal voltage $U_o = U_n /\sqrt{3}$, multiplied by the factor c.

The voltage factor c is introduced to take account of voltage variations depending on time and place, changing of transformer taps and other considerations. The values of c for the worst case condition in a low voltage installation are given in Table 7, derived from IEC 60909, Table 1.

The factor c is not intended to take account of the fault impedance. This Technical Report assumes zero fault impedance.

Voltage factor c

Nominal voltage	Voltage factor c	
	C_{max}	C_{max}
100V to 1000V	1.1	0.95

C_{max} is recommended for use when estimating maximum fault currents and C_{min} when estimating maximum disconnecting times.

However a designer knowing his voltages may be higher or lower than these figures may be wise if he uses his own best estimate.

For example when calculating maximum loop impedances for Tables 41.2 to 41.6 as appropriate a supply voltage of C_{min} $U_n/\sqrt{3}$, that is $0.95.430/\sqrt{3} = 219.4$V. Whereas Appendix 2 of BS 7671 advises that the permitted voltage range in the UK is 230V +10%/- 6% that is 216.2 to 253.0 V respectively.

6.3.3.2 Calculation

The maximum prospective fault current at the origin of the installation shown in Figure 6.7 is given by considering a three-phase-to-neutral/earth fault at the origin of the installation. Here the neutral/earth return has no significance:

$$I_{sc} = \frac{U_o}{Z_x + Z_D}$$

The designer may wish to use the factor "c" described above that is:

$$I_{sc} = \frac{C_{max} U_o}{Z_x + Z_D}$$

However if the line to neutral/earth open circuit voltage will exceed C_{max} U_o the open circuit voltage should be used.

The minimum prospective fault current is given by considering a single-phase-to-earth fault at the extremity of the circuit:

$$I_{ef} = \frac{U_{oc}}{Z_x + Z_D + Z_1 + Z_2 + Z_{PEN}}$$

or

$$I_{ef} = \frac{C_{max} U_o}{Z_x + Z_D + Z_1 + Z_2 + Z_{PEN}}$$

(For TN-S systems Z_{PEN} is replaced by the impedance of the supply system protective conductor.)

This is also the current to be used in determining the maximum disconnection times for shock protection. Examples of calculations are given in Chapter 18.

6.3.3.3 Measurement

Earth loop impedance and prospective fault current testers are readily available and widely used particularly for measuring the earth fault loop impedance of final circuits. From this loop impedance, by simple calculation, the fault current can be determined as $I_f = U_{oc}/Z_s$. These loop impedance test instruments may also have an earth fault current range. The two readings are obviously interchangeable by the formula above.

These test instruments will measure earth fault currents and prospective short circuit currents (see section 14.1.10 of Chapter 14).

The loop impedance is generally measured by charging up a capacitor on the first half-cycle, connecting a load from within the test instrument and discharging on the second half-cycle. The charge remaining in the capacitor is used to calculate the loop impedance. Loop impedance test instruments are specifically designed for a particular purpose and care has to be taken when using them in circumstances for which they were not designed, e.g. for very high fault levels or for earth loops with high impedance, say 500 Ω or more.

Typically, such test instruments determine the fault current flowing for a single-phase-to-earth fault. As can be seen from Figure 6.7, this current is approximately only half the value of that for a three-phase-to-earth fault. Therefore, if they are being used to make a quick assessment of the fault current on a three-phase installation, the estimate of fault current should be at least doubled. This presumes that the neutral/earth return path impedance equals the line conductor impedance.

Instruments are not readily available for measuring high earth fault current levels such as will be found close up to 11 000/433 V transformers. The designer will need to obtain the impedance values from the transformer owner and calculate the fault levels.

Expert's comment

Where the supply is provided at low voltage the local distribution network operator (DNO) will own the HV/LV transformer. Where supply is provided at high voltage, the transformer will be owned by the customer and the local DNO will not have any records of the privately owned plant. Also, note that there are now an increasing number of independent distribution networks embedded within the traditional DNOs' networks. Independent distribution network operators (IDNOs) take supply from the local DNO at HV or LV and distribute it on to customers within their own network. In this case the designer will need to ask the customer to obtain the name of the IDNO from his electricity energy supplier, who will be paying Distribution Use of System (DUoS) to the IDNO.

6.3.3.4 Ascertained by enquiry
General

The Regulations make specific reference to ascertaining by enquiry as a method of determining the fault current. Regulation 28 of the Electricity Safety, Quality and Continuity Regulations 2002 (as amended) states:

28. Information to be provided on request

A distributor shall provide, in respect of the existing or proposed consumer's installation which is connected or is to be connected to his network, to any person who can show reasonable cause for requiring the information a written statement of –

(a) the maximum prospective short-circuit current at the supply terminals;

(b) for low voltage connections, the maximum earth loop impedance of the earth fault path outside the consumer's installation;
(c) the type and rating of the distributor's protective device or devices nearest to the supply terminals;
(d) the type of earthing system applicable to the connection; and
(e) the number of phases, frequency and voltage,

which apply, or will apply, to that installation.

6.3.3.5 Ascertained by enquiry – single-phase supplies up to 100 A

Distributors will generally provide an estimate of the maximum prospective short-circuit current at the electricity supplier's cut-out, based on a declared level of 16 kA (0.55 p.f.) at the point of connection of the service line to the LV distribution cable. The fault level will be this high only if the installation is close to the distribution transformer. However, over the lifetime of an installation, changes may very well be made to the distribution network and consequently designers must install equipment suitable for the highest fault levels that might occur. Attenuation or reduction of these fault levels is allowed for service lines on the customer's premises. These allowances are made on the assumption that, while the distribution network on the public highway might change, the service line on the customer's premises will remain unchanged, or at least not be changed without the occupant's knowledge. Therefore, attenuation is allowed only for cable on the customer's premises. If it is known that the distribution cable is on the far side of the road, this allowance cannot be made, since at a later date the service cable may be rejointed to a distribution cable on the nearside of the road. Within these constraints, attenuation as per Table 6.6 can be made.

There are some inner city locations, particularly in London, where the maximum prospective short-circuit current on the distributing main exceeds 16 kA. The local Distribution Network Operator (or Independent Distribution Network Operator in the case of embedded networks) will need to be consulted in these cases.

6.3.3.6 Ascertained by enquiry – three-phase supplies

Distributors will provide estimated maximum prospective short-circuit currents (p.s.c.c.) at the cut-out of three-phase supplies based on either:

(a) a declared level of 25 kA (0.23 p.f.) at the point of connection of the service line to the busbars in the board's distribution substation, see Table 6.7 or
(b) a declared level of 18 kA (0.5 p.f.) at the point of connection of the service line to the lower voltage distribution main, see Table 6.8.

Knowledge of the power factor (p.f.) as well as the p.s.c.c. enables a more accurate calculation of p.s.c.c. downstream of the supply to be carried out. The p.s.c.c. can be resolved into resistive and reactive components.

Obviously, knowledge of the source of the supply is necessary to use the attenuation tables and Tables 6.6 to 6.8 are provided for this. Again, attenuation can only be allowed for the length of service line on the customer's premises. Figure 6.8 gives some guidance as to how this may be done.

Table 6.6 230 V single-phase supplies up to 100 A (estimated maximum prospective short-circuit current at the distributor's cut-out based on declared level of 16 kA (0.55 p.f.) at the point of connection of the service line to the LV distribution main)

Length of service line (m)	Up to 25 mm² Al or 16 mm² Cu service cable or overhead line		35 mm² Al or 25 mm² Cu service cable or overhead line (looped service)	
	p.s.c.c. (kA)	p.f.	p.s.c.c. (kA)	p.f.
0	16.0	0.55	16.0	0.55
1	14.8	0.63	15.1	0.61
2	13.7	0.69	14.3	0.66
3	12.6	0.74	13.5	0.70
4	11.7	0.78	12.7	0.74
5	10.8	0.82	12.0	0.77
6	10.1	0.84	11.4	0.79
7	9.4	0.86	10.8	0.82
8	8.8	0.88	10.3	0.83
9	8.3	0.89	9.7	0.85
10	7.8	0.91	9.3	0.86
11	7.4	0.92	8.8	0.88
12	7.0	0.92	8.4	0.89
13	6.6	0.93	8.1	0.90
14	6.3	0.94	7.7	0.91
15	6.0	0.94	7.4	0.91
16	5.7	0.95	7.1	0.92
17	5.5	0.95	6.9	0.92
18	5.3	0.96	6.6	0.93
19	5.1	0.96	6.4	0.93
20	4.9	0.96	6.2	0.94
21	4.7	0.96	6.0	0.94
22	4.5	0.97	5.8	0.95
23	4.4	0.97	5.6	0.95
24	4.2	0.97	5.4	0.95
25	4.1	0.97	5.3	0.95
26	3.9	0.97	5.1	0.96
27	3.8	0.98	5.0	0.96
28	3.7	0.98	4.8	0.96
29	3.6	0.98	4.7	0.96
30	3.5	0.98	4.6	0.96
35	3.1	0.98	4.0	0.97
40	2.7	0.99	3.6	0.98
45	2.5	0.99	3.3	0.98
50	2.2	0.99	3.0	0.98

Values taken from Table 1 of Engineering Recommendation P25/1 1996 and Table 2 of Engineering Recommendation P26 1985. Updated information can be found in the new ENA document.

Table 6.7 400 V three-phase supplies, 25 kA p.s.c.c. (estimated maximum p.s.c.c. at the distributor's cut-out based on declared levels of 25 kA (0.23 p.f.) at the point of connection of the service line to the busbars in the distributor's distribution substation)

Length of service line (m)	Service line cross-sectional area									
	95 mm² Al		120 mm² Al		185 mm² Al		240 mm² Al		300 mm² Al	
	p.s.c.c. (kA)	p.f.	p.s.c.c. (kA)	p.f.	p.s.c.c. (kA)	p.f.	p.s.c.c. (kA)	p.f.	p.s.c.c. (kA)	p.f.
0	25.0	0.2	25.0	0.2	25.0	0.2	25.0	0.2	25.0	0.2
5	23.1	0.4	23.3	0.3	23.7	0.3	23.8	0.3	23.9	0.3
10	21.1	0.5	21.7	0.4	22.4	0.4	22.7	0.3	22.8	0.3
15	19.2	0.6	20.1	0.5	21.1	0.4	21.6	0.4	21.8	0.3
20	17.5	0.6	18.6	0.6	20.0	0.5	20.5	0.4	20.9	0.4
25	16.0	0.7	17.2	0.6	18.9	0.5	19.6	0.4	20.0	0.4
30	14.6	0.7	16.0	0.7	17.9	0.5	18.7	0.5	19.2	0.4
35	13.5	0.8	14.9	0.7	16.9	0.6	17.8	0.5	18.4	0.4
40	12.4	0.8	13.9	0.7	16.1	0.6	17.1	0.5	17.7	0.5
45	11.6	0.8	13.0	0.7	15.3	0.6	16.3	0.5	17.1	0.5
50	10.8	0.8	12.3	0.8	14.6	0.6	15.7	0.6	16.4	0.5

Values taken from Table 1 of Engineering Recommendation P26 1985. Updated information can be found in the new ENA document.

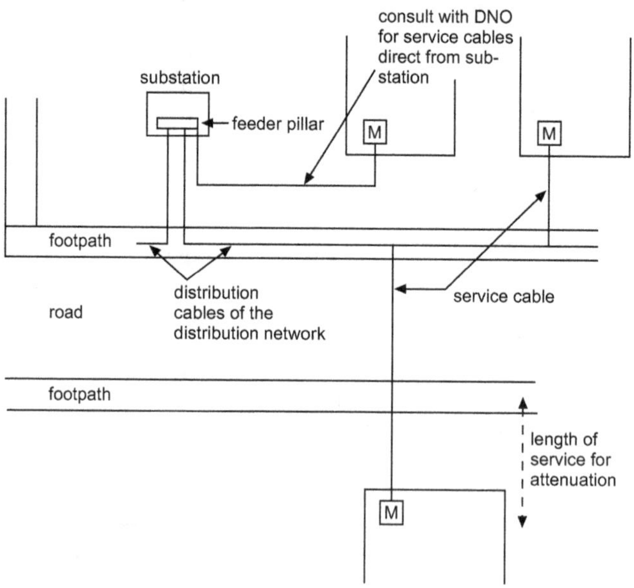

Figure 6.8 Distribution and service cables

Table 6.8 400 V three-phase supplies, 18 kA p.s.c.c. (estimated maximum p.s.c.c. at the distributor's cut-out based on declared levels of 18 kA (0.5 p.f.) at the point of connection of the service line to the LV distribution main)

Length of service line (m)	Service line cross-sectional area													
	Up to 35 mm^2 Al		70 mm^2 Al		95 mm^2 Al		120 mm^2 Al		185 mm^2 Al		240 mm^2 Al		300 mm^2 Al	
	p.s.c.c. (kA)	p.f.	p.s.c.c. (kA)	p.f.	p.s.c.c. (kA)	p.f.	p.s.c.c. (kA)	p.f.	p.s.c.c. (kA)	p.f.	p.s.c.c. (kA)	p.f.	p.s.c.c. (kA)	p.f.
0	18.0	0.5	18.0	0.5	18.0	0.5	18.0	0.5	18.0	0.5	18.0	0.5	18.0	0.5
5	14.8	0.7	16.2	0.6	16.6	0.6	16.8	0.6	17.1	0.5	17.2	0.5	17.3	0.5
10	12.2	0.8	14.5	0.7	15.3	0.6	15.7	0.6	16.2	0.6	16.5	0.5	16.6	0.5
15	10.2	0.8	13.1	0.7	14.1	0.7	14.7	0.6	15.4	0.6	15.8	0.6	16.0	0.6
20	8.8	0.9	11.9	0.8	13.0	0.7	13.7	0.7	14.7	0.6	15.1	0.6	15.4	0.6
25	7.6	0.9	10.8	0.8	12.1	0.7	12.9	0.7	14.0	0.6	14.6	0.6	14.9	0.6
30	6.8	0.9	9.9	0.8	11.3	0.8	12.2	0.7	13.4	0.7	14.0	0.6	14.4	0.6
35	6.0	0.9	9.1	0.9	10.6	0.8	11.5	0.7	12.9	0.7	13.5	0.6	13.9	0.6
40	5.5	0.9	8.5	0.9	9.9	0.8	10.9	0.8	12.3	0.7	13.0	0.6	13.5	0.6
45	5.0	0.9	7.9	0.9	9.3	0.8	10.3	0.8	11.8	0.7	12.5	0.6	13.1	0.6
50	4.6	0.9	7.4	0.9	8.8	0.8	9.8	0.8	11.4	0.7	12.1	0.7	12.7	0.6

6.3.3.7 Supply loop impedances

When electricity distribution network operators are asked for maximum prospective short-circuit current levels and maximum earth loop impedances, they may very well respond, for a PME supply, that the maximum fault level is 16,000 A and the loop impedance is typically 0.35 Ω. Clearly these two figures are not compatible. If an open-circuit supply voltage of 250 V and a loop impedance of 0.35 Ω are assumed, a fault current of 714 A is quickly determined, considerably smaller than the figure for maximum fault current provided. In a similar way that a distributor will provide the maximum fault level that is likely to arise, it will also provide the highest earth fault loop impedance. Fault levels and earth fault loop impedances of the supply network may change during the lifetime of an installation. When selecting switchgear ratings, the highest fault levels must be presumed; but, when using knowledge of the distributor's loop impedance for determining protection against electric shock, or the low fault current withstand of electric cables, the highest loop impedance must be presumed.

Values of earth loop impedance 'quoted' in the Energy Networks Association Engineering Recommendation P23-1, 'Customers Earth Fault Protection for Compliance with the IEE Wiring Regulations for Electrical Installations', are typically as follows for single-phase supplies:

(i) TN-C-S supplies: 0.35 Ω
(ii) TN-S supplies: 0.8 Ω
(iii) TT supplies: 20 Ω.

(i) and (ii) above for TN-C-S and TN-S supplies are loop impedances at the origin of the installation; (iii) for TT installations is effectively the resistance of the source transformer earth electrode.

Expert's comment

These are *typical* maximum values and not the absolute maximum that will be experienced during the course of an installation's lifetime.

DNOs are under no legal obligation to provide or maintain any particular value of earth loop impedance in any standard or regulation.

During the lifetime of an installation a number of circumstances may alter the earth fault loop impedance either permanently or temporarily.

- The DNO may make permanent alterations to the local LV network or change the local distribution transformer to a larger or smaller size to meet ongoing network requirements. This may lower or raise the earth loop impedance for some years into the future.
- The DNO may need to 'back-feed' the LV network from another source to maintain supplies while maintenance or repairs are carried out. This alternative source may be another, but more remote, distribution transformer via an extended length of LV network, or it may be from a mobile generator. In either case the earth fault loop impedance is likely to change significantly for the duration of the work, which may be hours or up to several weeks.
- The Electricity Safety, Quality and Continuity Regulations 2002 now permit customers to install small-scale power generation that will feed into the local LV network. This may take the form of combined-heat-and-power units, Stirling engine central-heating boilers, fuel cells, wind turbines, photovoltaic units etc. The effect on local networks of these units is not yet fully understood. However, they will have the effect of altering the p.s.c.c. and EFLI values as their generation outputs vary throughout the day/year.

From the above it is clear that a value of EFLI obtained by measurement, calculation or enquiry is valid only at the point in time it was obtained and may not continue for the lifetime of the installation. Indeed, an EFLI measurement made when the in-feed from local embedded generation is at a high level may result in a lower value being recorded than would occur when the generation is not providing any output.

This leaves the designer with the problem of selecting a value of EFLI for the earth fault protection calculations that will need to remain valid throughout the lifetime of the installation.

One DNO has provided the following information:

- New housing estates that have their own substations – Most DNOs design these networks to have an EFLI of less than 0.25 ohm under normal running conditions. This is primarily to ensure low levels of flicker when large single-phase switched loads such as instantaneous showers are operated.

During temporary back-feeding this may rise to 0.5 ohm or more depending on the design of the adjacent LV network.

- Infill housing, where single or small groups of houses are built within existing established housing. The original network will not have been designed to current methodologies and may exhibit higher values of EFLI than the larger new developments referred to above.

It may be useful for the designer to refer to Table 1 of IEC/TR 60725 'Consideration of reference impedances and public supply network impedances for use in determining disturbance characteristics of electrical equipment having rated current <75 amps per phase'.

This table publishes the results of an international survey of residential premises' complex supply impedances carried out in 1980.

The results for the UK are summarized below:

98 per cent of premises 0.46 + j0.45 ohms (= 0.644 ohm impedance)

90 per cent of premises 0.25 + j0.23 ohms (= 0.34 ohm impedance)

From these values one can surmise that the typical value of 0.35 ohm referred to in Energy Networks Association Engineering Recommendation P23-1 may refer to the conditions pertaining to 90 per cent of UK properties under normal feeding conditions.

Since 1980 little load-related LV reinforcement has taken place in established residential areas due to stabilization of load growth. Therefore, up to 10 per cent of premises may still exceed 0.35 ohm. However, the number of LV Networks with EFLI in the order of 0.644 ohm have been reduced by reinforcement projects resulting from voltage complaints.

The 17th Edition of BS 7671 requires the use of RCD protection in significantly more circumstances than in previous editions in order to provide enhanced protection to occupiers who may damage cables embedded in plaster, for example while carrying out DIY activities. The net result being that most circuits will require RCD protection regardless of the value of external EFLI.

Consumer units are now readily available with split busbars that enable RCD protection to be applied separately to one, two or more bus sections. Furthermore, RCBOs can now be installed within consumer units to protect high-reliability circuits e.g. to freezers and fire alarms.

This enables designers to readily provide earth fault protection, compliant with BS 7671, at any value of external EFLI that may be encountered on a DNO LV network even under abnormal feeding arrangements.

There remains the issue of protecting the circuit from the meter to the consumer unit busbars. Regulations 433.3.1 (iii) and 434.3 (iv) allow the designer to utilise the DNO's main fuse for this purpose, provided the DNO agrees that the characteristics of the device are suitable.

For short-circuit protection a 100 amp BS 88-3 (formerly BS 1361) fuse would provide adequate protection provided the meter tails are a minimum of 25 mm^2 and no longer than 3 m from cut-out to consumer unit.

For earth faults, Regulation 411.3.2.6 requires supplementary equipotential bonding to be provided as protection against electric shock where the five-second automatic disconnection time required in Regulation 411.3.2.3 cannot be achieved.

At the consumer unit, the connection of the installation's protective conductors to the main earth bar achieves the objective of supplementary bonding. That is to say, the potential differences that can occur between items of equipment within the installation during the occurrence of an earth fault between the meter and the consumer unit busbars are limited to safe values.

Protection from electric shock is therefore still maintained where the DNO cut-out fuse's earth fault disconnection time exceeds five seconds.

6.3.4 *Rated short-circuit capacity* Regulation 434.5.1

The breaking capacity of each protective device shall be not less than the prospective short-circuit current or earth fault current, whichever is the greater, at the point at which the device is installed. The calculation of the prospective short-circuit current is dealt with in Chapter 18. Table 6.9 provides the rated short-circuit capacities for a range of devices.

A lower breaking capacity is allowed if another protective device having the necessary breaking capacity is installed on its supply side. In this situation, it is necessary that the characteristics of the devices be coordinated, so that the energy let-through of the upstream device does not exceed that which can be withstood without damage by the device on the load side. A cartridge fuse might be used to back up circuit-breakers, as shown in Figure 6.9.

If the circuit-breakers in the figure are rated at 9 kA, and the prospective short-circuit current for a three-phase-to-earth fault is 20 kA, a general-purpose BS 88 fuse can provide backup protection. Figure 6.10 (see image on p. 130) shows the superposition of a backup fuse characteristic F' and F'' on that of the circuit-breaker (c).

For overloads the circuit-breakers will interrupt the fault before the fuse operates. In the event of a fault, the circuit-breaker will clear the fault if the fault current I does not exceed the selectivity limit I_s. The circuit-breaker manufacturer's advice on the selection of the fuse must be taken. The takeover current I_0 must not exceed the rated short-circuit capacity I_{cn} of the circuit-breaker.

BS EN 60898-1 *Specification for circuit breakers for overcurrent protection for household and similar installations* requires (Annex D3) that, on request, the manufacturer of the circuit-breaker shall state the type and characteristics of the fuses to be used in conjunction with the circuit-breaker, and state the maximum prospective short-circuit current for which the combination of fuse and circuit-breaker is suitable. Obviously, the preferred approach is that the manufacturer's advice be taken. BS EN 60898 does state that in some practical cases it may be sufficient to compare the operating characteristics of the circuit-breaker and the fuse. However, it also says that special attention must be paid to the I^2t values of the circuit-breaker

Table 6.9 Rated short-circuit capacities

Device type	Device designation	Rated short-circuit capacity (kA)	
Semi-enclosed fuse to BS 3036 with category of duty	S1A	1	
	S2A	2	
	S4A	4	
Cartridge fuse to BS 1361 Type I		16.5	
Type II		33.0	
General purpose fuse to BS 88-2		50 at 415V	
Cartridge fuse to BS 88-3 Type I		16	
Type II		31.5	
General purpose fuse to BS 88-6		16.5 at 240V 80.0 at 415V	
Circuit-breakers to BS 3871	M1	1	
	M1.5	1.5	
	M6	6	
	M9	9	
Circuit-breakers to BS EN 60898* and RCBOs to BS EN 61009		I_{cn}	I_{cs}
		1.5	(1.5)
		3.0	(3.0)
		4.5	(4.5)
		6	(6.0)
		10	(7.5)
		15	(7.5)
		20	(10.0)
		25	(12.5)

* Two rated short-circuit capacities I_{cn} and I_{cs} may be quoted for circuit-breakers.
I_{cn} = rated ultimate short-circuit capacity (marked on the device).
I_{cs} = rated service short-circuit breaking capacity. This is the maximum level of fault current operation after which further service is assumed without loss of performance.

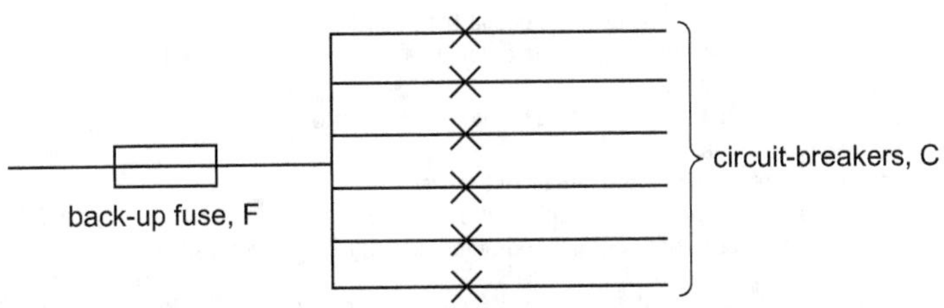

Figure 6.9 Backup fuse protecting circuit-breakers

and of the fuse during the break time, and the effects on the circuit-breaker of the peak value cutoff current of the fuse. Figure 6.10, as well as plotting the time/current characteristics of the devices, also, in the lower figure, shows the energy let-through I^2t plotted against the fault current I. Generally, the advice is given that the takeover current I_0 should be no greater than 80 per cent of I_{cn}, the rated short-circuit capability of the circuit-breaker, unless the manufacturer has carried out specific tests to ensure that a closer relationship is acceptable.

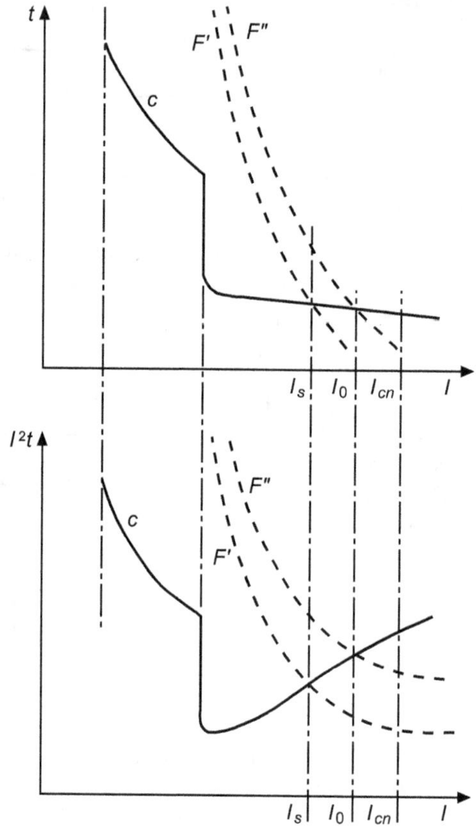

where:

I = prospective short-circuit current
I_{cn} = rated short-circuit capacity of the circuit-breaker alone
I_s = selectivity limit current; for currents below this the circuit-breaker will operate leaving the fuse intact
I_0 = take over current; this must not be greater than the rated short-circuit capacity of the circuit-breaker
F' = minimum pre-arcing characteristic of the fuse
F'' = maximum operating characteristic of the circuit-breaker
C = maximum time/current I^2t current characteristics of the circuit-breaker

Figure 6.10 Coordination of circuit-breaker and backup fuse characteristics

6.3.5 *Omission of protection against overload (protection against fault current only)* Regulations 433.3, 434.5.2

If overload protection is not required because, for example, the load is fixed, and the overcurrent device cannot provide overload protection, because its rating I_n exceeds that of the cable I_z, then a calculation must be made to ensure that the circuit conductors are large enough to carry the fault currents without damage until the fault current device operates. Alternatively, this can be looked at as saying that the overcurrent device must interrupt the current in sufficient time to prevent damage to the cable. Regulation 434.5.2 provides an equation for this time t:

$$t = \frac{k^2S^2}{I^2}$$

or

$$S \geq \frac{I\sqrt{t}}{k}$$

where:

t = duration of the fault in seconds.
S = nominal cross-sectional area of the conductor in mm^2.
I = is the effective fault current, in amperes, expressed for a.c. as the rms value, due account being taken account of the current limiting effect of the circuit impedances.
k = is a factor taking account of the resistivity, temperature coefficient and heat capacity of the conductor material, and the appropriate initial and final temperatures, shown in Table 43.1.

k can be obtained from Table 43.1 or, for protective conductors, from Tables 54.2 to 54.6 of BS 7671. The fault current I is determined by calculation, knowing the impedance of the faulted circuit, and t is determined from the time/current characteristics of the devices given in Appendix 3 of BS 7671.

The calculated value of S must be less than the actual cross-sectional area of the conductor of the circuit.

Whenever the overcurrent device is providing protection against fault current, but not overload, this check must be made. Such a calculation is described in Chapter 18.

6.3.6 *Characteristics of protective devices* Regulation 434.5.2

For circuit-breakers to BS EN 60898 – that is type B, C or D – there is a requirement that for a fault of very short duration (less than 0.1 s):

$$k^2S^2 > I^2t$$

where:

S = nominal cross-sectional area of the conductor in mm^2
k = factor from Table 43.1
I^2t = energy let-through quoted for the class of device in BS EN 60898, BS EN 61009 or as quoted by the manufacturer.

Clause 4.6 of BS EN 60898-1 states that circuit-breakers may be classified by energy-limiting class. Table 6.10 tabulates the energy-limiting classes with the maximum energy let-through as given in Table ZA of BS EN 60898-1 and endorsed with the minimum conductor csa (S) allowed by these energy let-throughs (I^2t) calculated using the formula:

$$S > \frac{\sqrt{\text{energy let through } (I^2t)}}{k}$$

These minimum conductor cross-sectional areas apply both to line and protective conductors.

This limitation on the csa of conductor requires, particularly for cables with reduced-section protective conductors, not only the use of energy-limiting Class 3 circuit-breakers, but for high fault levels either the use of circuit-breakers from the manufacturer quoting maximum energy let-through (I^2t) much less than those required by the standard, or, rather, larger csa protective conductors than is the norm. See section 11.4.2.3 for more details.

6.3.7 *Busbar trunking fault ratings*
Regulation 434.5.3

The regulation requires either of the following.

(i) The rated short-time withstand current (I_{cw}) and the rated peak withstand current of a busbar trunking system or powertrack system shall be not lower than the rms value of the prospective fault current and the prospective fault peak current value, respectively. The maximum time for which the I_{cw} is defined for the busbar trunking system shall be greater than the maximum operating time of the protective device.

(ii) The rated conditional short-circuit current of the busbar trunking system or powertrack system associated with a specific protective device shall be not lower than the prospective fault current.

The terms used are not defined in BS 7671, but are defined in BS EN 61439-1 as follows:

(i) Rated short-time withstand current I_{cw}
r.m.s value of short-time current, declared by the ASSEMBLY manufacturer, that can be withstood under specified conditions, defined in terms of a current and time.

(ii) Rated peak withstand current I_{pk}
value of peak short-circuit current, declared by the ASSEMBLY manufacturer, that can be withstood under specified conditions.

Table 6.10 Minimum conductor csa (S) by energy limiting class for type B and C circuit-breakers to BS EN 60898-1 [Source: Table ZA of BS EN 60898-1]

Table ZA. 1 – Permissible I^2t (let-through) values for class 3 circuit-breakers

Rated short-circuit capacity (A)	rated current up to and including 16A				rated currents 20, 25 and 32A			
	B type I^2t max (A^2s)	Minimum S (mm^2)	C type I^2t max (A^2s)	Minimum S (mm^2)	B type I^2t max (A^2s)	Minimum S (mm^2)	C type I^2t max (A^2s)	Minimum S (mm^2)
3000	15000	1.06	17000	1.67	18000	1.2	20000	1.23
4500	25000	1.4	28000	1.45	32000	1.6	37000	1.67
6000	35000	1.6	40000	1.74	45000	1.8	52000	1.98
10000	70000	2.3	80000	2.46	90000	2.6	100000	2.75

Table ZA.2 – Permissible I^2t (let-through) values for circuit-breakers with rated current exceeding 16A up to and including 32A (Note 1)

Rated short-circuit capacity (A)	rated current 40A				rated currents 50 and 63A			
	B type I^2t max (A^2s)	Minimum S (mm^2)	C type I^2t max (A^2s)	Minimum S (mm^2)	B type I^2t max (A^2s)	Minimum S (mm^2)	C type I^2t max (A^2s)	Minimum S (mm^2)
3000	21600	1.2	24000	1.3	28000	1.4	30000	1.5
4500	34800	1.6	45000	1.8	48000	1.9	55000	2.0
6000	54000	2.0	63000	2.2	65000	2.2	75000	2.3
10000	108000	2.8	120000	3.0	135000	3.1	145000	3.3

S is the conductor csa in mm^2 for k = 115.

Table 6.11 Minimum conductor csa determined from limits in BS EN 60898-1 for energy-limiting Class 3, Type B and C devices (note 1)

Energy-limiting Class 3 device rating	Fault level (kA)	Protective conductor csa (mm^2) (note 2)	
		B type	C type
Up to and including 16A	3	1.0	1.5 (1.13)
	6	2.5 (1.6)	2.5 (1.74)
Over 16 up to and including 32A	3	1.5 (1.2)	1.5 (1.23)
	6	2.5 (1.8)	2.5 (2.0)
40A	3	1.5 (1.3)	1.5 (1.34)
	6	2.5 (2.0)	2.5 (1.74)
50 and 63A	3	1.5 (1.45)	1.5 (1.51)
	6	2.5 (2.22)	2.5 (2.38)

Notes:

1. For other device types consult manufacturer's data
2. Smallest standard size conductor is listed; the csa in brackets is the limit in the standard.

Manufacturers' catalogues will provide this data, as Table 6.12.
With respect to conditional rating this is defined:

(iii) Rated conditional short-circuit current I_{cc}
value of prospective short-circuit current, declared by the ASSEMBLY manufacturer, that can be withstood for the total operating time (clearing time) of the short-circuit protective device (SCPD) under specified conditions.

Note: The short-circuit protective device may form an integral part of the ASSEMBLY or may be a separate unit.

Table 6.12 Example of manufacturers' permissible currents for busbar trunking

Rated current I_{th} (1)	200	315	400	630	800
Rated peak current kA	26	48	52	76	82
Permissible rated short-time current	7.2	18	19	31	35

6.3.8 Energy let-through calculation

Manufacturers of circuit-breakers provide energy let-through data for their devices that allow the adequacy of protective conductors to be checked. The particular manufacturer's data must be used for the particular device (frame size and rating), as they differ from manufacturer to manufacturer, for the standard is not sufficiently specific for standard tables to be prepared.

Data will be presented in a similar manner to Table 6.13.

Table 6.13 Manufacturer's energy let-through data for circuit-breakers

I^2t Energy let-through Amperes^2 seconds $\times 10^6$								
Circuit-breaker frame	Circuit-breaker rating (A)	Prospective fault current of the installation						
		10 kA	20 kA	25 kA	30 kA	36 kA	40 kA	50 kA
CD	16–100	0.28	0.42	0.47				
CN	125–250	0.52	0.70	0.71	0.72	0.73		
CH	16–250	0.52	0.70	0.71	0.72	0.73	0.74	0.75
SMA	300–800		8.6	12	15	21	25	36

Using the formula:

$$S = \frac{\sqrt{I^2 t}}{k}$$

and looking up k in Tables 54.2 to 54.6 of BS 7671, the minimum cross-sectional area can be calculated. For example, consider a 300 A circuit-breaker (frame type SMA above) is installed in a location where the fault level is 30 kA. What is the minimum size of copper protective conductor? From Table 6.13, the I^2t energy let-through would be 15×10^6 ampere^2 seconds.

From Table 54.3 k is 115

$$\text{and } S = \frac{\sqrt{I^2t}}{k} = \frac{\sqrt{15 \times 10^6}}{115}$$

$$S = 33.7 \text{ mm}^2$$

6.3.9 *Protection against overload and short-circuit*
Regulation 435.1

Where an overcurrent protective device is able to provide overload protection, and is capable of interrupting the maximum prospective short-circuit current at the point of installation, no further calculations need to be carried out on the ability of the circuit conductors to carry both the overloads and the fault currents, provided that the csa of neutral and protective conductors equals (or exceeds) that of the line conductor.

6.4 Calculation of k
Regulation 434.5, Tables 43.1, 54.2 to 54.6

Should a designer wish to calculate the value of k for materials other than those listed in BS 7671, or for different initial and final temperatures, use can be made of the generalized form of the adiabatic temperature rise formula from BS 7454 as follows:

$$I^2_{AD}t = \frac{\sigma_c(\beta+20)\times10^{-12}}{\rho_{20}} \ln\left(\frac{\theta_f+\beta}{\theta_i+\beta}\right)S^2 \qquad \text{(i)}$$

where:

I_{AD} = short-circuit current (rms over duration) calculated on an adiabatic basis (A)
t = duration of short-circuit (s)
σ_c = volumetric specific heat of the current-carrying component at 20°C (J/Km^3)
ρ_{20} = electrical resistivity of the current-carrying component at 20°C (Ωm)
S = cross-sectional area of the current-carrying component (mm^2). For conductors and metallic sheaths it is sufficient to take the nominal cross-sectional area (in the case of screens, this quantity requires careful consideration).
θ_f = final temperature (°C)
θ_i = initial temperature (°C)
β = reciprocal of temperature coefficient of resistance of the current-carrying component at 0°C (°C)
ln = $\log_e$ (natural logarithm).

The equation of Regulation 434.5.2 and Table 43.1 etc. can be rearranged as $I^2_{AD}t = k^2S^2$, hence it can be seen that:

$$k = \sqrt{\frac{\sigma_c(\beta+20)\times10^{-12}}{\rho_{20}} \ln\left(\frac{\theta_f+\beta}{\theta_i+\beta}\right)} \qquad \text{(ii)}$$

Table 6.14 Cable constants

Material	β* (k)	σ_c† (J/Km^3)	ρ*$_{20}$ (Ωm)
Copper	234.5	3.45×10^6	1.7241×10^{-8}
Aluminium	228	2.50×10^6	2.8264×10^{-8}
Lead	230	1.45×10^6	21.4×10^{-8}
Steel	202	3.80×10^6	13.8×10^{-8}

* Values taken from Table I from IEC 287:1991 (withdrawn).
† For further information see *Electra*, No. 24, October 1972.

The constants used in the above formula are given in Table 6.14.

Consider the k value for copper conductors with 70 °C thermoplastic in Table 43.1 of BS 7671, $\sigma_c = 3.45 \times 10^6$, $\beta = 234.5$, $\rho_{20} = 1.7241 \times 10^{-8}$, $\theta_f = 160$, $\theta_i = 70$ °C then from equation (ii):

$$k = \sqrt{\frac{3.45\times10^6(234.5+20)\times10^{-12}}{1.7241\times10^{-8}}\ln\left(\frac{160+234.5}{70+234.5}\right)}$$

$$= 114.79$$

the value given in Table 43.1 (i.e. 115) for 70 °C thermoplastic insulation.

6.5 Non-adiabatic calculations

Regulations 434.5 and 543.1.3

The heading to Table 43.1 states that the data are applicable only for disconnection times up to 5 s. The equation $S = \frac{\sqrt{I^2t}}{k}$ assumes that no heat is lost from the cable during the fault, i.e. the equation is *adiabatic* (from the Greek, meaning impassable to heat). The temperature rise can be assumed to be adiabatic for disconnection times of up to 5 s. If it is used for larger disconnection times it errs in giving larger conductor sizes than are strictly necessary to keep the temperature rise within the limits specified, as heat is lost from the conductor during the fault. This loss is not significant for short disconnection times.

If greater accuracy is required, the calculation may be made in accordance with the method in BS 7454 *Method for calculation of thermally permissible short-circuit currents, taking into account non-adiabatic heating effects.* Regulation 543.1.3 allows the csa of protective conductors to be calculated in this way. The method makes allowance for heat loss from the cable so that the calculation is non-adiabatic and more accurate particularly for long disconnection times.

The permissible non-adiabatic short-circuit current is given by:

$$I = \varepsilon \times I_{AD}$$

where:

I = permissible short-circuit current

I_{AD} = short-circuit current calculated on an adiabatic basis

ε = factor to allow for heat loss into the adjacent components. For adiabatic calculations $\varepsilon = 1$.

The general form of an empirical equation for the non-adiabatic factor is:

$$\varepsilon = \sqrt{1 + FA\sqrt{\frac{t}{S}} + F^2B\left(\frac{t}{S}\right)}$$

where:

F = factor to account for imperfect thermal contact between conductor or wires and surrounding or adjacent non-metallic materials – 0.7 is recommended (1.0 for oil-filled cables)

A, B = the empirical constants based on the thermal properties of the surrounding or adjacent non-metallic materials

$$A = \frac{C_1}{\sigma_c}\sqrt{\frac{\sigma_t}{\rho_t}}\,(\text{mm}^2/\text{s})^{1/2} \text{ where } C_1 = 2464\,\text{mm}/\text{m}$$

$$B = \frac{C_2}{\sigma_c}\left(\frac{\sigma_t}{\rho_t}\right)(\text{mm}^2/\text{s}) \text{ where } C_2 = 1.22\,\text{Km mm}^2/\text{J}$$

and

σ_c = volumetric specific heat of the current-carrying component (J/Nkm^3)

σ_i = volumetric specific heat of the surrounding or adjacent non-metallic materials (J/km^3)

ρ_i = thermal resistivity of the surrounding or adjacent non-metallic materials (K.m/W).

(Suggested values for these material constants are set out in IEC 949 or BS 7454.)

The general formula can be simplified for common combinations of materials as follows:

$$\varepsilon = \sqrt{1 + X\sqrt{\frac{t}{S}} + Y\left(\frac{t}{S}\right)}$$

where X and Y, incorporating the thermal contact factor of 0.7 (1.0 for oil-filled cables), are given in Table 6.15.

The designer will, only in the most carefully considered circumstances, have disconnection times exceeding 5 s. Consequently, the use of non-adiabatic equations should generally be restricted to flexible cables.

Non-adiabatic protective conductor characteristics are shown in Figures 6.11 and 6.12 together with adiabatic characteristics. The use of the non-adiabatic

Table 6.15 Constants for use in the simplified non-adiabatic formula for conductors and spaced wire screens

	Constants for copper		Constants for aluminium	
Insulation	X $[(mm^2/s)^{½}]$	Y (mm^2/s)	X $[(mm^2/s)^{½}]$	Y (mm^2/s)
PVC: ≤3 kV	0.29	0.06	0.40	0.08
>3 kV	0.27	0.05	0.37	0.07
XLPE	0.41	0.12	0.57	0.16
EPR: ≤3 kV	0.38	0.10	0.52	0.14
>3 kV	0.32	0.07	0.44	0.10
Paper: oil-filled	0.45	0.14	0.62	0.20
others	0.29	0.06	0.40	0.08

Note: The thermal contact factor used is 0.7 except for oil-filled cables when 1.0 is used.

characteristics of a cable will increase allowable loop impedances and as a consequence allowable cable lengths. For disconnection times of 5 s or less, the difference is small and, the quicker the disconnection time, the smaller the difference.

6.6 Protection of conductors in parallel

Regulations 433.4, 434.4, 523.8, App. 10, see also 9.6

BS 7671 allows cables run in parallel to be protected by a single device. The requirement is fulfilled provided the conductors are the same cross-sectional area, conductor material, length, disposition and have no branch circuits; it is usually to be assumed that the current will be shared equally among them, but see Regulation 523.7 for single-core cable limitations.

The use of conductors that are not of the same cross-sectional area is allowed by Regulation 433.4.2, if the suitability of the particular arrangement is verified. For conductors up to 120 mm^2, smaller conductors have relatively greater current-carrying capacity relative to their cross-sectional area than larger conductors, so that for conductors up to 120 mm^2 (see Appendix 10), all other factors being equal, load is shared proportionally to cross-sectional area, and the smaller conductor will not be damaged by load currents.

However, for larger conductors, the inductance of the cable – being fairly constant, whatever the csa of the cable – plays an increasingly important part as csa increases.

There is an equation in Appendix 10 for the current in a particular cable run in parallel with other cables, all of different cross-sectional areas.

It is perhaps more readily understood as:

$$I_k = \frac{I_b Z_t}{Z_k}$$

where:
I_b is the total current for all the parallel conductors
I_k is the design current for conductor k
Z_k is the impedance of conductor k
Z_t is the impedance of the conductors in parallel

$$\frac{1}{Z_t} = \frac{1}{Z_1} + \frac{1}{Z_2} + \ldots\frac{1}{Z_m}$$

Appendix 10 of BS 7671 provides much guidance for conductors in parallel. Figure 10A, 'Circuit in which an overload protective device is provided for each of the m conductors in parallel', seems to be an arrangement with some disadvantages. For similar cables there is no benefit in a protective device on each cable, since, if one is overloaded, all will be overloaded and for dissimilar cables the operation of one device will result in load transfer to the other cables and overloading of them, resulting in cascade tripping.

An adiabatic calculation must be carried out on conductors in parallel for a fault at the extremity, to ensure that live and protective conductors of the smaller cable meet the adiabatic equation. This calculation must always be carried out if the current-carrying capacity of a conductor is less than that of the nominal rating I_n of the protective device (see paragraph 6.3.4; see also section 9.6).

6.7 Flexible cables and cords

For fixed installation design, although Regulation 543.1.3 allows the use of non-adiabatic calculations, it is not usual to do so and the benefits are small. However, when determining the earth fault current capabilities of flexible cables, use is often made of the non-adiabatic equations when determining minimum flex sizes and maximum flex lengths.

Figures 6.11 and 6.12 show non-adiabatic and adiabatic conductor characteristics plotted over BS 88 fuse characteristics. The adiabatic characteristics are straight lines; the non-adiabatic curves are asymptotic with the adiabatic characteristics (as time decreases ε of the adiabatic equation approaches 1). These curves can be used to determine the limiting lengths of flexible cable for particular overcurrent devices.

Table 6.16 has been prepared from Figures 6.11 and 6.12. It is assumed that the circuit to the flex connection point is designed such that 0.4s disconnection is achieved for faults at the extremity of the fixed part of the circuit. The flex length is determined by limiting the loop impedance of the fixed circuit plus the flex, either:

(i) within the non-adiabatic limit of the flex, or
(ii) to ensure disconnection within 5 s.

The figures in brackets are the flex length allowed if not limited by 5 s disconnection. There is no particular reason why flex lengths should be limited such

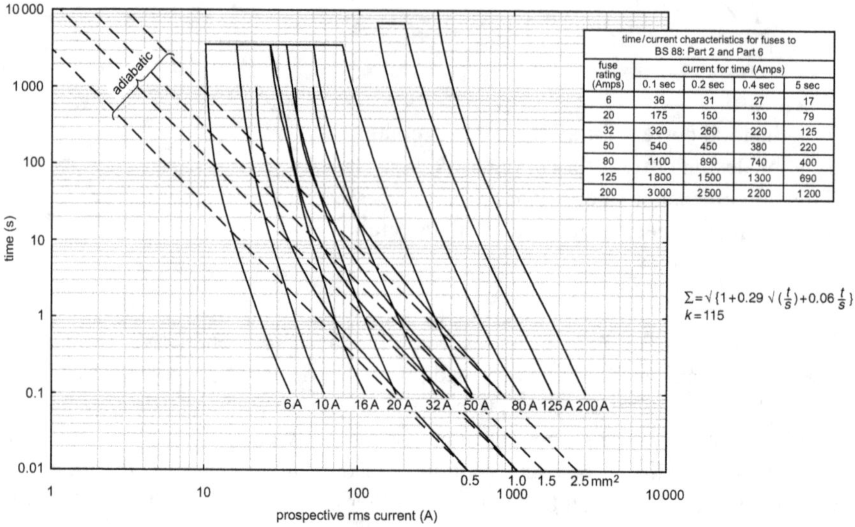

fuse rating (Amps)	current for time (Amps) 0.1 sec	0.2 sec	0.4 sec	5 sec
6	36	31	27	17
20	175	150	130	79
32	320	260	220	125
50	540	450	380	220
80	1100	890	740	400
125	1800	1500	1300	690
200	3000	2500	2200	1200

time/current characteristics for fuses to BS 88: Part 2 and Part 6

Figure 6.11 70 °C PVC flexible cord non-adiabatic characteristics

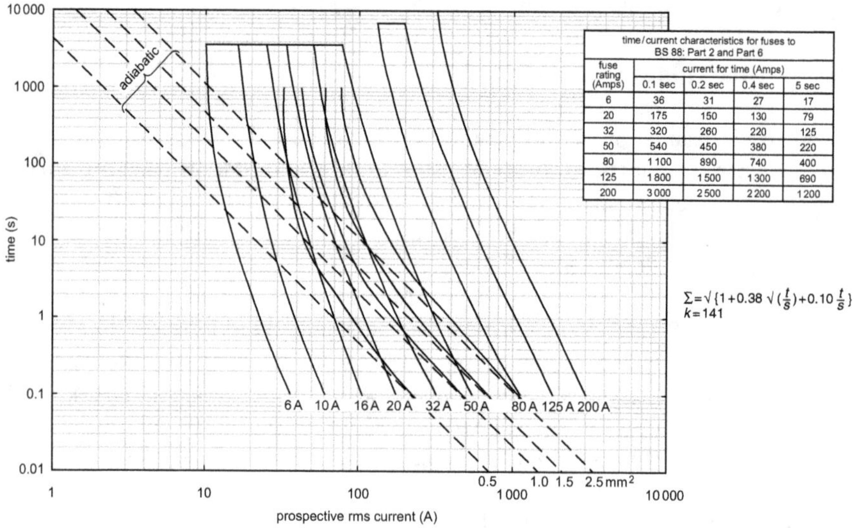

fuse rating (Amps)	current for time (Amps) 0.1 sec	0.2 sec	0.4 sec	5 sec
6	36	31	27	17
20	175	150	130	79
32	320	260	220	125
50	540	450	380	220
80	1100	890	740	400
125	1800	1500	1300	690
200	3000	2500	2200	1200

time/current characteristics for fuses to BS 88: Part 2 and Part 6

Figure 6.12 60 °C rubber flexible cord non-adiabatic characteristics

Table 6.16 Maximum lengths of flexible cables when used in circuits protected by BS 88 fuses (circuits designed for 0.4 s disconnection at the flex outlet or socket)

BS 88 fuse rating (A)	Conductor size (mm^2)	PVC cable (thermoplastic)		Rubber cable	
		Maximum non-adiabatic fault current, I_s (A)	Maximum cable length, Note 1 (m)	Maximum non-adiabatic fault current, I_s (A)	Maximum cable length, Note 1 (m)
16	0.50	75	4	60	13
	0.75	60	19	Note 3	24(∞)(2)
	1.00	50	33(43)(2)	Note 3	33(∞)(2)
20	0.75	120	2	100	8
	1.00	95	14	Note 3	26(∞)(2)
	1.25	70	32(42)(2)	Note 3	32(∞)(2)
32	1.00	250	NP	200	2
	1.25	200	3	175	7
	1.50	170	10	110	26(∞)(2)
	2.50	95	43(75)(2)	Note 3	43(∞)(2)

Notes:

1. Cable length calculated assuming 240 V open-circuit voltage and conductor impedance from Table 41.1 of BS 7671 at 60 °C.
2. Where a 5 s disconnection time at the end of the flex limits the length, this limited length is tabled. The length in brackets is the non-adiabatic limiting length. In some circumstances the length is infinite when the conductor characteristics do not intersect the device characteristics.
3. Limited only by the fault capability of the fuse.

NP Not permitted.

that disconnection occurs in 5 s, other than that some limit has to be set, and 5s has been adopted for fixed installations.

$$\text{Maximum flex length for } 5s \text{ disconnection} = (Z_{5s} - Z_{0.4s})\frac{1000}{(R_1+R_2)\times C_t}$$

where:
Z_{5s} = maximum loop impedance for 5 *s* disconnection
$Z_{0.4s}$ = maximum loop impedance for 0.4 *s* disconnection
C_t = factor to correct ($R_1 + R_2$) from 20 °C to operating temperature.

$$\text{Maximum flex length to meet non-adiabatic limits} = (Z_A - Z_{0.4})\frac{1000}{(R_1+R_2)\times C_t}$$

where: Z_A = maximum loop impedance to meet non-adiabatic requirements Z_A is determined from the intersection of the non-adiabatic conductor characteristics with the device characteristic.

When circuit-breakers protect the final circuit supplying a flexible cable, a different approach needs to be taken as there is no difference in loop impedance for 0.4 or 5 *s* disconnections for the characteristics in Appendix 4 of BS 7671. The design maximum value of loop impedance for such a circuit will have to be reduced by an amount to allow for the loop impedance of the cable, and it is suggested that 3 m of flexible cord could be presumed.

Manufacturers' characteristics may show a difference in loop impedance for 0.4 and 5 *s* disconnection; however, the approach described above is probably still the most appropriate.

6.8 Appliance flexible cords

It is a requirement of the Plugs and Sockets etc. (Safety) Regulations 1994 that most, if not all, appliances sold to the public be fitted with a standard UK BS 1363 plug (or come fitted with an approved conversion plug for use in conjunction with a non-UK plug which complies with the safety provisions of IEC 884-1).

It is standard practice to fit either a 3 A or a 13 A fuse to these plugs, appropriate to the size of the flex and the load. The rating of the plug fuse I_n must exceed the rating of the appliance.

Concern is sometimes expressed at the general guidance given that, if the rating of the device is less than 3 A (or 720 W at 240 V), a 3 A fuse be fitted, and, for a load in excess of 3 A, a 13 A fuse.

The role of the plug fuse is to protect the flex. Figure 6.13 shows that a 0.5 mm^2 flex is protected by a 3 A fuse whatever the flex length, and a 1.0 mm^2 PVC flex is protected whatever its length by a 13 A fuse. The 1.0 mm^2 flex non-adiabatic curve just clips the upper edge (worst case) of the operating zone of the 13 A fuse. This is clearly very useful practice for appliances with long flexes, such as vacuum cleaners and lawn mowers.

6.9 Motor starters

Regulations 435.2 and 536.5.2

Quite commonly, motor contactors and starters are backed up by a short-circuit protective device, often the fuse at the origin of the motor circuit. Regulation 435.2 specifically allows the type of coordination described in BS EN 60947-4-1. This is the part of the British Standard for low voltage switchgear and controlgear, which specifies requirements for electromechanical contactors and motor starters. This standard describes two types of coordination:

Type 1: Coordination requires that, under short-circuit conditions, the contactor or starter shall cause no danger to persons or installation and may not be suitable for further service without repair and replacement of parts.

Type 2: Requires that, under short-circuit conditions, the contactor or starter shall cause no danger to persons or installation and shall be suitable for further use. The

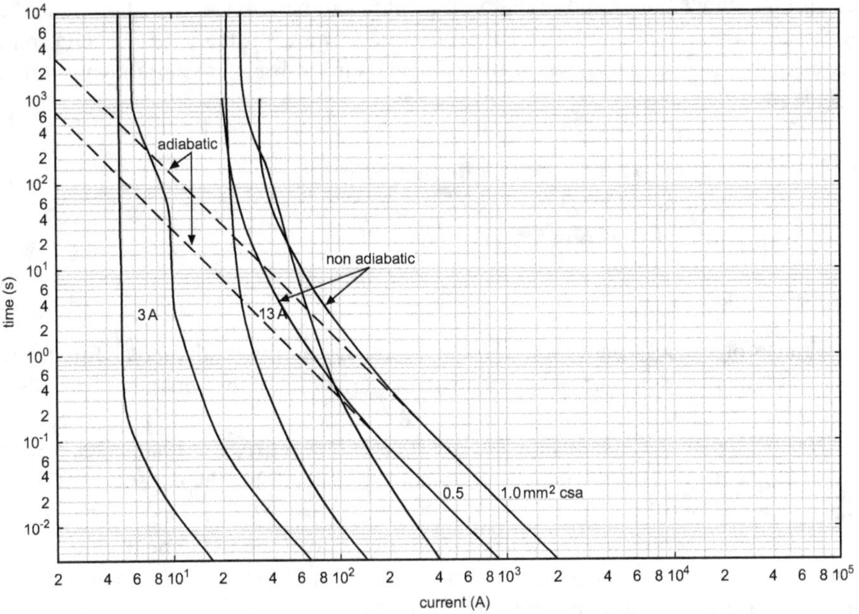

Figure 6.13 70 °C PVC flexible cord non-adiabatic characteristics plotted on the fuse characteristics of a 13 A fuse to BS 1362

risk of contact welding is recognized, in which case the manufacturer shall indicate the measures to be taken as regards maintenance of the equipment.

The reason for the statement in the second paragraph of Regulation 435.2 can now be understood, as it does allow Type 1 coordination, which, while safe, can render the contactor or starter unfit for future use. This would contravene the first requirement of Regulation 435.1.

Designers, clearly, must be aware of the differences between Type 1 and Type 2 coordination and, if they require the motor or starter to be operable after a fault, they must specify Type 2 coordination to BS EN 60947-4-1:1992.

6.10 Coordination of overload current and fault current protection Regulations 434.5.1, 435.2 and 536.1

Regulation 435.2 allows a device with a lower breaking capacity than the prospective short-circuit current to be installed, if it is backed up by another protective device having the necessary breaking capacity on its supply side. This situation normally arises where one device such as a circuit-breaker is providing protection against overload protection, and the backup device, a fuse, is providing protection against fault current. The requirement of Regulation 435.2 is a necessary one, that the backup device (normally a fuse) should limit the energy let-through under fault conditions to that which can be withstood without damage by the overload device. Discrimination between, and coordination of, overcurrent devices fuse-to-fuse, fuse to circuit-breaker and circuit-breaker to circuit-breaker is discussed in Chapter 10.

6.11 Harmonics and overcurrent protection
Regulations 431.2.3, 523.6.1-.3, 551.5.2, Appendix 4 sections 5.5 and 5.6

6.11.1 Introduction

Fault protection is determined in the same way as for fundamental frequency loads, as faults are determined by circuit characteristics, not load characteristics. For overload protection, this is not so.

Normal – that is, fundamental 50 Hz 3-phase – load currents, if balanced, cancel out in the neutral. This is a natural consequence of the 120° displacement of each phase. See Figure 6.14.

However, the third, and other triple, harmonics combine in the neutral to give a neutral current that has a magnitude equal to the sum of the third harmonic content of each phase. The heating effect of this neutral current can have a noticeable effect on the temperature at which a cable operates, particularly where four-core cables are used.

The tabulated current-carrying capacities of cables given in Appendix 4 of BS 7671 are identical for both three- and four-core cables. The presumption is that,

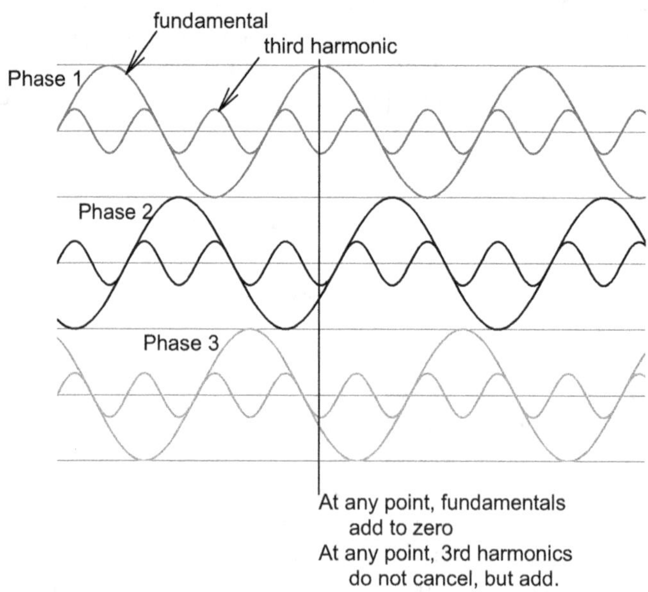

Figure 6.14 Third harmonics

when the neutral conductor in a multicore cable carries current as a result of unbalanced line currents, the heat generated by the neutral current is offset by the reduction in heat generated by line currents. This is not so for triple harmonics.

6.11.2 Overcurrent protection
Regulation 523.6.1-.3, Appendix 4 sections 5.5 and 5.6

The usual formula

$$I_z \geq I_n \geq I_b$$

is applicable for triple harmonic content up to 15 per cent (see Regulation 523.6.1). For greater harmonic content, selection can be made as follows:

for triple harmonic content	0–15%	$I_z \geq I_n$
for triple harmonic content	15–33%	$I_z \geq I_n/0.86$
for triple harmonic content	33–45%	$I_Z = I_n \frac{3h}{86}$
for triple harmonic content and $I_n > I_b$	above 45%	$I_Z = I_n \frac{3h}{100}$

where: I_n is the rating of the overcurrent device in the line conductors
h is the percentage triple harmonic
I_z is the current-carrying capacity of the cable under particular installation conditions.

Overcurrent protection may be provided by devices in the line conductor. However, it may be appropriate to fit overcurrent detection in the neutral, which must disconnect the line conductors, but not necessarily disconnect the neutral (see Regulation 431.2.3).

For PEN conductors in TN-C or TN-C-S systems, the PEN conductor must not be switched (Regulation 537.1.2). It may be appropriate to fit an overcurrent detection device in the neutral. However, this must actuate disconnection of the lines. Overcurrent devices in the line conductors will not necessarily provide overload protection to the neutral unless carefully selected with knowledge of the harmonic content.

Chapter 7
Protection against overvoltages

(Advisers: Lindsay Moody, N.G. Bailey, Tony Haggis I.Eng MIET, Tony Haggis Consulting Ltd; W.J.S. Rogers C.Eng, MIET; Nigel Foster Cu Technical Services Ltd)

7.1 Introduction
Chapter 44

There are four main sections in Chapter 44 of IEC 60364:

- Section 442: Protection of low voltage installations against temporary overvoltages due to earth faults in the high voltage system and due to fault in the low voltage system;
- Section 443: Protection against overvoltages of atmospheric origin or due to switching;
- Section 444: Protection against electromagnetic disturbances;
- Section 445: Protection against undervoltage.

7.2 Protection against faults between high voltage systems and Earth
Section 442 (IEC 364-4-442; HD 384.4.442; IEC 61936-1)

7.2.1 The Electricity Safety, Quality and Continuity Regulations (ESQCR)

Regulation 8(2) of the ESQC Regulations requires that, in respect of any high voltage installation, the earthing shall be designed, installed and maintained so as to prevent danger in any low voltage network occurring as a result of any fault in the high voltage network. The advice in the Guidance on the ESQC Regulations is:

> Duty holders must ensure that persons are not at risk of danger from low voltage networks due to the rise in potential of the earthing system caused by the release of earth fault current from the high voltage system. In practice, duty holders will either interconnect the earthing conductors connected to high voltage equipment and those connected to the low voltage system where the combined resistance to earth is very low or, alternatively, operate separate earth electrodes, in which case the effect of overlapping resistance areas should be such as to prevent the transfer of hazardous potentials from the high to the low voltage system.

The Electricity Supply Regulations (replaced in 2003 by the Electricity Safety, Quality and Continuity (ESQC) Regulations (2002)) required that, where in a substation the HV equipment earth and the LV neutral earth were common, the resistance to earth must not exceed 1 Ω. For most substations this value was provided by the uninsulated protective sheaths of the older types of cable in use and normally was sufficient to lower the impedance of these cable sheaths to ensure sufficiently low earth potential rise (EPR) for general combinations of HV and LV earth systems, even with very high earth fault current. (Earth potential rise is given the designation fault voltage (U_f) in BS 7671.)

However, modern cables have insulated protective sheaths, and earthing characteristics of networks have changed: 1 Ω earth systems or electrodes are not a guarantee of acceptable earthing conditions for many situations. The 1 Ω limit is not repeated in the new ESQC Regulations and therefore a decision must be made on actual earth potential rise of an earth system based on effective impedance and current flowing into the ground at that point.

In urban high voltage distribution networks where cables with uninsulated metallic sheaths still predominate and insulated sheath cables form a relatively small proportion of the total length, the interconnected nature of the HV and LV earthing systems together with the bonding of metallic gas and water pipes results in a low overall earth resistance. Also, the mutual inductance between phase core and screen encourages the majority of HV earth fault current to return to the source substation via the cable sheath/screen rather than through the substation earth electrodes. This 'reduction factor' further reduces EPR. (See ENA Engineering Recommendation S34 and BS 7354, BS EN 50522 Annex I). This ensures a low EPR and is termed a 'Global Earthing System'.

However, the overall earth resistance will be considerably higher than older urban networks, for example where old cables have been replaced with insulated sheath cables and for new networks constructed entirely of insulated sheath cables. In these scenarios, which are becoming more common, the overall resistance to Earth relies mainly on the earth electrodes installed at substations, additional earths installed at the remote ends of LV PME cable systems, and sometimes other fortuitous earths such as bonded steel reinforced building foundations. Furthermore, the beneficial effect of phase/screen mutual inductance is considerably reduced in the modern single core unarmoured XLPE cable compared to traditional Paper Insulated Lead Covered Steel Wire Armoured (PILCSWA) cables.

For example: referring to the formulae in Table 7 of BS 7354;

For PILCSWA C = 76 and for unarmoured single-core XLPE C = 450. When applied to a 1 km length of 185 mm^2 11 kV cable with R_A = 1 ohm and R_B = 10 ohms.

Per unit ground current is:

ß = 0.02 for PILCSWA, meaning only 2% of the total earth fault current returns via the earth electrode

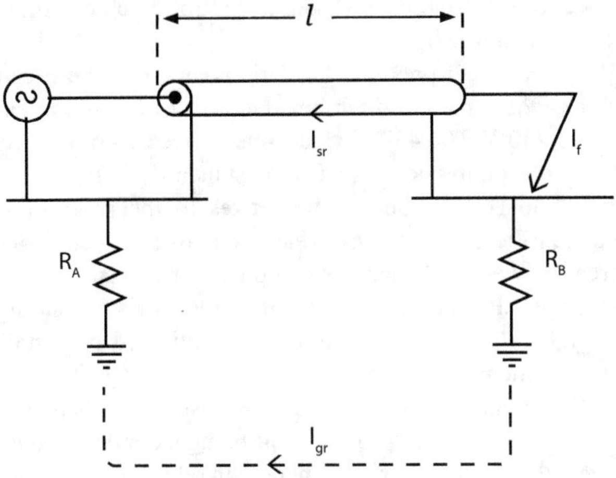

Figure 7.1 Common HV and LV earthing (from table 7 of BS 7354)

A is the conductor csa in mm^2
E is the system voltage in kV
L is the length in km
I_{gr} is the ground return current in A
I_{sr} is the sheath return current in A
I_f is the fault current in A
R_A is the resistance of the earthing arrangement of the exposed-conductive-parts of the equipment of the low voltage installation
R_B is the resistance of the earthing arrangement of the low voltage system neutral, for low voltage systems in which the earthing arrangements of the transformer substation and of the low voltage system neutral are electrically independent.
$R_{AB} = R_A + R_B$

ß = 0.13 for XLPE, meaning that 13% of the total earth fault current returns via the earth electrode.

As it is the current in the earth electrode R_B that results in EPR it can be seen that modern cable systems present increased earthing challenges, particularly where the earth fault current is relatively high.

This challenge may also occur within Global Earthing Systems where a new building's substation is supplied directly from the source substation by XLPE cable rather than by direct connection to the local network in the street. The building may not be bonded to metallic gas and water pipes (as plastic pipes now prevail) or to the multiple earthed LV network. Thus the building and substation become an isolated earth electrode system with the possibility of a high EPR. The earthing design may take into account the resistance of the building's foundations to augment the substation electrode. Care must be taken to control touch potentials both within and around the outside of the building.

BS 7430:1998 supported previous regulations and also the Electricity Supply Regulations with a requirement for combined earth systems not to exceed 1 Ω and EPR not to exceed 430 V. The 430 V value was derived from the requirements of the International Telecommunications Union standards detailed in Engineering Recommendation S36-1 (ER S36-1), and relates to the protection of telecoms equipment from the EPR transferred from electricity distribution systems. The main concern is for remotely earthed telecoms equipment and circuits that terminate in a substation with a high EPR, or terminating within the area surrounding a substation affected by the high EPR. The hazard is usually mitigated by installing isolation units where the equipment is terminated.

Engineering Recommendation S36-1 defines substations where the EPR exceeds 430 V or 650 V as 'Hot', with the 430 V limit being extended to 650 V where the operating voltage is 33 kV or higher and where high reliability circuit protection will clear the earth fault within 200 ms. Substations with EPR values lower than these limits are defined as 'Cold'.

The electricity supply industry earthing standard Energy Networks Association Engineering Recommendation 41-24, and BS 7430:1998, use the same 430 V and 650 V limits to determine whether or not the LV neutral earth system can be combined with the HV earth system. The electricity supply industry has therefore used the same terminology as defined in ER S36-1 to refer to substations with combined HV and LV neutral earths as 'Cold' and 'Hot' where they are separated. It has therefore been generally accepted that a transfer potential of up to 430 V from the HV to the LV system is acceptable, irrespective of the final touch potential values seen on the LV system.

Regulation 8(2)(b) of the Electricity Safety, Quality and Continuity Regulations states that:

(2) *A generator or distributor shall, in respect of any high voltage network which he owns or operates, ensure that—*

- *(a) the network is connected with earth at, or as near as is reasonably practicable to, the source of voltage but where there is more than one source of voltage in that network, the connection with earth need only be made at one such point;*
- *(b) the earth electrodes are designed, installed and used in such a manner so as to prevent danger occurring in any low voltage network as a result of any fault in the high voltage network; and . . .*

DNO substations supply public LV TN-C networks which employ Protective Multiple Earthing (PME) in which the neutral earth (PEN) conductor is earthed at the end of each main in addition to the neutral earth electrode at the substation. In BS EN 50522 the permissible touch potential (Figure 7.4) is doubled where the LV system has multiple earths (i.e. F factor = 2). By taking into consideration the fact that in most situations it is expected that shoes will be worn, the higher permissible touch potential values in Figure 7.5 can be applied, which raises the allowable touch potential limit to 233 V for fault durations of 1 second. Multiply this by the F factor of 2, giving a value of 466 V, means that DNOs can still use the existing 'Hot' and 'Cold' criteria for determining whether the HV and LV earths can be combined. This, together with the protective equipotential bonding required in buildings by BS 7671, ensures that dangerous touch potentials are not experienced due to HV earth fault potentials transferred from the DNO's HV system.

However, the ESQC Regulations do not permit the TN-C system to be used in consumer's installations and so the installation must have separate neutral and protective conductors. As the LV neutral is only earthed at one point (i.e. the transformer star point) the BS EN 50522 F factor remains at 1. In both cases the touch potential limits in Figures 7.4 and 7.5 must be observed.

The designer must now look to BS 7671 and more particularly to BS EN 50522:2010 *Earthing of power installations exceeding 1 kV a.c.* for the limits to earth fault voltage rise (EPR).

7.2.2 BS 7671: 2008 Amd 3 2014

BS 7671 simply advises that fault voltages (see figure 7.2) as calculated using Table 44.1 of BS 7671, shall not exceed a dangerous level (Regulation 442.2.1). It is noted in BS 7671 that for global earthing systems where the HV earth (R_E) and LV earth (R_B) are connected together (figure 7.2) as is usual in urban distribution networks the requirements may be met.

A global earthing system is defined in BS EN 50522 as:

> **3.4.19 global earthing**
>
> equivalent earthing system created by the interconnection of local earthing systems that ensures, by the proximity of the earthing systems, that there are no dangerous touch voltages.
>
> *NOTE 1* Such systems permit the division of the earth fault current in a way that results in a reduction of the earth potential rise at the local earthing system. Such a system could be said to form a quasi equipotential surface.
>
> *NOTE 2* The existence of a global earthing system may be determined by sample measurements or calculation for typical systems. Typical examples of global earthing systems are in city centres and urban or industrial areas with distributed low- and high-voltage earthing.

When designing an installation that includes an HV supply, reference might be made to BS 7430: *Code of practice for protective earthing of electrical installations*. However section 5 of this standard refers the reader to other standards as follows:

> Specific requirements that should be followed for high voltage installations including substations are found in BS EN IEC 61936-1:2011 and BS EN 50522:2010. Requirements that should be followed for the protection of LV installations against temporary overvoltages due to earth faults in an HV supply system are given in BS 7671:2008+3, Section 442.

7.2.3 *HV fault voltages: the hazard*

A fault to Earth on the HV system will result in a ground current I_E in the HV substation earthing arrangement (e.g. local earth grid or electrode) (see Figures 7.2 and 7.3). This will be greatest if the HV fault occurs on cables or equipment in or close to the substation supplying the affected LV system. The fault current flowing through the HV earthing system will raise the potential of the substation metalwork relative to true earth. This earth potential rise (EPR) or fault voltage (U_f) will be impressed on any metalwork connected to the HV earthing, including the LV earthing if there is a common HV/LV earthing system. This voltage will be transmitted to the earth connections of all LV equipment via the PE conductors. This voltage may be above the touch voltage limits in BS EN 50522 Figure 7.4 or 7.5 except in a Global Earth System. Additionally, faults in nearby substations operating with higher voltages may transfer ground current and EPR to an LV system. It is therefore necessary to establish the prospective EPR and, where necessary, control touch potentials to safe values by protective equipotential bonding or segregating the HV earth from the LV earth or a combination of both.

7.2.4 *Substations with separated earths*

For substations in TN systems with separated earth systems, see Figure 7.2, fault voltage will not be directly impressed on the LV neutral and earth systems of the associated supplies with reference to remote earth. However, voltage will be transferred from the HV electrodes to the LV electrodes via the soil, the value depending on the proximity of the LV electrodes to the HV electrodes. A series of concentric voltage contours will occur around the HV electrode system, *on the surface of the surrounding ground*, which gradually falls to zero over some distance, which varies according to the area of the HV electrode system and the local soil resistivity. The value of U_f will depend on the potential of the voltage contour where the LV electrode is installed, its equivalent source resistance and the combined earth impedance of the LV electrodes and other earth connections, such as the resistance of the steel reinforcement in building foundations bonded to the building earth. In a correctly designed system, U_f will be low. However, a stress voltage (U_1) will exist on the LV equipment located inside and around the substation.

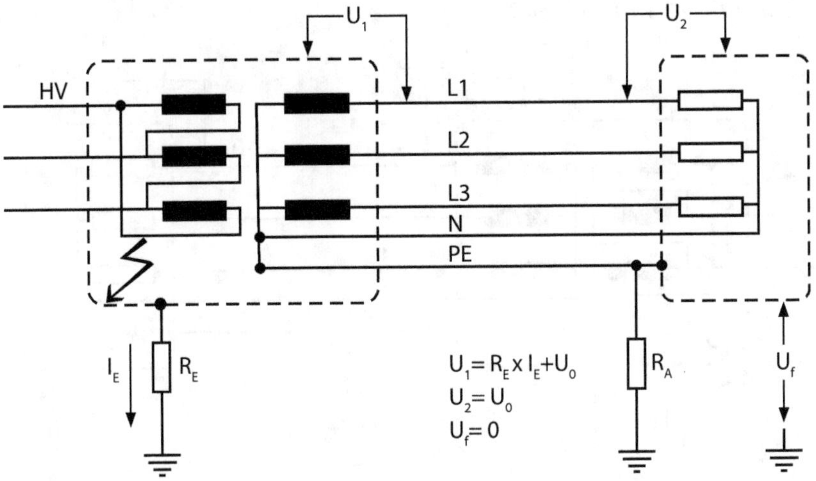

where:

I_E = is the part (component) of the earth fault current in the high voltage system that flows through the earthing arrangement of the transformer substation ground

R_E = is the resistance (or impedance) of the earthing arrangement of the transformer substation

R_A = is the resistance (or impedance) of the earthing arrangement of the exposed-conductive-parts of the equipment of the low voltage installation supplied from the substation

R_B = is resistance of the substation earthing arrangement of the low voltage system neutral, for low voltage systems in which the earthing arrangements of the transformer substation and of the low voltage system neutral are electrically independent

U_0 = is nominal a.c. rms line voltage of the low voltage system to Earth

U_f = is power frequency fault voltage that may affect premises supplied from the low voltage system between exposed-conductive-parts and earth for the duration of the fault

U_1 = is power frequency stress voltage between the line conductor and the exposed-conductive-parts of the low voltage equipment of the transformer substation during the fault

U_2 = is power frequency stress voltage between the line conductor and the exposed-conductive-parts of the low voltage equipment of the low voltage installation during the fault

Note: It is recommended that the minimum separation between the LV and HV electrodes be calculated to confirm the separation is sufficient. Many DNOs use a standard separation distance which varies from company to company irrespective of the actual requirements. In many cases these standard distances are more than required so the transferred touch potentials on the LV system are well within acceptable levels; however in high resistivity ground conditions the separation distance can be excessive, with the standard distances resulting in touch potentials on the LV system that exceed permissible limits.

Figure 7.2 Fault voltages U_f and stress voltages U_1 and U_2 arising from HV faults – TN systems with separated HV and LV earths

7.2.5 Substations with common HV/LV earth

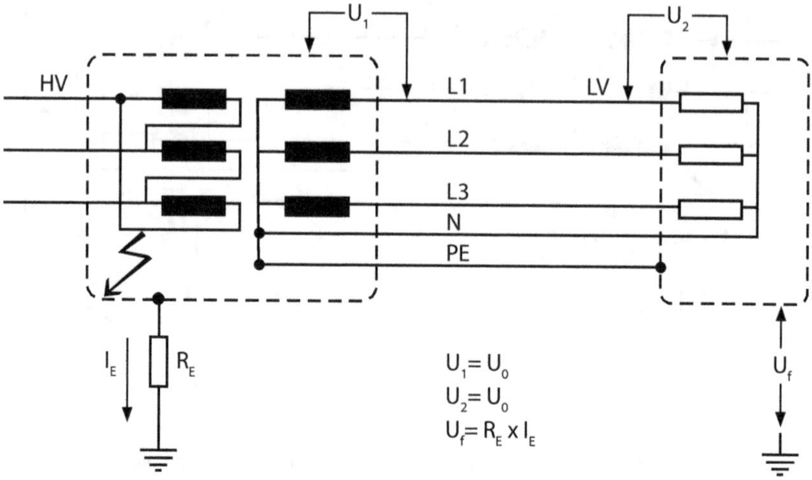

where:

I_E = is the part (component) of the earth fault current in the high voltage system that flows through the earthing arrangement of the transformer substation ground

R_E = is the resistance (or impedance) of the earthing arrangement of the transformer substation

R_A = is the resistance (or impedance) of the earthing arrangement of the exposed-conductive-parts of the equipment of the low voltage installation supplied from the substation

R_B = is resistance of the substation earthing arrangement of the low voltage system neutral, for low voltage systems in which the earthing arrangements of the transformer substation and of the low voltage system neutral are electrically independent

U_0 = is nominal a.c. rms line voltage of the low voltage system to Earth

U_f = is power frequency fault voltage that may affect premises supplied from the low voltage system between exposed-conductive-parts and earth for the duration of the fault

U_1 = is power frequency stress voltage between the line conductor and the exposed-conductive-parts of the low voltage equipment of the transformer substation during the fault

U_2 = is power frequency stress voltage between the line conductor and the exposed-conductive-parts of the low voltage equipment of the low voltage installation during the fault

Figure 7.3 Fault voltages U_f and stress voltages U_1 and U_2 arising from HV faults – TN systems with common HV/LV earthing system

In TN systems derived from substations with a common HV/LV earth system, a power frequency fault voltage (U_f) will arise between all metalwork connected to the substation earth system, including the connected LV supply cables and true earth for the time it takes the HV protection to operate.

For TN systems and a combined HV and LV earth system the LV installation fault voltage (U_f) will be given by:

$$U_f = I_E R_E$$

The fault voltage is defined as voltage between a given point of fault and reference earth resulting from an insulation fault.

In a global earthing system with protective equipotential bonding there will be minimal risk of electric shock. In all situations designs will have to be carried out in accordance with BS EN 50522.

7.2.6 BS EN 50522 Earthing of power installations exceeding 1 kV a.c.

In all substations exceeding 1kV the substation fault voltage (prospective touch voltage) must be kept within the limits prescribed in BS EN 50522.

The UK requirements (in BS EN 50522) are different from the rest of Europe as on the advice of the UK Health and Safety Executive (HSE), consensus was reached in the relevant BSI committees that HV earthing systems must be designed according to tolerable voltages based on body impedances exceeded by 95 per cent of the population, rather than the 50 per cent values adopted in EN 50522. Also worth noting has been the consensus among PEL/99 and GEL/600 to move away from using the tolerable body current curve 'c1' to curve 'c2' from DD IEC/TS 60479-1:2005, again, based on advice from the UK HSE. This marks a departure from the very strict deterministic limits observed previously under ENATS 41-24. However, the reduction in values of IEC published body impedances means that the resultant values of tolerable voltages are not greatly affected and certainly not reduced.

Expert's Note

At the time of writing, a decision is to be made on the correct body impedance values to use, as those used in BS EN 50522, taken from IEC/TS 60479-1 do not apply the correction factors detailed in 60479-1, which if applied correctly result in lower body impedance values, which in turn results in lower permissible touch potential values.

A proportion of the substation power frequency fault voltage will appear in premises supplied from the network connected to the substation. Reasonable equipotential conditions will be maintained in the premises by the main protective bonding conductors and this will reduce risk of electric shock. However, services and cables into or from the protected zone suffer a stress voltage with risk of damage.

For TN systems and a combined HV and LV earth system, the LV installation fault voltage $U_f = I_E R_E$.

BS 7671 requires that the fault voltage (U_f) shall not exceed a dangerous level.

A designer will need to make reference to BS EN 50522 and may need to take advice from an expert.

The basic outline of the requirements is given below.

The fault voltage (U_f), or in BS EN 50522 terms the earth potential rise (U_E),does not exceed twice the permissible touch voltage (U_{TP}) (dependent on the clearance time of the HV protection).

U_E less than or equal to U_{TP} or $U_E \leq 2U_{TP}$

UTP is given by Figure NA1 or NA.2 of BS EN 50522, see figures 7.4 and 7.5 below.

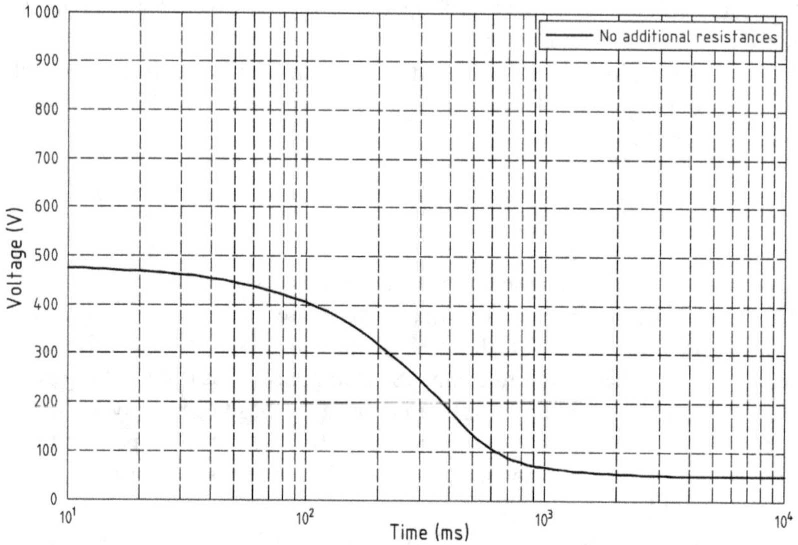

Figure 7.4 Permissible touch voltages – no additional resistances, (figure NA1 of BS EN 50522)

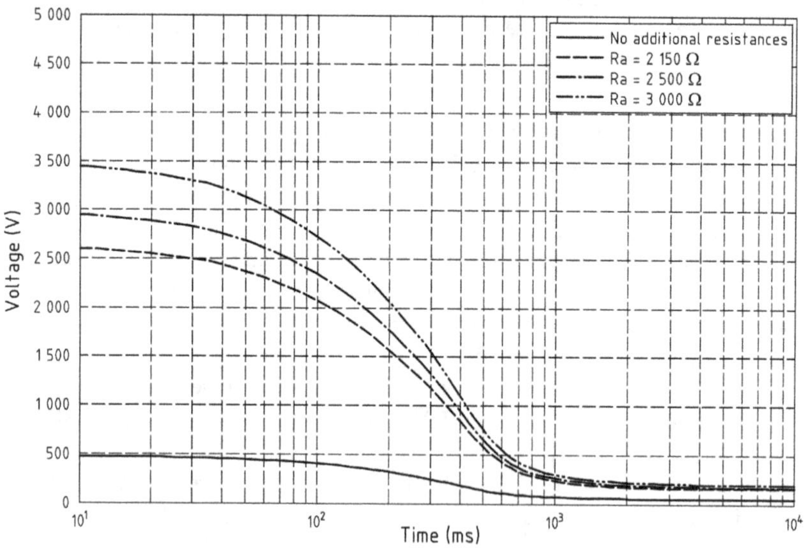

Figure 7.5 Permissible touch voltages – no additional resistances, (figure NA2 of BS EN 50522)

If this cannot be met specific measures or additional measures as described in BS EN 50522 must be taken. It is also acceptable to allow for shoe resistance and additional resistances, depending upon the circumstances pertaining to the particular installation under consideration such as a layer of chippings, in which case higher permissible touch potential values from figure NA.2 and from table NA.1 may be applicable. Allowing for such additional resistance values can significantly reduce the size and cost of installation and make positioning of the substation less problematic, as well as reducing the required separation distance between separate HV and LV earth systems.

7.3 Power frequency stress voltages

Depending on the decision to combine or separate HV and LV earth systems, faults on HV equipment may impose stress voltages on low voltage equipment both at the transformer substation (U_1) and at the LV equipment (U_2) (see Figures 7.2 and 7.3).

Where the HV and LV earths are combined the EPR must be maintained within the limits of BS 7671 Table 44.2 for the disconnection time of the HV supply.

DNO protection systems employed with Neutral Earth Resistors or Reactors (NER) on the primary substation transformer star points will normally clear earth faults within 5 seconds and so the 1200 V EPR limit is applicable. It should be noted that, for HV supplies from overhead lines, EPR may be significantly higher than 1200 V unless a very low earth resistance can be achieved.

DNOs may instead employ tunable Arc Suppression Coil (ASC or Petersen Coil) star point earthing, which will hold earth faults for periods in excess of 5 seconds and possibly several hours as is the practice in many mainland European utilities. This makes the maximum allowable EPR 250 volts. However, achieving an EPR below 250 volts is not as onerous as it first seems. Because the inductance of the ASC is tuned to equal and oppose the capacitance of the overhead line conductors, only a few amps of earth fault current flows which in turn produces very low EPR as it passes to earth via the substation's earth electrode.

Where the HV and LV earths are separate the stress voltage applied to the transformer LV winding becomes $U_1 = R_E \times I_E + U_o$, which may be in the order of several kV.

DNOs are aware of risks from separating neutrals and the unavoidable risk of high stress voltage and purchase transformers built to Energy Networks Association Technical Specification 35-1. The LV AC withstand levels are specified to suit the values of EPR and earth fault clearance times encountered on public distribution networks:

- Ground-mounted transformers to ENATS 35-1 have an LV winding AC withstand of 3 kV rms, which is sufficient for an EPR of up to 2.75 kV. Even when supplied from an overhead line, a ground-mounted transformer benefits from both its own local earth electrode and the earth electrode of the cable termination at the overhead/underground transition pole. Whilst these combined resistances may not ensure a 'Cold' site, the resulting EPR will normally be within the 3 kV capability of the transformer LV winding with the HV and LV earths segregated.

- Pole-mounted transformers are usually specified with the ENATS 35-1 option of 10 kV rms LV winding AC withstand. As the transformer steelwork is inaccessible from ground level high touch potentials can be tolerated and a typical EPR may be up to 5.5 kV with a simple earth electrode system of less than 40 ohms. HV and LV earths are always segregated.

However, BS 7671 Table 44.2 prescribes a stress voltage limit of 1200 V + U_0. Note that in BS 50522 Table 2 (the source for BS 7671 Table 44.2) note c states *"Limit may be increased if appropriate LV equipment is installed or EPR may be replaced by local potential differences based on measurements or calculations".*

Therefore the full transformer LV winding AC withstand of 3 kV can be utilised where the HV and LV earths are separate and the (EPR + U_0) stress does not appear on the LV switchgear.

Where HV and LV earths are to be combined it could be possible to specify LV switchgear with a stress voltage in excess of (1200 V + U_0) volts and therefore tolerate an EPR above 1200 V. However, this would be inadvisable. In commercial/industrial installations it is very difficult to achieve safe touch potentials where the EPR is above 1.2 kV and so the earthing system has to be designed to keep well below this value in any case.

Expert's comment

Substations installed in urban areas

The majority of urban areas will behave as a Global Earth, due to the highly interconnected nature of the existing underground cables and other metallic services. Commercial/industrial estates often consist of steel-framed buildings which contribute significantly to lowering the overall earth resistance.

Where new substations are connected directly to the existing DNO HV cable network in the street then they can normally be considered to be part of a Global Earth and the HV & LV earth combined.

However, where a new substation is supplied directly by XLPE cable over some distance from the DNO primary substation there is a real possibility that the substation could have an EPR high enough to exceed the EPR and touch voltages in BS EN 50522. Whilst DNO substations supplying a multiply earthed public LV network can apply an 'F' factor of 2 to the touch voltage in NA.1 and NA.2 an 'F' factor of 1 is applicable to private networks without multiple LV earthing. The permissible touch voltages are therefore half that of DNO LV networks and this should be borne in mind when seeking advice from the DNO.

An earthing specialist must be employed to evaluate the EPR and design the earthing system. Reference will need to be made to ENA Engineering Recommendation S34, BS 7354 and BS EN 50522 Annex 1 to determine the 'reduction factor' for the proportion of earth fault current entering the local electrode system (producing EPR) compared to the earth fault current returning via the cable screens.

Making accurate earth resistance measurements in urban areas is often impractical using the standard Fall of Potential, Slope or even the Star-Delta methods, due to lack of unmade surfaces and the presence of metallic services (see BS EN 50522 NC5.1, NC5.2, NV 5.4). Computer modelling is often employed to design the system using soil resistivity values based geological knowledge and/or measurements taken at the closest available open space. Often, the only way to confirm the resistance of the newly installed earth installation is to connect the earth tester between the new earth and the existing network cables at the time cable laying/jointing takes place. Alternatively the Clamp Method (NC.9) can be used after the systems are interconnected. These methods compare the resistance of the new installation to the much lower resistance of the existing interconnected network being used as the reference.

Substations in non-urban Areas

Isolated factories and commercial/industrial estates supplied via overhead lines will usually have the highest EPR and are the most challenging to design.

A combined HV/LV earth system is the preferred solution. However, due to the local soil resistivity and size of site it may or may not make it possible to keep touch potentials within BS EN 50522 limits even when including the steel frames and foundations of buildings in the overall assessment.

Separating the HV and LV earths of a commercial/industrial installation is difficult to achieve in practice and earthing needs to be considered in the site's design at the earliest opportunity. There are several options available:

Most DNOs can provide an LV metered supply for loads up to 1000 kVA. This takes the form of a standard DNO compact unit substation fitted with an LV ACB instead of distribution feeder cable cabinet. The customer installs and owns the LV cables from the ACB outgoing terminals to the LV switchroom. The HV substation (and earth) can be placed on the site boundary as preferred by DNOs for operational access. The LV earth can be installed at the LV switchroom, thus providing significant electrical separation between the two earth systems. Many customers prefer this option as they do not have the ongoing responsibility of owning HV equipment. The downside is that the Distribution Use of System (DUoS) charges are higher for LV metered supplies compared to HV metered supplies.

For supplies over 1000 kVA, or where the customer wishes to pay HV DUoS charges, then it is possible to place the HV substation and customer's transformer at the site boundary and again run the LV cables to the LV switchroom where the LV neutral earth is installed.

Where loads are such that for LV volt drop and/or cable cost considerations it is necessary to site the transformer closer to the load then segregating the two earthing systems becomes more challenging. Often this entails insulating the switchgear and transformer from the substation floor and running an insulated HV earth cable to a remote HV earth electrode. Floor insulation has to be installed and maintained for operator safety. Specialist advice is essential.

On large sites with multiple transformers it becomes impractical to separate HV and LV earths. Fortunately the presence of large buildings means that their foundations make a significant contribution to keeping the EPR low. Careful attention to bonding may be necessary, particularly in areas of high soil resistivity, to ensure touch voltages are controlled to safe values both inside and outside the buildings. It may prove expedient to use TT earthing to lighting columns and entry barriers etc. with the steelwire armours 'gapped' where cables leave the buildings. Alternatively, Class 2 equipment could be used outside the building.

LV winding earth fault on segregated HV/LV earth systems

A consequential risk arising from separation of 11 kV and LV earth systems is the possibility of long-term standing voltage on transformer tanks and 11 kV and LV earth systems if an LV earth fault occurs on the transformer. This might be a winding fault or an LV bushing flashover, occurring as a result of increased transient voltages caused by the earth systems separation. An LV earth fault is produced passing current to the 11 kV tank, to ground via 11 kV electrodes, and back into the LV neutral conductor via the LV electrodes. Most cases of this type of fault involve pole-mounted transformers but very occasionally occur with ground-mounted types.

As an example, should HV earth resistance be 1 ohms and the LV be of a similar order of resistance, then the resulting current flowing would be about 250 V /2 = 125 amps This would introduce about 125 V standing, as an EPR on each earth system. T × U_f and LVU_f will both equal 125 V.

The transformer HV protection could not detect this condition and would not operate to isolate the transformer and it would await reporting as a source of electric shock. See Figure 7.6.

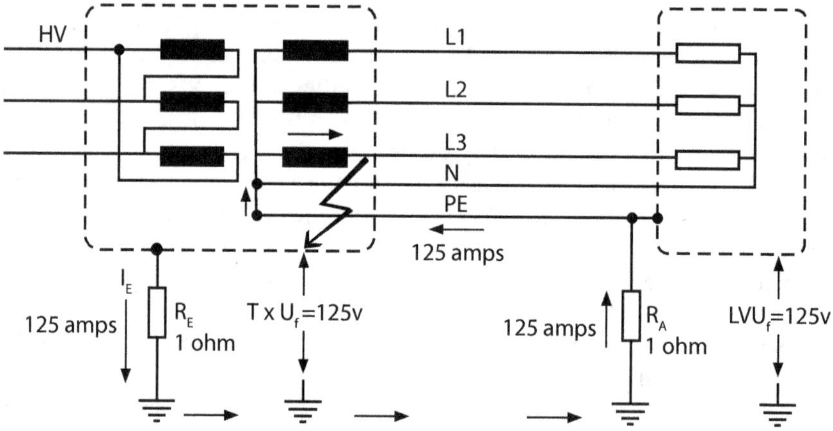

Figure 7.6 LV earth fault to transformer tank

The touch voltage hazard on the transformer tank and HV switchgear is readily addressed by the substation earthing which will be designed to cope with the value of EPR arising from HV earth faults. Persons entering a substation will be wearing footwear and so the touch voltage values in Figure NA2 for $R_A = 2150\ \Omega$, $2500\ \Omega$ or $3000\ \Omega$ can be applied. However, it must be remembered that this touch voltage may remain for a long period and so the +10 second (10^4 ms) part of the curve must be used rather than the 3 second (3×10^3 ms) value that would be used for an HV fault EPR.

Whilst the standing touch voltage of 125 V $T \times U_f$ may be safely controlled within the substation it must be remembered that the LV touch potential LVU_f of 125 V will be transferred to all parts of the LV system via the PE conductors. This is more difficult to control as persons may or may not be wearing footwear and the circuit may extend outside the building to external equipment such as lighting columns, entry barrier systems etc. In some installations 125 V may be an unacceptable risk.

However, in the previous example both the HV and LV earths had the same resistance.

Now consider the following example:

HV resistance R_E is 1 ohm and the LV R_A is 0.25 ohms. The current flow now becomes 250 V/1.25 ohms = 200 amps.

200 amps passed through R_E of 1 ohm – $T \times U_f = 200 \times 1 = 200$ volts

200 amps passed through R_A of 0.25 ohm – $LVU_f = 200 \times 0.25 = 50$ volts

Thus, it can be seen that the value of LVU_f depends on the ratio of the HV to LV earth resistance values.

The incidence of this type of transformer fault and subsequent standing voltage is extremely rare on ground-mounted equipment and control measures are not normally considered necessary.

However, in particularly sensitive installations, adjusting the ratio of R_A to R_E could be done to ensure that the majority of the touch voltage appears in the substation where it can be more easily be controlled.

Target HV earth resistance.

It can be seen from the previous discussion of transformer LV winding faults that from a standing voltage point of view it is desirable to have an HV earth resistance higher than the LV earth resistance when it is necessary to separate HV/LV earths. This infers that a high HV earth value is desirable, which is counter intuitive.

Certainly, in order to combine the HV and LV earths, a low resistance is necessary to keep the EPR and touch voltage below BS EN 50522 safe values.

However, where a small site has to be supplied via overhead lines the HV and LV earths will often have to be separated. It is a common mistake to install an extensive HV earth electrode system in an attempt to keep the HV substation EPR as low as possible. In fact a 20 ohm substation earth will raise sufficient earth fault current to operate DNO earth fault protection within 1 second and this value is often used as standard in DNO substations.

Consider Fig 7.7.

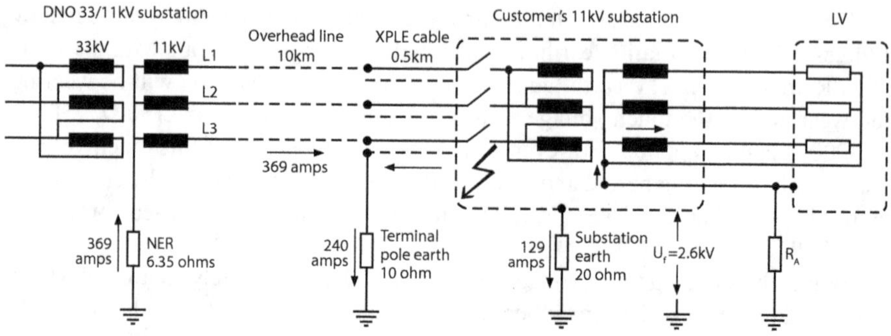

Figure 7.7 Customer's 11 kV substation with separate HV and LV earthing

A customer's 11 kV substation with separate HV and LV earths is supplied via overhead line and cable from a DNO primary 33/11 kV substation.

The DNO's 33/11 kV transformer has a 6.35 ohm Neutral Earth Resistor. At the customer's substation the effective resistance is the combination of 10 + 20 ohms in parallel (6.67 ohms).

The combined resistance of the cable and overhead line is 4.18 ohms and the voltage driving the earth fault current is

$$\frac{11000}{\sqrt{3}} = 6350\ V$$

By simple Ohm's Law,

$$I_f = \frac{6350}{6.35 + 4.18 + 6.67} = 369\ amps$$

This 369 amps of earth fault current would readily cause the DNO earth fault protection to operate within 1 second.

Applying the formula from BS 7354 we can determine the amount of earth fault current passing through the substation's 20 ohm earth.

Per unit ground current

$$\beta = \frac{I_{gr}}{I_f} = \frac{\frac{C}{(a+9E)}}{\sqrt{\left\{\left(\frac{C}{a+9E} + \frac{R_{AB}}{l}\right)^2 + 0.6\left(\frac{\rho}{aE}\right)^{0.1}\right\}}} = 0.35$$

Of the 369 amps earth fault current, only 35% (129 amps) passes through the substation earth resulting in an EPR of 2.6 kV.

Should it be necessary to ensure that in the event of a transformer LV winding earth fault the standing voltage LVU_f is below 50 V the ratio of HV to LV earth resistance needs to be 5:1 which is 6.67/5 = 1.33 ohms.

If additional earth electrodes were installed to reduce the substation earth from 20 ohms to 5 ohms then the total fault current would increase from 369 amps, to 458 amps of which 313 amps would now pass through the substation's 5 ohm earth.

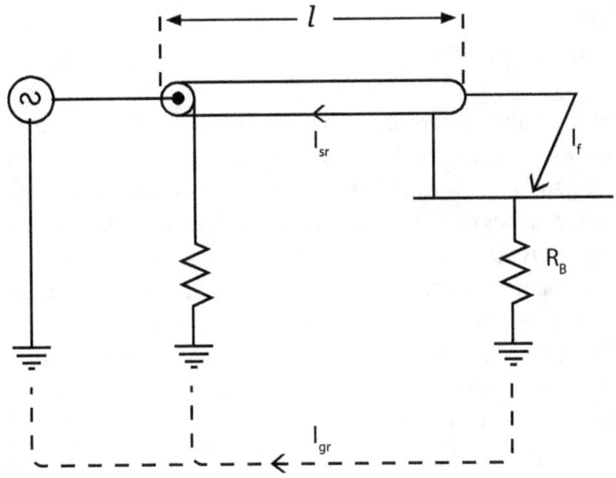

Figure 7.8 Circuit for separate HV and LV earthing (from table 7 of BS 7354)

The new EPR would become 1.56 kV, which would still require the HV and LV earths to be separated. However, to get the 5:1 ratio the LV earth would need to be less than 0.66 ohm which could prove difficult in poor soil resistivity on a small site. In any case, if 0.66 ohm could be achieved for the LV earth then the EPR may be low enough to combine the HV and LV earths.

Note that the above calculations are very simplistic and only consider the principal resistances by simple Ohms Law. To accurately calculate earth fault current the reactive impedance of all elements has to be considered and Symmetrical Component theory applied, which can result in considerable calculation. Sample calculations have shown that the simple Ohm's Law approach results in values of earth fault current about 5 to 10 per cent higher than those calculated by Symmetrical Components. Overestimating earth fault current results in slightly higher values of EPR being calculated. This in turn results in slightly conservative measures being put in place to control touch voltages thus adding a further measure of safety. The simple Ohm's Law approach is usually used except at the highest voltages.

The main problem with attempting to obtain low values of HV earth electrode resistance in unfavourable soil is that an extensive earth electrode system covering a large area may be necessary. This will cause the voltage contours around the HV electrode to be significantly extended, causing additional HV EPR to be impressed upon the LV earth electrode. The separation distance between the HV and LV earths would need to be increased significantly which could prove unattainable on small sites. Furthermore, the increased voltage contours can export HV EPR further from the site and impinge upon other parties.

The earthing design specialist has to model these effects in earthing design software and iterate the design several times until a safe and cost-effective design is attained.

7.4 Protection against overvoltages of atmospheric origin or due to switching
Section 443

The frequency of thunderstorms in the UK is so low as not to warrant special protective measures generally to be taken against lightning strikes (see Regulation 443.2.2). Lightning strikes and switching cause overvoltages on the system. All equipment is required to have a degree of withstand against such voltages. BS 7822-1:1995 (IEC 60664-1:1992) *Insulation Coordination for Equipment Within Low-Voltage Systems* provides for coordination of impulse voltage withstands. Switchgear at the origin of the installation (Category IV) is required to have a higher rated impulse withstand than equipment used in the fixed installation (Category III) and greater than that in the energy-consuming equipment (Category II). Table 7.1 lists the rated impulse voltages for each overvoltage category. It is to be noted that, in setting the rated impulse voltages for each voltage category, it is presumed that the installation will comply with Section 443 (IEC).

Regulation 443.2.1 states that where an installation is supplied only by underground cables (contains no overhead lines), and equipment has withstand levels as per Table 7.1 (Table 44.3 of BS 7671), no further protection against overvoltages of atmospheric origin is required. However, a note was added in 2008 to say protection against overvoltage may be necessary where a higher reliability or higher risks (e.g. fire are expected). Where installations are supplied by overhead line and there are fewer than 25 thunderstorm days per year (AQ1), again, no additional protection is required if the equipment complies with Table 7.1.

Table 7.1 Rated impulse voltage for equipment energized directly from the low voltage mains

Nominal Voltage of the supply (V)		Voltage line to neutral derived from nominal voltages a.c. or d.c. up to and including (V)	Rated impulse voltage (Note 1)			
			Overvoltage category (Note 2)			
Three-phase	Single-phase		I	II	III	IV
		50	330	500	800	1500
		100	500	800	1500	2500
	120–240	150	800	1500	2500	4000
230/400 277/480		300	1500	2500	4000	6000
400/690		600	2500	4000	6000	8000
1000		1000	4000	6000	8000	12000

Notes:

1. Equipment with these rated impulse voltages can be used in installations in accordance with Section 443.
2. Category I is equipment for connection to circuits in which measures are taken to limit transient overvoltages to an appropriately low level.
   Category II is energy-consuming equipment to be supplied from the fixed installation.
   Category III is equipment in fixed installations and for cases where the reliability and the availability of the equipment is subject to special requirements.
   Category IV is for use at the origin of the installation.

[Expert's comment: Some of the above voltages are not directly energized from LV mains.]

Figure 7.9 shows the thunderstorm days per year throughout the world. The thunderstorm days per year in the UK are between five and twenty, so additional protection is not required, as discussed above. This is not the case in many parts of the world, where protection against overvoltage due to lightning is necessary even if the equipment complies with Table 7.1 (see Regulation 443.2.3).

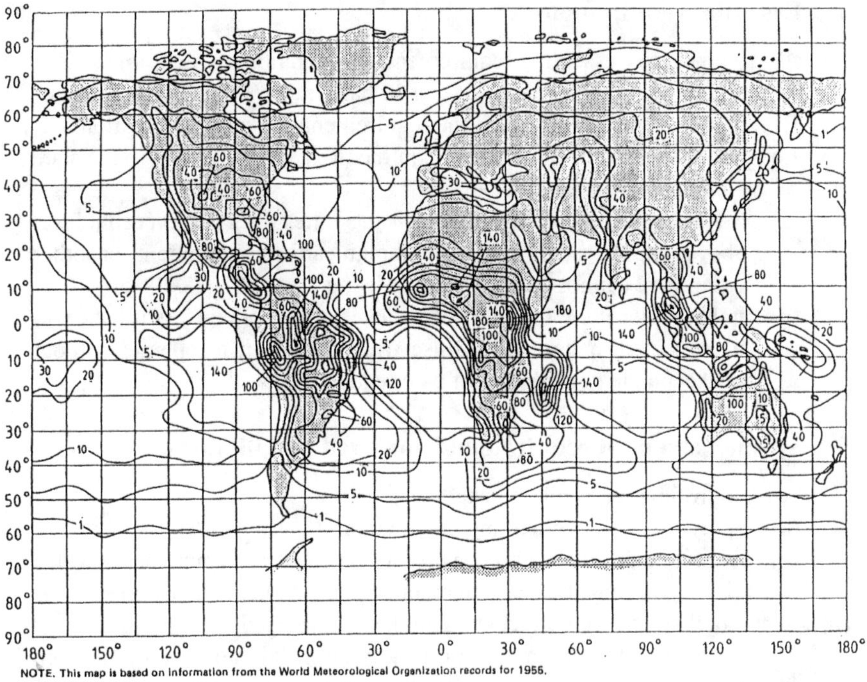

Figure 7.9 *Map showing thunderstorm days per year throughout the world* [Expert's comment: Experience with the EATL plotting system is that, generally, in the immediate vicinity of the coast, with tall structures, there will be elevated risk.]

7.5 Electromagnetic compatibility

Section 444, Regulation 512.1.5

7.5.1 The Electromagnetic Compatibility Regulations

The Electromagnetic Compatibility (EMC) Directive 2004/108/EC was transposed into UK law by the EMC Regulations 2006 (SI 2006 No. 3418), which came into force on 20 July 2007. These Regulations replaced and repealed the EMC Regulations 2005.

The intention (Regulations 4 and 5) is that, in its intended use, equipment should not interfere with radio and telecommunication as well as other equipment and that equipment should have sufficient immunity to electromagnetic emissions/disturbances that are normally present to work properly when used as intended.

The Regulations require that all electrical and electronic apparatus marketed in the UK, including imports, satisfy the requirements of the EMC Directive. Regulations 4 and 5 are copied below with permission of the EMC Directive.

4. Essential requirements

(1) A reference to 'essential requirements' in relation to equipment is a reference to the requirements set out in paragraph (2) and in the case of fixed installations shall include the requirements set out in regulation 5.

(2) Equipment shall be designed and manufactured, having regard to the state of the art, so as to ensure that –

(a) the electromagnetic disturbance it generates does not exceed a level above which radio and telecommunications equipment or other equipment cannot operate as intended; and

(b) it has a level of immunity to the electromagnetic disturbance to be expected in its intended use which allows it to operate without unacceptable degradation of its intended use.

5. Specific essential requirements for fixed installations

(1) A fixed installation shall be installed –

(a) applying good engineering practices; and

(b) respecting the information on the intended use of its components, with a view to meeting the essential requirements set out in regulation 4.

(2) Such good engineering practices shall be documented.

(3) The responsible person in relation to a fixed installation shall hold such documentation at the disposal of the enforcement authority for inspection purposes for as long as the fixed installation is in operation.

This legislation applies to equipment forming part of an electrical installation and presumably to the equipment as installed. A CE mark on equipment is a statement by the manufacturer/supplier that the equipment complies with all relevant directives including the EMC directive (see section 2.5).

7.5.2 Scope

Regulations 444.1

The scope advises that currents due to lightning, switching operations, short-circuits and other electromagnetic phenomena can cause overvoltages and electromagnetic interference. These effects arise where large metal loops are unintentionally formed by cabling and bonding arrangements in the installation, where different electrical wiring systems are installed in different routes, e.g. power cables being separate from data transmission cabling. Power and data cables need to follow common routes, but with adequate separation/segregation, so that interference between systems is reduced without creating aerial loops, which will receive electromagnetic input.

The values of the induced voltages depend on the rate of change of the lightning or other currents and the size of the loops. Within power cables, high rates of change of current can arise with the starting currents of motors, e.g. lift motors, and this can result in overvoltage in cables of information technology systems, which can interfere with or even damage the equipment.

It is recommended that all electrical equipment meet the appropriate EMC requirements, and be in accordance with the relevant standards. The emission and immunity standard is BS EN 61000 series. Equipment should be selected to comply with these standards or with standards for specific types of equipment where these have been published.

7.5.3 Sources of electromagnetic disturbance
Regulation 444.4.1

Equipment sensitive to EMI should preferably not be installed close to possible sources of Electromagnetic interference (EMI) including:

- switching devices (for inductive loads in particular);
- electric motors;
- fluorescent lighting;
- welding machines;
- computers;
- rectifiers;
- choppers;
- frequency converters/regulators;
- lifts;
- transformers;
- switchgear; and
- power distribution busbars.

7.5.4 Measures to reduce Electromagnetic Interference (EMI)
Regulation 444.4.2.1

The precautions to be taken against electromagnetic disturbances required are what has always been sensible and good Practice:

1. Separate power and information technology (IT) cabling, taking particular care with high current equipment and equipment resulting in causing rapid fluctuations in current e.g. lift motors and motors generally, welding equipment, discharge lighting, power transformers.
2. Ensure that the live, neutral and protective conductors of power circuits are installed in very close proximity e.g. same cable or conduit, avoiding ariel

loops that can be created by the circuit conductors (these will transmit and receive magnetic radiation).

3. Not installing any combined neutral and protective (PEN) conductors in the building as they will result in neutral currents circulating in the conductive parts of the building.
4. Installing equipotential bonding, for example on each floor of a multi-storey building. Additionally BS 7671 requires "consideration" of:
5. measures to prevent power circuit fault currents flowing in the screens of IT equipment including cable screens.
6. the use of surge protective devices.

The general measures listed in IEC 60364-4-44 are:

(a) the installation of surge protective devices and/or filters;
(b) metal sheaths of cables, bonded to a common protective equipotential and functional earthing system;
(c) cables and, in particular, protective conductors should not form loops that will emit and receive interference by the use of common routes for power, signal and data circuit wiring;
(d) power and signal cables should be kept separate and should, wherever practicable, cross each other at right angles;
(e) cables with concentric conductors should be used to reduce currents induced into the protective conductor;
(f) symmetrical multicore cables (e.g. screened cables containing separate protective conductors) should be used for the electrical connections between converters and motors, which have frequency controlled motor drives;
(g) signal and data cables should be used according to the EMC requirements of the manufacturer's instructions;
(h) where a lightning protection system is installed:
    - power and signal cables should be separated from the down conductors of lightning protection systems (LPS) by either a minimum distance or use of screening; the minimum distance should be determined by the designer of the LPS in accordance with BS EN 62305-3;
    - metallic sheaths or shields of power and signal cables should be bonded in accordance with the requirements for lightning protection given in BS EN 62305-3 and BS EN 62305-4;
(i) where screened signal or data cables are used, care should be taken to limit the fault current from power systems flowing through the screens and cores of signal cables, or data cables, that are earthed. Additional conductors may be necessary, e.g. a bypass equipotential bonding conductor for screen reinforcement (see Figure 44.R1).
(j) where screened signal cables or data cables are common to several buildings supplied from a TT-system, a bypass equipotential bonding conductor should be used (see Figure 44.R2); the bypass conductor should have a minimum cross-sectional area of 16 mm^2 Cu or equivalent, and the equivalent cross-sectional area should be dimensioned in accordance with Regulation 544.1 of BS 7671;

(k) the impedance of equipotential bonding connections should be kept as low as possible:
- by being as short as possible;
- by having a cross-section shape that results in low inductive reactance and impedance per metre of route, e.g. a bonding braid with a width-to-thickness ratio of five to one;

(l) where an earthing busbar is intended (according to IEC 60364-4-444.5.8) to support the equipotential bonding system of a significant information technology installation in a building, it may be installed as a closed ring.

7.5.5 System selection for EMC Regulation 444.4.3

TN-C systems are not recommended, which is not a problem in the UK, as they are not allowed by regulation 8(4) of the ESQC Regulations. For TN-C supply systems (PME), separated neutral and earth is recommended in the installation, that is a TN-C-S system. TN-S systems are preferred to avoid load currents flowing in the PEN conductors and in particular in loops and parallel paths. The separation of neutral and earth should be arranged as near to the origin as practicable.

7.5.6 Multiple power sources and EMC Regulation 444.4.6

For multiple power sources, e.g. two transformers or a generator, the star points should for EMC purposes be earthed at one point (see Figure 7.10).

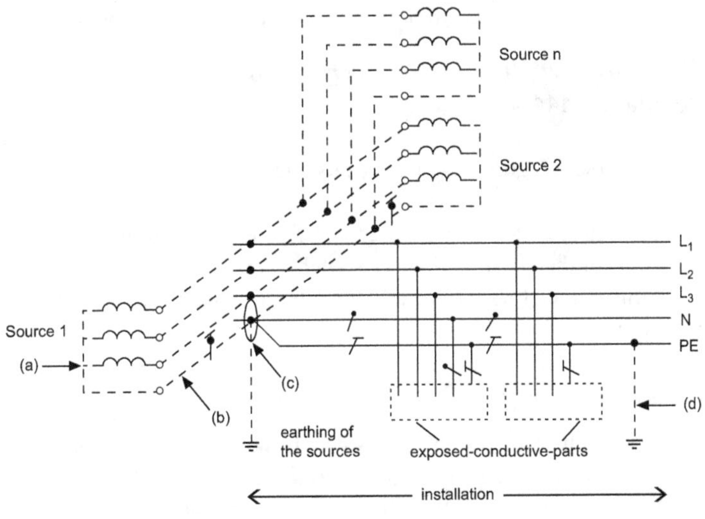

Figure 7.10 TN multiple-source power supplies to an installation with connection to Earth of the star point at one and the same point [Source: IEC 603644-44, Fig. 44.R7B]

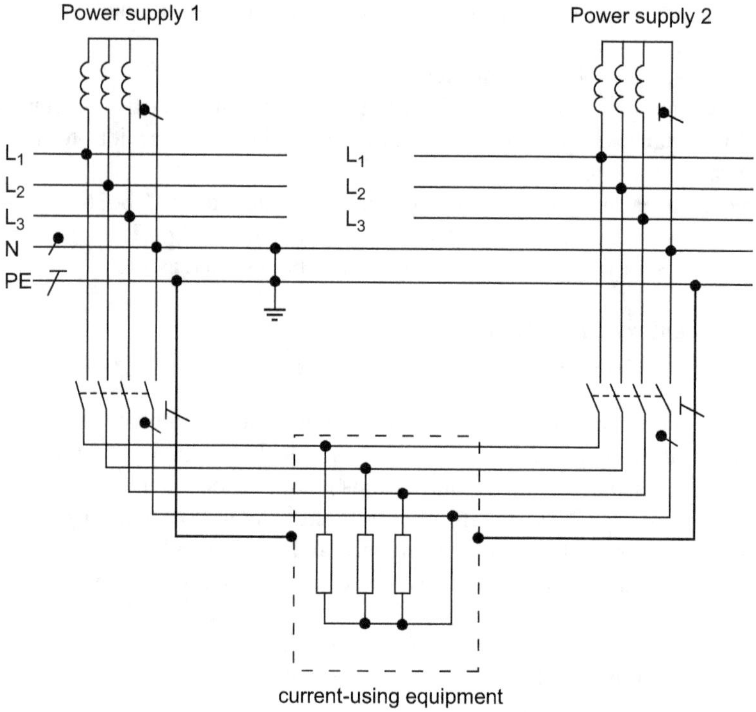

Figure 7.11 Three-phase alternative power supply with four-pole switch [Source: IEC 60364-4-44, Figure 44.R9A]

7.5.7 Alternative supplies, four-pole changeover switch Regulation 444.4.7

This method (see Figure 7.11) prevents electromagnetic fields due to stray currents in the main supply system of an installation. The sum of the currents within one cable must be zero. It ensures that the neutral current flows only in the neutral conductor of the circuit, which is switched on. The third harmonic (150 Hz) current of the line conductors will be added with the same phase angle to the neutral conductor current.

7.5.8 Existing buildings

Figure 7.12, taken from IEC 60364-4-44, shows an existing installation where steps have been taken to reduce EMI effects in an existing building. Loops are avoided in the wiring systems, the phase, neutral and earth conductors are kept together, and any additional functional earths similarly follow the same route as the phase, neutral and safety earths. This reduces emissions of electromagnetic waves due to the self-cancelling effects of balanced currents. The electromagnetic field generated by the phase current is cancelled by the fields generated by the neutral and protective conductor currents if the three conductors are physically close. Induced

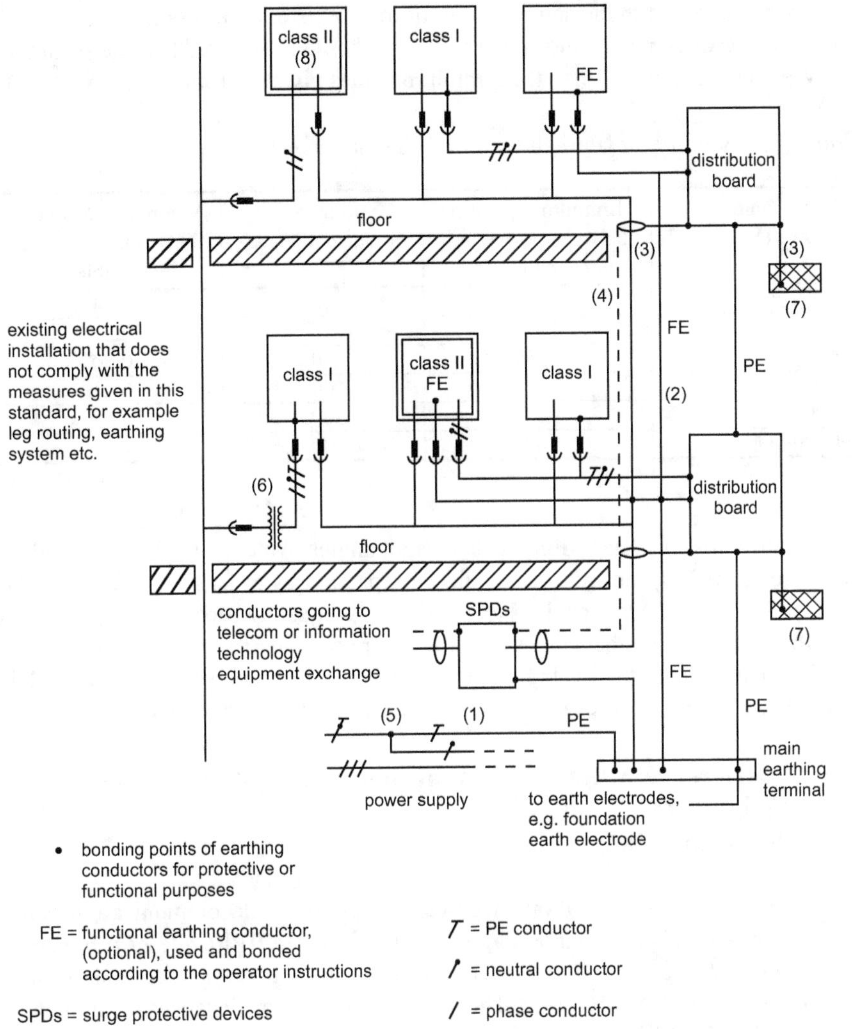

Figure 7.12 Illustration of measures of protection against EMC in an existing building (Figure 44.R11 of IEC 60364-4-44)

voltages in conductors due to EMC emissions from other equipment will then tend to be self-cancelling. The system is TN-S.

7.6 IET publication: 'Electromagnetic interference: report of the Public Affairs study group'

An IET publication, 'Electromagnetic interference: report of the Public Affairs study group', advises that it is important to ensure the correct routing of cables in order to minimize coupling effects. In particular, power and signal cables should not be grouped together in the same conduit or tray. Where, of necessity, parallel cable runs exist between power and signal cables, it should be ensured that the separation between them is not less than the minimum values given in Table 7.2.

Table 7.2 Separation between power and signal cables

Power cable voltage (V)	Minimum separation between power and signal cables (m)	Power cable current (A)	Minimum separation between power and signal cables (m)
115	0.25	5	0.24
240	0.45	15	0.35
415	0.58	50	0.50
3 300	1.10	100	0.6
6 600	1.25	300	0.85
11 000	1.40	600	1.05

Ideally, crossover should be made at right angles. Antenna feeders, particularly those connected to transmitters, should be separated by at least 10 cm from other cables. Small pulse, data cables etc. should be adequately separated (typically not less than 5 cm) from other cables and, wherever possible, from each other.

In environments where high interference levels are present, it may well be helpful to use optical fibres in preference to coaxial cables or twisted pair cables.

7.7 Interference with telecommunications and data-processing equipment
Regulation 528.2

Guidance on the separation of power cables from telecommunication cables is given in a number of standards including BS 6701:2010 *Code of practice for installation of apparatus intended for connection to certain telecommunication systems* and also BS EN 50174-2. The minimum separation distances recommended are reproduced in Table 7.3.

Telecommunication systems used in any way in connection with a close proximity to high voltage equipment require special attention, and reference must be made to BS 6701.

7.8 Interference with other electricity users

The Energy Networks Association Engineering Recommendation P28, 'Planning limits for voltage fluctuations caused by industrial, commercial and domestic equipment in the UK', provides guidance on the acceptability of load-change rates with respect to the supply, e.g. guidance on the type of motor starting necessary for a particular frequency of starting.

Table 7.3 Separation distances between power and telecommunication circuits

Power cable voltage to earth (V)	Separation
Exceeding 600V a.c. or 900V d.c.	1. a distance of not less than 150 mm; or 2. a distance of not less than 50 mm, effected by a divider meeting the requirements of BS 7671
Exceeding 50V a.c. or 120V d.c., but not exceeding 600V a.c. or 900V d.c.	*Outside the building* (a) a distance of not less than 50 mm; or (b) a divider meeting the requirements of BS 7671. *Inside the building* (a) a distance of not less than 50 mm; or (b) a divider meeting the requirements of BS 7671; unless one or more of the following conditions is met (Note 1: if one or more of these conditions is met then separation is not deemed to be necessary): (1) the electricity supply cables are enclosed in a separate conduit or trunking which, if metallic, is earthed in accordance with BS 7671; (2) the electricity supply cables are of a mineral-insulated type; (3) the electricity supply cables are of an earthed armoured construction; (4) the electricity supply cables are of a flexible double-insulated type (e.g. 'kettle leads' supplying 230V mains power to telecommunications equipment in cabinets). (Note 2: Standard electrical 230V 'twin and earth type' cabling is not flexible double-insulated.)

Chapter 8
Common rules for the selection and erection of equipment

(Adviser: K. Morton BSc C.Eng MIEE, Health and Safety Executive)
Chapter 51

8.1 Compliance with standards
Section 511, Appendix 1

8.1.1 Harmonized standards

BS 7671 requires every item of equipment (which includes cables and luminaires) to comply with the relevant requirements of the applicable British Standard or Harmonized Standard appropriate to the intended use of the equipment. A harmonized standard is a European standard formally presented by CENELEC (European Committee for Electrotechnical Standardization) to the European Commission in response to a mandate under a New Approach Directive, and covering its essential requirements, or one drawn up by CENELEC with reference to the Low Voltage Directive.

It can be seen that a harmonized standard has a very particular meaning. We will initially consider those harmonized standards drawn up by CENELEC with reference to the Low Voltage Directive.

The Low Voltage Directive (which is not a New Approach Directive – more on this later) is implemented in the UK by the Electrical Equipment (Safety) Regulations 1994, and includes in its requirements that electrical equipment offered for sale be safe (Regulation 5(1)).

Expert's comment

The Low Voltage Directive applies to all apparatus operating on any a.c. supply between 50 V and 1000 V or d.c. supply between 75 V and 1500 Vs. Recent guidance from the European Commission has made it clear that the voltage limits referred to are the supply voltage or any voltage generated that appears at an output terminal, and not voltages generated within and confined internally to the equipment.

Electrical equipment that complies with the appropriate CENELEC harmonized standard can be taken to comply with the Electrical Equipment (Safety) Regulations 1994.

8.1.2 European Norms (ENs)

CENELEC European standards are given the designation 'EN' (European Norm). All member countries of CENELEC (and this includes all the European National Committees), are obliged to implement European standards by publishing them unchanged as a national standard, and to withdraw any conflicting national standards. In the UK such standards have the prefix 'BS EN'.

This is of particular importance to specifiers and installers. Electrical equipment complying with a CENELEC electrical safety standard can be considered to meet the requirements of the Electrical Equipment (Safety) Regulations 1994 and BS 7671. It will also meet Europe-wide requirements for electrical safety.

8.1.3 Harmonized documents
Preface (to BS 7671)

CENELEC also publishes HDs (harmonized documents). These are not to be confused with harmonized standards; however, HDs are also standards, but they carry with them less onerous implementation requirements than ENs. CENELEC member countries must implement an HD by public announcement of the HD number and title, and withdraw any conflicting national standards. An HD does not have to be published unchanged, as does an EN.

BS 7671 implements the HDs in the 384 and 60364 series (the 384 series is gradually being replaced by the 60364 series to align with IEC numbering). These are listed in the Preface to BS 7671. BS 7671 is not obliged to be identical to these HDs, but it must include their technical intent and not conflict with them. It is to be noted that BS 7671 includes the requirements of many IEC 60364 parts for which there is as yet no HD. This is particularly true of Part 7.

8.1.4 CE marking

CE marking is required on all electrical equipment sold in the UK by the Electrical Equipment (Safety) Regulations. It is to be applied after a manufacturer has assessed that his product complies with the essential health-and-safety requirements of relevant regulations implementing applicable European Directives.

CE marking is fundamentally to prevent barriers to trade so that a manufacturer in one EU member state can claim compliance with the relevant directive, and the goods can be sold in all EU states without hindrance, unless non-compliance can be shown.

Relevant regulations may include those implementing the EMC Directive (the EMC Regulations), the Construction Products Directive and the Machinery Safety Directive (the Supply of Machinery (Safety) Regulations 2008 as amended), as well as the Low Voltage Directive (the Electrical Equipment (Safety) Regulations 1994).

The Equipment and Protective Systems Intended for Use in Potentially Explosive Atmosphere Regulations 1996 implement one of the two ATmosphere EXplosive Directives and are relevant to some aspects of low voltage installations.

The various regulations require the manufacturer to have third-party assessment or prepare a technical construction file demonstrating compliance; this is to be made available should conformity be challenged. If an independent or third-party assessment is required, the approval mark of one of the approval bodies nominated by the IEC should be looked for. Some typical UK electrical equipment approval marks are shown in Figure 2.1.

The New Approach Directives, which include the EMC Directive, the Construction Products Directive and the Machinery Directive, are complied with by meeting the requirements of harmonized standards. The new approach is that only general requirements are specified in the directive, and compliance is effected by complying with harmonized standards.

BS 7671 does allow foreign national standards based on International Electrotechnical Commission (IEC) standards to be used, provided that the designer or other persons responsible for specifying have verified that any differences between the foreign standard and the corresponding British Standard or Harmonized Standard do not result in a lesser degree of safety. In practice, such an assessment needs to be made by experts in such matters, e.g. those test houses listed in the directory of accredited laboratories prepared by the United Kingdom Accreditation Service (UKAS).

8.2 Voltage
Regulation 512.1.1

Nominal voltages for low voltage, public electricity supply systems

HD 472 was implemented in the UK on 1 January 1995, when the nominal voltage in the UK was changed to 230 V + 10% – 6% from the previous nominal voltage of 240 V±6%. This change was effected by an amendment to the Electricity Supply Regulations 1988 and is now a requirement of the ESQC Regulations 2002. This change was a step towards the harmonization of voltages within Europe. Those countries with nominal voltages of 240 V were obliged to move to 230 V +10% – 6% and those with nominal voltages of 220 V changed to 230 V – 10% + 6%. The original intent was that, by the year 2005, all voltages would be harmonized at 230 V ± 10%. However, representations by the IET and others that the appliance and equipment standards had not been amended to reflect the wider supply voltage range were heeded by the then DTI, who then became BERR, and in mid-2009 became known as BIS, and there has as yet been no further widening of the supply voltage range.

The harmonization of voltage is not actually being effected by adopting common supply voltages, but by requiring equipment manufacturers to make products that will work over a much wider voltage range. It is fortunate that, within the European Union, all countries had already adopted a frequency of 50 Hz for a.c. supplies. This frequency is also widely used in the Middle and Far East, Africa and South America. The other frequency adopted around the world is 60 Hz, as in the United States.

It is most likely that supply voltages in the UK will remain the same as they were before the nominal change from 240 V to 230 V, as voltages supplied within the range 240 V ± 6% will lie within the range of 230 V + 10% – 6%. It is unlikely that a supply company would want to reduce its voltage, as it would then reduce its revenue (power being proportional to the square of the voltage), although there may be situations where power companies will wish to keep the demand down (load management) in order to defer reinforcements of the system. It may, in particular circumstances, be cost-effective for this course of action to be taken, and there is no reason why this should not be done, provided the requirements of the ESQC Regulations continue to be met.

This is significant for installation design. A reduced voltage could, in extreme cases, result in circuit-breakers not operating instantaneously, with repercussions with respect to shock protection and the adiabatic constraints on circuit conductors. Similarly, the time for disconnection of a fuse will be increased, but, as there are no step changes in the characteristics of a fuse, a reduction in voltage would result only in a proportional increase in the disconnection time.

Many new large installations are being fitted with autotransformers to reduce the supply voltage, and the suppliers claim energy consumption is reduced as a consequence. This is interesting. Installations with their own transformer could achieve the same effect by altering the transformer tap setting to reduce the voltage by say 2.5 per cent.

Care must obviously be taken to ensure installed equipment will operate satisfactorily and it may be necessary to reduce voltage drop in the installation to reflect reduced supply voltage. Some interesting calculations can be made. Reducing voltage will increase current for a given power output and increase copper losses.

8.3 Compatibility
Regulation 512.1.5

8.3.1 General

Regulation 512.1.5 requires that every item of equipment be selected and erected so that it will neither cause harmful effects to other equipment nor impair the supply during normal service, including switching operations. Harmful effects to which a designer will need to give consideration include:

(i) emission of heat;
(ii) vibration;
(iii) earth leakage currents;
(iv) electromagnetic interference to the installation in general;
(v) interference with telecommunication and data-processing equipment;
(vi) interference with other electricity users, including harmonic currents, high switching rates, signals transmission etc.; and
(vii) flicker effects.

8.3.2 Cable conductor operating temperature
Regulation 512.1.5

The second paragraph added in 2011 reminds the designer that most switchgear is designed for cable conductor temperatures not exceeding 70 °C. This is necessary as the thermosetting cable ratings given in Appendix 4 of BS 7671 assume a conductor operating temperature of 90 °C. If cable conductor cross sectional areas are selected on this basis (90 °C) it may result in overheating and failure of the switchgear. Thermosetting cables terminating in switchgear will need to be derated to ensure the conductor operating temperature does not exceed 70 °C.

8.3.3 Switchgear enclosures, effect on rating

When a device is installed within an enclosure with other devices e.g. as in a switchboard the devices will be derated by the heat emission from the other devices in the enclosure. Section 10.3 of this book provides advice.

In principle this is analogous to cable ratings; a cable has a clipped direct/free air rating, so does a protective device. Group the device with other loaded devices and all the devices are de-rated. Thus diversity factors and enclosed current ratings come into effect.

Another important fact is that a device in an enclosure can have a significantly different current rating to that marked on it including incoming switch and outgoing devices.

8.4 External influences
Regulation group 512.2

The requirement that every item of equipment should be appropriate to the situation in which it is to be used is an obvious one, but one that needs stating. An example of this is the ambient temperature in household premises, which is unlikely to exceed the 30 °C assumed for cable ratings, and the 20 °C ambient assumed for accessories. However, there are situations where these temperatures will be exceeded, and then the designer must take note of them, for example in airing cupboards, or in the vicinity of heating appliances.

8.5 Accessibility
Regulations 132.12, 513.1

8.5.1 General

Except for permanent joints in cables as allowed by Regulation 526.3, every item of equipment is required to be installed so as to facilitate its operation, inspection and maintenance, and to provide access to each connection. This is a very important requirement for the users of an installation, and it is one that is backed by the Electricity at Work Regulations 1989. Switchgear and controlgear should not be placed in any situation that will adversely affect their operation, e.g. too high, too low or in insufficient space.

8.5.2 Operating and maintenance gangways

CENELEC has agreed requirements for operating or maintenance gangways specifying spacing between switchgear panels and minimum distances from walls to facilitate operational access in HD 60364-7-729: *Low-voltage electrical installations – Part 7-729: Requirements for special installations or locations – Operating or maintenance gangways*. There is a UK special national condition not allowing unprotected live parts and making reference to the Electricity at Work Regulations and Health and Safety Executive Guidance Note HSR 25, the memorandum of guidance on the Electricity at Work Regulations. This has now been included in BS 7671 as Section 729, see Chapter 15 of this Guide.

Annex B of HD 60364-7-729 advises the following:

United Kingdom

Clause 729.410.3.7: In the UK where the gangway has unprotected live parts refer to UK government guidance on the Electricity at Work Regulations 1989, Statutory Instrument 1989, number 635, guidance document HSR25.

In the Health and Safety Executive's view, work activities in a gangway of such construction (unprotected live parts) would be counter to the requirements of EWR (Regulations 4, 7 and 14). It is important to note that the HSE objection (and consequently the UK note) relates to gangways with exposed live parts, not those that have the live parts enclosed and inaccessible in normal operation to persons in the gangway. For this reason information on the recommendations for gangways only in installations where the protective measure of barriers or enclosures applies is repeated below.

8.6 Identification and notices
Section 514

The general requirements for identification and notices are important to those responsible for the operation and maintenance of an installation. Except where there is no possibility of confusion, a label or other suitable means of identification shall be provided to indicate the purpose of each item of switchgear and controlgear. This is another requirement that is reinforced by the Electricity at Work Regulations 1989, and is essential for safe operation. Regulation 514.1.2 goes further and states that, as far as is reasonably practicable, wiring shall be so arranged or marked that it can be identified for inspection, testing, repair or alteration of the installation. This means that cable runs from switchboards should be grouped together, with cables and conductors running neatly and logically on cable trays etc.

8.6.1 Conduit colour
Old Regulation 514.2.1

Interestingly Regulation 514.2 of the 2008 regulations was deleted in 2011. The requirement was that where conduit is to be distinguished from a pipeline or another service, orange is the colour to be used. The reason is probably that it lead to disputes as to whether BS 7671 required conduit to be coloured orange. It is clear now that it does not.

BS 1710: *Specification for identification of pipelines and services* makes reference to this colour for electrical services for use where there is a need to identify pipes by colour. The colour identifiers for other services are shown in Table 8.1, and the safety colours in Table 8.2. Application of the basic colours and safety colours is shown in Figure 8.1.

Table 8.1 Identification colours: Table 1 of BS 1710 Specification for identification of pipe colours

Service	Colour	BS 4800:2000 colour code
Basic identification colours		
Water	Green	12 D 45
Steam	Silver-grey	10 A 03
Oils – mineral, vegetable or animal	Brown	06 C 39
Combustible liquids		
Gases in either gaseous or liquefied condition (except air)	Yellow ochre	08 C 35
Acids and alkalis	Violet	22 C 37
Air	Light blue	20 E 51
Waste effluents	Black	00 E 53
Electrical services and ventilation ducts	Orange (note 3)	06 E 51
Safety colours		
Fire	Red	04 E 53
Water from a public supply	Auxiliary blue	18 E 53
Water from any other source	Flint grey	00 A 09
Warning	Yellow	08 E 51
Code and other colours		
Refrigeration	Blue	18 E 51
Medical gases and refrigeration	Golden brown	06 D 45
Building services and refrigeration	Crimson	04 D 45
Building services and refrigeration	Emerald green	14 E 53
Building services	Salmon pink	04 C 33
Building services and medical gases	Primrose	10 E 53
Building services and medical gases	White	00 E 55
medical gases	French grey	12 B 21
medical gases	French blue	20 D 45
Refrigeration	Sea green	16 C 37
Refrigeration	Dark mauve	02 C 37

Note: The colours reproduced in BS 1710 and the names of colours are given for guidance only. When ordering, only the BS 4800 reference number should be specified, as manufacturers often attribute different names to these colours.

Table 8.2 Safety colours

Safety colour (1)	BS colour reference, BS 4800	Purpose
Red	04 E 53	Firefighting
Yellow	08 E 51	Warning
Auxiliary blue	18 E 53	With basic identification colour green – fresh water (2)

Notes:

1. The colour names given in column 1 are only included for guidance since different colour names may be used by different manufacturers for the same colour reference.
2. Potable or non-potable.

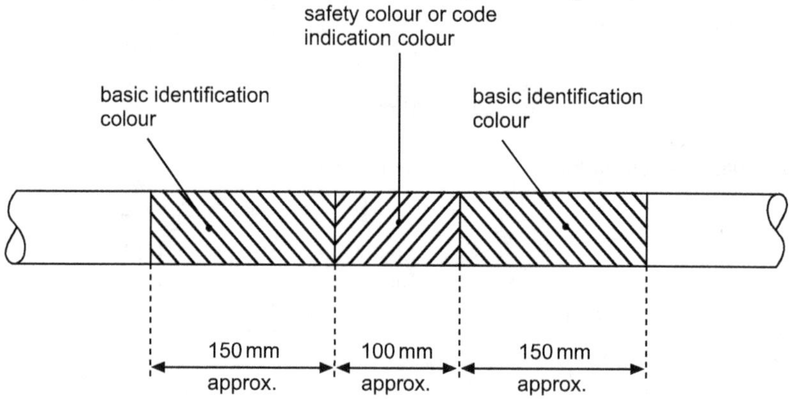

Figure 8.1 Application of safety colours and code indication colours

8.6.2 Identification of utility ducts, pipes and cables

The National Joint Utility Group (NJUG) have agreed colours for ducts, pipes and cables laid in the public highway (roads, footpaths etc.) – see Table 9.3 of Chapter 9.

8.7 Identification of conductors

Regulation group 514.3, Appendix 7

Appendix 7 of BS 7671 advises that the means of identification of conductors (cable core colours and the alphanumeric marking) has been harmonized with:

- HD 384.5.514 *Identification,* including 514.3 *Identification of conductors;*
- HD 308 S2:2001 *Identification of cores in cables and flexible cords;*
- BS EN 60445:2010 *Basic and safety principles for man-machine interface, marking and identification of equipment terminals and of terminations;*

HD 384.5.514 Identification

Harmonization document HD 384.5.514, *Selection and erection of electrical equipment – Common rules*, requires single-core cables and insulated conductors to be coloured as follows:

(i) protective conductor – bi-colour combination of green and yellow;
(ii) neutral or midpoint conductors – blue;
(iii) PEN conductors – green/yellow throughout their length and, in addition, light blue markings at the terminations or light blue throughout the length with, in addition, green/yellow markings at the terminations;
(iv) line conductors shall be identified throughout their length by the colours brown or black or grey; the use of one of these colours for all of the line conductors in a circuit is permitted.

BS EN 60445:2000 Basic and safety principles for man–machine interface, marking and identification of equipment terminals and of terminations

BS EN 60445 specifies requirements for marking by alphanumeric, that is, not by colour (see Table 8.3).

BS EN 60445 specifies requirements for marking by alphanumeric, that is, not by colour (see Table 8.3).

BS EN 60446 permits the following colours: black, brown, red, orange, yellow, green, violet, grey, white, pink, turquoise. Particular conductors are required to be coloured as follows:

(i) protective conductor – bi-colour combination of green and yellow;
(ii) neutral or midpoint conductors – blue;
(iii) PEN conductors – green/yellow throughout their length and, in addition, light blue markings at the terminations or light blue throughout the length with, in addition, green/yellow markings at the terminations;
(iv) a.c. phase – the colours brown or black are preferred.

BS EN 60204 Safety of machinery

The harmonized standard BS EN 60204 *Safety of machinery, electrical equipment of machines* has the following colour-coding requirements, which are in alignment with HD 384.5.51:

(i) protective conductor – bi-colour combination of green and yellow;
(ii) neutral conductor – light blue;
(iii) a.c. and d.c. power circuits – black;
(iv) a.c. control circuits – red;
(v) d.c. control circuits – blue;
(vi) interlock control circuits supplied from an external source – orange;
(vii) other conductors – black, brown, red, orange, yellow, green, violet, grey, white, pink, turquoise.

Table 8.3 Marking of equipment terminals intended for certain designated conductors and of terminations of these conductors

Designated conductor	Identification of conductors[a] and conductor terminations	Equipment terminal marking	Graphical symbols[f]
a.c. conductors			
Line 1	L1[d]	U	
Line 2	L2[d]	V[e]	
Line 3	L3[d]	W[e]	
Neutral conductor	N	N	
d.c. conductors			⎓
Positive	L+	+ or C	+
Negative	L−	− or D	−
Mid-point conductor[b]	M	M	
Protective conductor[b]	PE	PE	
PEN conductor[b]	PEN	PEN	
PEM conductor[b]	PEM	PEM	
PEL conductor[b]	PEL	PEL	
Functional earthing conductor[b]	FE	FE	
Functional equipotential bonding conductor[b c]	FB	FB	

[a] Identification by colour see IEC 60446 (colours described earlier).
[b] Definitions see IEC 60050-195.
[c] The former distinction between equipotential connection and frame or chassis connection no longer exists according to IEC 60050-195. The concept functional equipotential bonding conductor includes both concepts. The graphical symbol ^ (IEC 60417-5021) is deprecated.
[d] Numeral after 'L' is necessary only in systems with more than one phase.
[e] Necessary only in systems with more than one phase.
[f] The graphics shown correspond to the symbol numbers in IEC 60417.
Source: Table 1 of BS EN 60445:2000

8.8 Diagrams
Regulation 514.9.1

Regulation 514.9.1 requires that a legible diagram, chart, table or equivalent form of information shall be provided, indicating:

(i) the type and composition of each circuit;
(ii) the protective measure used for compliance with Regulation 410.3.2, i.e. provisions for basic protection and fault protection;

(iii) information necessary for the identification of each device performing the functions of protection, isolation and switching; and
(iv) any circuit or equipment vulnerable to the electrical tests required by Part 6 of BS 7671.

A typical schematic diagram to provide this information is shown in Figure 8.2. For simple installations, the information can be provided in the form of a schedule, and there is a requirement for a durable copy of the schedule relating to the distribution board to be provided within or adjacent to each distribution board. This would then require a schedule of installations similar to Table 8.4.

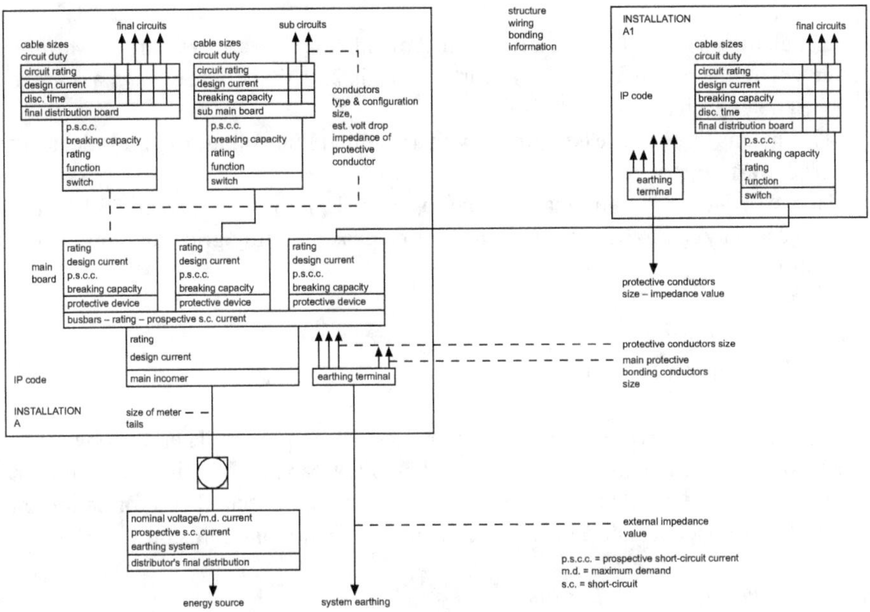

Figure 8.2 Installation diagram per Regulation 514.9

Table 8.4 Installation schedule for a distribution board

No.	Device type	Device rating (A)	Points served	Circuit description	Cable size (mm^2)	Type
RCD 1	RCD	30 mA				PVC/t + c.p.c.
Circuit 1	B	32	30	Ground-floor ring	2.5/1.5	PVC/t + c.p.c.
Circuit 2	B	32	20	First-floor ring	2.5/1.5	PVC/t + c.p.c.
Circuit 3	B	32	1	Cooker	6.0/2.5	PVC/t + c.p.c.
RCD 2		30 mA				
Circuit 4	B	10	15	Ground-floor lights	1.5/1.0	PVC/t + c.p.c.
Circuit 5	B	10	10	First-floor lights	1.5/1.0	PVC/t + c.p.c.
Circuit 6	B	10	1	Alarms	1.5/1.0	PVC/t + c.p.c.

Equipment vulnerable to testing: RCDs, heating programmer, lounge dimmer switch

8.9 Warning, etc. notices
Section 514.10

Clear notices suitably located are a very necessary requirement of any installation. It is to be remembered that those responsible for operating and maintaining an installation do not have the same knowledge of the installation possessed by the persons who designed or installed it.

8.9.1 Voltage
Regulation 514.10.1

Regulation 514.10.1 was clarified in Amendment No.1 to require a warning notice where the nominal voltage to earth exceeds 230 V and where such a voltage is not to be expected.

This means three phase equipment with a nominal line voltage of 400 V does not require a warning notice.

However where the nominal voltage to earth (U_0) does not exceed 230 V, but three-phase may not to be expected, say a three phase lighting grid, labelling would be helpful.

8.9.2 Isolation
Regulation 514.11.1

The warning with respect to isolation is particularly important. There are obviously considerable hazards with any equipment that is not capable of being isolated by a single device. The regulation requires not only a warning, but also information on the location of each isolator, unless there is no possibility of confusion.

8.9.3 Periodic inspection and testing
Regulation group 514.12

The requirement to install a notice advising of the recommended date of the next inspection is also important. The Regulations themselves do not specify what the intervals between inspections should be. There is a requirement in the electrical installation certificate for a recommendation to be given as to when the installation is to be further inspected and tested, and this information also must be marked on the periodic inspection notice of Regulation 514.12.2. The designer of the installation should have had in mind what frequency of inspection he was assuming when he made his selection of equipment and protective measures, bearing in mind the use and the nature of the installation.

8.10 Rubber mats

The question quite often arises as to whether there is a need for rubber mats in front of switchgear for the operator to stand on. The short answer is that there do not appear to be any HSE recommendations about their use. It could be argued that they assist in the creation of an earth-free area when working on switchgear, but, given the prevalence of earthed metalwork in most installations, this is of little advantage. If they are being used, the duty holder needs to identify the purpose of use so that a false sense of safety is not made for people working on the switchgear; and they also need to be maintained to ensure their continuing integrity – it is not unknown for the insulation properties of a mat to be compromised by foreign objects being embedded in the mat.

Their use was common in front of feeder pillars (fuse boards) with exposed live parts in indoor Electricity Board substations (in 1960/70ties) and they were of obvious use when working live (one hand behind your back). Compliance with the Electricity at Work Regulations of such working practices would be difficult to justify even for authorized personnel and would not receive HSE backing.

Chapter 9
Selection and erection of wiring systems

Chapter 52
9.1 Introduction
Sect 520

Cables and equipment, including trunking, conduit, track system and busbar trunking systems, are required to be selected to comply with the fundamental requirements of Chapter 13. Table 9.1 (see table on p.192) provides guidance on the selection of PVC cables to BS 6004, the cables most commonly used.

9.2 Electromagnetic effects
Regulations 521.5.1, 521.5.2

Single-core cables armoured with steel wire or tape are not allowed for a.c. circuits. The conductors of all a.c. circuits installed in ferromagnetic enclosures, such as conduit, trunking and switchgear, are to be arranged so that all the line conductors and the neutral conductor, if any, and appropriate protective conductor(s) are contained within the same enclosure. If a single-core cable carrying a current is encircled by an iron ring, eddy currents will be induced in that iron, and these will heat the enclosure. For this reason, if the individual conductors of a three-phase circuit enter an enclosure, such as a switch, via individual holes, those holes must be slotted (see Figure 9.1).

9.2.1 Electromechanical stresses
Regulation 521.5.1

The electromagnetic forces can become increasingly significant as fault currents increase above 15 kA rms. The cut-off characteristics of fault protective devices restrict the forces being produced. Information needs to be obtained from the manufacturer of the device.

9.2.1.1 Cables
It can be assumed that unarmoured multicore cables of all sizes are strong enough to withstand the electromagnetic bursting forces where protection is coordinated with cable size, in compliance with Regulation 435.1 or 434.5.2, when the fault current protective device is of the current-limiting type. However, cable manufacturers should

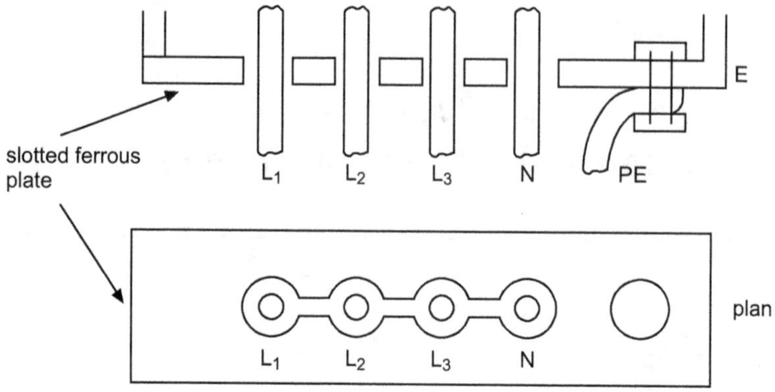

Figure 9.1 Slotting of end plates for single-core cables

be consulted where let-through currents are likely to be in the region of 25 kA rms or higher.

Similarly, it can be assumed that armoured multicore extruded-insulation cables are also strong enough to withstand bursting forces without damage, whatever type of protective device is employed. However, where currents may exceed 25 kA rms, it may be wise to consult the cable maker.

In the case of single-core cables, electromagnetic forces act to move the cables apart, and the force is dependent on the cable size, type and spacing. The closer the cables, the greater the forces are. Fault currents in excess of 20 kA peak may cause the cables to bend, and with higher currents there is the possibility of broken fixings. To reduce the risk of damage, the cables may be strongly bound together at frequent intervals, to resist the electromagnetic forces. It may well be appropriate to make provision for lateral movement and longitudinal relief by snaking. The cable manufacturer will provide information on the type of bindings and supports necessary.

The weakpoints with respect to longitudinal expansion forces and electromagnetic forces are on unsupported lengths of cable at joints and terminations. Electromagnetic forces are highest in the region of the crutch of multicore cable terminations, e.g. where the conductors are closest. Adequate support in the form of a suitable gland or binding is needed to reinforce the end of the cable and to support the cable at the crutch. Lateral support for the cables should be considered where tails are long enough to bend and impose strain on terminals. Advice is given in IET Guidance Note 6: *Protection Against Overcurrent.*

9.2.1.2 Busbar trunking

For fault withstand of busbar trunking see Regulation 434.5.3 and section 6.3.7 of this Commentary.

9.3 Selection and erection in relation to external influences
Section 522 and Appendix 5

9.3.1 Introduction

Appendix 5 of BS 7671 gives a classification and codification of external influences from CENELEC HD 60364-5-51. The use of this classification system is being progressively extended, and will, when it is more widely recognized within equipment standards, enable designers to assess rather more precisely the suitability of equipment for the particular external influences.

Cables must be appropriate for the external influences and method of installation. Table 9.1 provides guidance on unarmoured PVC insulated cables.

9.3.2 Ambient temperature
Regulation group 522.1

The requirements in this regulation deal with the suitability of cables for the highest and lowest local ambient temperatures, and not the effect of the ambient temperature on the cable rating. The later is discussed with respect to Section 523 (section 9.5).

The regulations remind designers that not only should cables be suitable for the range of ambient temperature to be expected, but that some cables should not be handled at low temperatures. For example, cables that are insulated and/or sheathed with general-purpose PVC should not be stored in refrigerated spaces or other situations where the temperature is consistently at or below 0 °C. It is also advised that such cables, or cables that are paper-insulated or have bituminous compound beddings, should not be installed when the ambient temperature is below 0 °C. Cable temperatures must be greater than 0 °C and should have been so for the previous 24 hours or so before being handled.

See Table 9.1 for cable types suitable for low temperatures.

9.3.3 Impact
Regulation group 522.6

9.3.3.1 Cables installed under floors or above ceilings
Regulation 522.6.201

The requirements in BS 7671 with respect to the laying of cables under floors or above ceilings are intended to ensure that cables are not installed in any position where they are likely to suffer damage by contact with the floor or ceiling or their fixings. Unarmoured cables passing through a joist are required to be installed at least 50 mm measured vertically from the top or bottom, as appropriate, of the joist or batten, or be enclosed in earthed steel conduit, trunking or ducting, or be incorporated in an earthed metallic sheath suitable for use as a protective conductor,

Table 9.1 Guide to the use of PVC insulated cables

BS 6004 Table No.	Harmonized code designation	Cable type and use	Comments
4a and 4b	4a Solid conductor: H07V-U Stranded conductor: H07V-R 4b H07V-K	*Single-core, non-sheathed general-purpose* Installation in surface mounted or embedded conduits, or similar closed systems.	Suitable for use in channels with cover. Suitable for fixed protected installation in or on light fittings and inside appliances, switchgear and controlgear, for voltages up to 1,000 V a.c. or up to 750 V to earth, d.c.
5	Solid conductor: H05V-U Stranded conductor: H05V-R Flexible conductor: H05V-K	*Single-core and twisted twin, non-sheathed, for internal wiring* Fixed, protected installation inside appliances and in or on light fittings.	Suitable for installation in surface-mounted or embedded conduits, only for signalling or control conduits.
6	None allocated (see note 6).	*Light PVC-sheathed* Fixed installation in dry or damp premises.	Unsuitable for outdoor use or embedding in concrete
7, 8 and 9	National type	*Single-core, flat twin and three-core, PVC-sheathed, with and without protective conductor* Fixed installation in dry or damp premises.	Suitable for installation in walls, on boards and in channels or embedded in plaster.
10a and 10b	10a Solid conductor: H07V3-U Stranded conductor: H07V3-R 10b H07V3-K	*Single-core, non-sheathed for low-temperature installation* As for Tables 4a and 4b, but suitable for installation at low temperatures.	Suitable for use in channels with cover. Suitable for fixed protected installation in or on light fittings and inside appliances, switchgear and controlgear, for voltages up to 1,000 V a.c. or up to 750 V to earth, d.c.
11a, 11b and 12	11 Solid conductor: H07V2-U Stranded conductor: H07V2-R Solid conductor: H05V2-U Stranded conductor: H05V2-R 12 Flexible conductor: H05V2-K	*Single core and twisted twin, heat-resisting non-sheathed* For internal wiring only.	Maximum conductor temperature in normal use 90 °C. Not to be used in contact with objects at temperatures higher than 85 °C. Not suitable for fixed installations in distribution systems.

13	H05WC4V5-K	*Oil-resisting, PVC-sheathed, screened* The interconnection of parts of machines used for manufacturing purposes, including machine tools, where some degree of protection against electromagnetic interference is required. After installation the cables may be moved for the repositioning, maintenance, adjustment and inspection of machines provided that the cable is not mechanically stressed during movement, but screened cables are not designed for continual flexing.	The cables are resistant to general-purpose mineral oils but are not designed for continuous immersion in oil. They are intended for use inside buildings. Contamination by hydrocarbons, acids and alkalis should be avoided and the cables should be protected against mechanical damage. Installation in conduits, trunking etc. is advised. Where contact with special oils is likely, advice should be sought from the manufacturer

Notes:

1. None of the cable types specified in this standard (BS 6004) is intended to be laid underground.
2. The cables specified in this standard are suitable for use where the combination of ambient temperature and temperature rise due to load results in a conductor temperature not exceeding 70 °C (90 °C for heat-resisting types specified in Tables 11 and 12 of BS 6004) and, in the case of a short-circuit (maximum allowable time 5 s), the maximum conductor temperature does not exceed 160 °C for conductor sizes up to and including 300 mm^2 and 140 °C for conductor sizes above 300 mm^2.
3. The short-circuit temperature is based on the intrinsic properties of the insulating material. It is essential that the accessories that are used in the cable system with mechanical and/or soldered connections be suitable for the temperature adopted for the cable.
4. Installation requirements and current ratings are detailed in BS 7671, or in individual appliance specifications.
5. An expanded guide to use for cables specified in this standard is given in BS 7540. Typical uses of the different types of cable specified in this standard are given in Table A.1. below.
6. This type of cable is included in CENELEC Harmonization Document HD 21.4 S2, but, owing to lack of agreement on a colour code for rigid multicore cables, this cable type cannot be considered as harmonized, and no harmonized code designation has been allocated.

Source: Table A.1 of BS 6004

or have mechanical protection against damage sufficient to prevent penetration of cable by nails, screws and the like. See Figure 9.2.

The alternative to providing mechanical protection sufficient to prevent penetration of the cable by nails, screws and the like is sometimes debated. Hammered nails have surprising penetrative power when they are steadied by being hammered

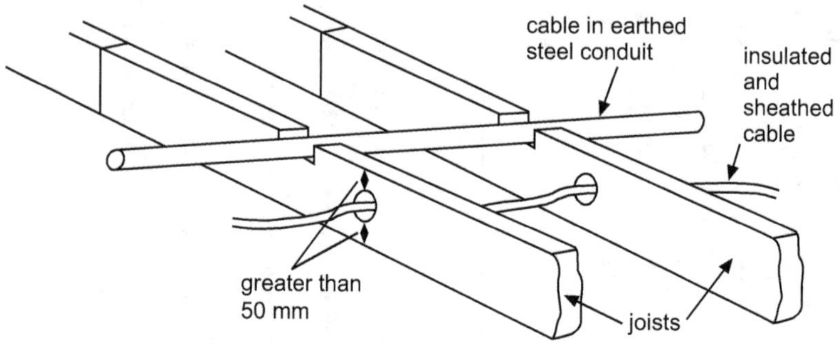

Figure 9.2 Cables in floors and ceiling

initially through a floorboard, and when the article they are about to penetrate is held by the joist, and self-tapping fixing screws will penetrate steel plate.

The National House Builders' Registration Council (now referred to as the NHBC) recommends that timber joists and studs should be notched and drilled only within the limits shown in Table 9.2 and Figure 9.3.

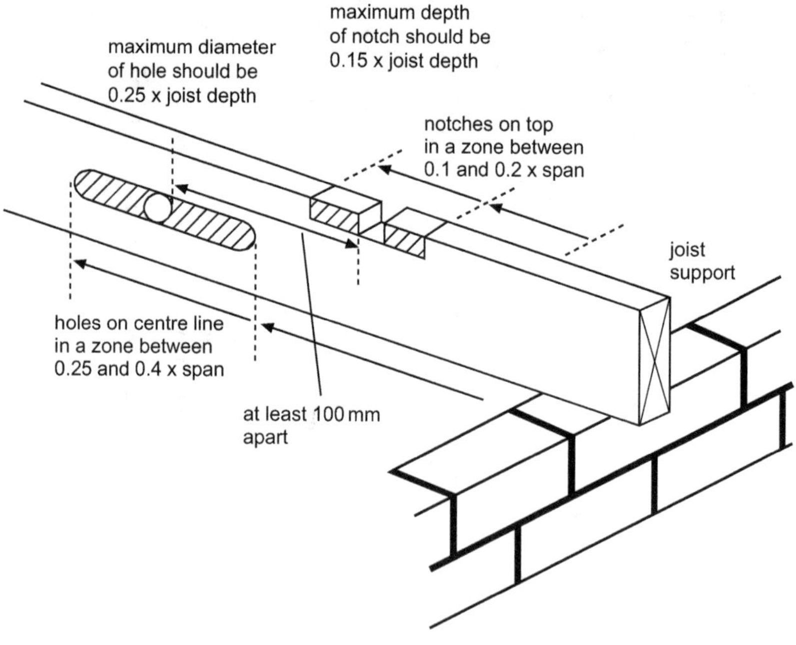

Figure 9.3 NHBC recommendations for joists

Notches and drilled holes in the same joist should be at least 100 mm apart horizontally (see Figure 9.3).

Table 9.2 Notching and drilling of joists

Item	Location	Maximum size
Notching joists up to 250 mm depth	0.07 to 0.25 of span	0.125 × depth of joist
Notching joists up to 250 mm depth	Within 0.1 to 0.2 of span	0.15 × depth of joist
Drilling joists up to 250 mm depth	Centre line 0.25 to 0.4 of span	0.25 × depth of joist
Drilling studs	Centre line 0.25 to 0.4 of height	0.25 × depth of joist

Note: Notches may be at the top or bottom of a joist, but not both at the same end.

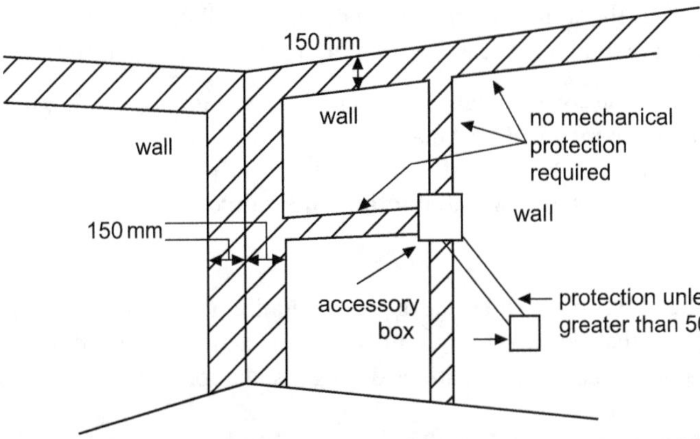

In domestic and similar premises 30 mA RCDs are required to protect cables installed at a depth of less than 5 cm unless mechanically protected by earthed steel conduit or similar.

Figure 9.4 Permitted cable routes in walls

9.3.3.2 Cables installed in walls or partitions
Regulation 522.6.201

Regulation 522.6.201 requires that a cable concealed in a wall or partition shall:

(i) be at least 50 mm from the surface, or
(ii) be mechanically protected (e.g. by conduit, trunking, or armouring), or
(iii) be installed in specified zones (Regulation 522.6.201 (i)) and protected by a 30 mA RCD.

The relaxation for installations under the supervision of an instructed or skilled person has been removed by Amendment 3. The reason for the change was probably that members of the committee now thought installing shallow cables in zones was insufficient as the lay person was not aware of them and the use of RCDs was now much more acceptable. Problems had been encountered with different understandings as to what under the supervision of a skilled or instructed person, meant.

9.3.3.3 Cables installed in metal-framed etc. walls or partitions
Regulation 522.6.203

An additional requirement (to those in Regulation 522.6.201) is included for cables installed in metal-framed etc. walls or partitions. Even if installed more than 50 mm from the surface they must:

(i) be protected by a 30 mA RCD, or
(ii) be mechanically protected sufficiently to avoid damage to the cable during construction of the wall or partition and during installation of the cable, or
(iii) be mechanically protected (e.g. by conduit, trunking, or armouring).

Again the relaxation for installations under the supervision of an instructed or skilled person has been removed by Amendment 3.

9.3.3.4 Cable capacities of conduit, ducting and trunking
Regulation group 522.8

This regulation group includes the general requirements for the installation of cables, requiring that the wiring system be so selected and erected as to avoid damage to the cables and their terminations during installation, use and maintenance (Regulation 522.8.1).

The 15th Edition contained a number of appendices that provided advice on these matters. These included:

(i) notes on the selection of types of cable and flexible cords, particularly uses and external influences;
(ii) notes on methods of support for cables, conductors and wiring systems; and
(iii) cable capacities of conduit and trunking.

These appendices are no longer included in BS 7671. The guidance provided is, however, included in Guidance Note 1: *Selection & Erection* and in the *On-Site Guide*, both published by the IET.

The guidance given for determining the cable capacities of conduit and trunking allows the designer or installer to estimate the size of conduit or trunking necessary for the cables to be installed. Reference is made for trunking to a 45 per cent space factor. This does not provide spare capacity within the trunking; it is simply the space remaining when round cables are laid in place. When the maximum number of cables allowed by the guidance are installed, then no more can be accommodated. It is presumed cables will be laid from the top neatly. If there are 'T's or conduit take-offs the capacity will be further reduced.

When it is proposed to install additional cables in existing conduit or trunking, thus increasing the number of cables in the group, care must be taken to ensure that this action does not damage the existing installation. The cables in the existing trunking may very well have been sized on the basis of a particular group rating factor, and if further cables are added, although there is space in the trunking, the existing cables may well be overloaded by the addition. If spare capacity is requested by the client, the designer must take account of the ultimate rating factor. Clearly, at first sight, it may seem to be wise to leave spare capacity, but it will increase the cost of an installation, and it is unlikely to be cost-beneficial unless the additions to the installation are imminent. It is not necessarily the most cost-effective method to install a large number of cables grouped together in one trunking; it may very well be more cost-effective to have two or more smaller groups.

The designer must also give detailed attention to the stresses and strains that might be imposed on the building by the cables. This is of particular importance with respect to suspended cables, which can result in considerable longitudinal stress on the supporting structures. There is considerable movement within cables as they heat up under load and cool down again when the load is switched off, or when the ambient temperatures are low; this movement can produce very high longitudinal forces, additional to the weight of the cable.

9.4 Buried cables
Regulation 522.8.10

The requirements for buried cables are general. They must be mechanically protected by earthed armour or metal sheath or both, unless enclosed in a conduit or duct that provides equivalent protection. Cables are required to be marked by cable covers, or a suitable marking tape, or by suitable identification in a conduit or duct. This leaves the designer with much freedom as to how the position of the cable is marked. The requirement for depth is simply that it be sufficient to avoid cables being damaged by any disturbance of the ground reasonably likely to occur. The general guidance given is to comply with the requirements of the ESQCR and those of NJUG (see 9.4.1 and 9.4.2).

9.4.1 Buried cables: Electricity Safety, Quality and Continuity Regulations

The requirements of the ESQCR must be complied with by anyone who owns or operates a network supplying consumers. Particularly, they must be complied with if cables are laid in roads or footpaths and good guidance must be provided for other locations. Regulation 14 requires:

14. Excavations and depth of underground cables

(1) Every underground cable shall be kept at such depth or be otherwise protected so as to avoid, so far as is reasonably practicable, any damage or danger by reason of such uses of the land which can be reasonably expected.

(2) In addition to satisfying the requirements of paragraph (1), an underground cable containing conductors not connected with earth shall be protected, marked or otherwise indicated so as to ensure, so far as is reasonably practicable, that any person excavating the land above the cable will be given sufficient warning of its presence.

(3) The protection, marking or indication required by paragraph (2) shall be made by placing the cable in a pipe or duct or by overlaying the cable at a suitable distance with protective tiles or warning tape or by the provision of such other protective or warning device, mark or indication, or by a suitable combination of such measures, as will be likely to provide an appropriate warning.

There is an old DTI publication entitled *Guidance on the Electricity Safety, Quality and Continuity Regulations 2002* that supplements the regulations. The guidance on Regulation 14 is as follows:

Regulation	Guidance
14(1)	This requirement places a continuous duty on generators and distributors, subject to regulation 12 General restriction on the use of underground cables, to install and keep underground cables at a suitable depth or otherwise protect the cables (e.g. by steel plates in shallow trenches) in order to prevent danger, so far as is reasonably practicable. Duty holders should decide on the depth or other protection in the light of the use of the land at installation and other likely uses of the land in the future. Generators and distributors may not necessarily be held to account for breach of this regulation if they are not aware of an increased risk of danger due to the actions of other parties (see comments for regulation 12 General restriction on the use of underground cables). Useful guidance on the depth and relative position of underground services is available from the National Joint Utilities Group. Safety guidance on how to avoid underground services when carrying out excavation works is available from the HSE.

14(2) Marking or protection of underground cables is required for all voltages, i.e. high voltage as well as low voltage (including service cables), in order to offer warning to persons excavating in the vicinity of live cables (contractors working in the street as well as consumers digging in the garden).

In the case of low voltage cables, this requirement need only apply to cables installed after commencement of the Regulations (see regulation 2(6)). If any material alteration is made to a low voltage network then a mark or indication should be applied whether to new or to pre-commencement cables (see regulation 2(8)). For example, if after commencement of the Regulations a distributor exposes a low voltage cable installed without marker tape in 1995, he must install a mark or indication at the new excavation to comply with the requirements of this regulation.

14(3) The methods that duty holders should employ to demonstrate compliance with this regulation and thereby reduce the risk of injury to contractors or members of the public are listed in order of preference as follows:-

(i) Cable installed in a duct with marker tape above.
(ii) Cable installed in a duct only.
(iii) Cable laid direct and covered with a protective tile.
(iv) Cable laid direct and covered with marker tape.
(v) Some other method of mark or indication.

In consideration of the methods by which cables should be marked or protected, duty holders should make allowance for the environment in which the cables are installed and the risks to staff who may need to expose and work on the cables in future.

9.4.2 Buried cables: positioning and colour coding of underground utilities' apparatus

The NJUG has agreed colours, depths of cover and positions for ducts, pipes, cables and marker/warning tapes when laid in the street. These are reproduced in Tables 9.3a and b, and in Figure 9.5 and form useful guidance to designers as to the depth at which cables might reasonably be expected to be laid generally as well as in public paths and highways.

The advice for depth of cover for carriageway crossings is that it will be determined by the carriageway construction, the method of excavation and the need to minimize the effects of the loading and vibration generated by vehicular traffic. In normal circumstances apparatus should be installed below the carriageway construction. Failure to lay apparatus beneath the carriageway construction may significantly increase the damage due to distributed traffic loads, and both damage to apparatus and disruption may occur when apparatus or highway is repaired or renewed. Where practicable, pipes and cables that cross the carriageway should be laid in ducts for ease of future maintenance and the avoidance of congestion.

Table 9.3a NJUG Guidelines on the Positioning and Colour-Coding of Underground Utilities' Apparatus: Table 1 – Recommended Colour Coding of Underground Utilities Apparatus

All depths are from the surface level to the crown of the apparatus

Utility	Duct	Pipe	Cable	Marker systems	Recommended minimum depths (mm)	
					Footway/Verge	Carriageway
Electricity HV (high voltage)	Black or red tile	N/A	Red or Black	Yellow with black legend	450–600	750–1,200
Electricity LV (low voltage)	Black or red	N/A	Black or red	Yellow with black legend	450	600
Gas	Yellow	see note 1 below	N/A	Yellow with black legend	600 (footway) 750 (verge)	750
Water non Potable & Grey Water	N/A	Black with green stripes	N/A	N/A	600–750	600–750
Water – Firefighting	N/A	Black with red stripes or bands	N/A	N/A	600–750	600–750
Oil/fuel pipelines	N/A	Black	N/A	Various surface markers Marker tape or tiles above red concrete	900 All work within 3 metres of oil fuel pipelines must receive prior approval	900 All work within 3 metres of oil fuel pipelines must receive prior approval
Sewerage	Black	No distinguishing colour/material (e.g. Ductile Iron may be red; PVC may be brown)	N/A	N/A	Variable	Variable
Communications	Grey, white, green, black, purple	N/A	Black or light grey	Various	250–350	450–600
Water	Blue or Grey	Blue polymer or blue or uncoated Iron/GRP. Blue polymer with brown stripe (removable skin revealing white or black pipe)	N/A	Blue or blue/ black	750	750 minimum
Water pipes for special purposes (e.g. contaminated ground)	N/A	Blue polymer with brown stripes (non-removable skin)	N/A	Blue or blue/ black	750	750 minimum

Note 1

*** PE – Up to 2 bars – yellow or yellow with brown stripes (removable skin revealing white or black core pipe). – Between 2 to 7 bar -orange. Steel pipes may have yellow wrap or black tar coating or no coating. Ductile Iron may have plastic wrapping Asbestos & Pit/ Spun Cast Iron – No distinguishable colour

Reproduced with permission from NJUG. Note that these documents are best practice and guidelines offered by NJUG to its members and the wider community. No responsibility/liability can be held by NJUG for any changes, omissions or other deviations from the advice given.

Table 9.3b NJUG Guidelines on the Positioning and Colour-Coding of Underground Utilities' Apparatus: Table 2 – Recommended Colour-Coding of Other Underground Apparatus (at the time of publication the following were current examples of known highway authority apparatus colour coding)

Asset owner	Duct	Pipe	Cable	Marker systems	Recommended minimum depths* (mm)	
					Footway/ Verge	Carriageway
			Street lighting			
England and Wales	Black or orange (consult electricity company first)	N/A	Black	Yellow with black legend	450	600
Scotland	Purple	N/A	Purple	Yellow with black legend or purple	450	450
Northern Ireland	Orange	N/A	Black or Orange	Various	450	450
			Other			
Traffic control	Orange		Orange	Yellow with black legend		
Street furniture	Black	N/A	Black	Yellow with black legend	450	600
Telecoms	Purple/orange	N/A	Black	Various	Various	
			Motorways and trunk roads – England and Wales			
Communications	Purple	N/A	Grey	Yellow with black legend	450	
Communications power	Purple	N/A	Black	Yellow with black legend		
Road lighting	Orange	N/A	Black	Yellow with black legend		
			Motorways and trunk roads – Scotland			
Communications	Black or grey	N/A	Black	Yellow with black legend		
Road lighting	Purple	N/A	Purple	Yellow with black legend		

* All depths are from the surface level to the crown of the apparatus.

These guidelines describe utility industry practice. However, it should not be assumed that all apparatus will conform to the recommendations for positioning and colour coding contained in this publication.

Reproduced with permission from NJUG. Note that these documents are best practice and guidelines offered by NJUG to its members and the wider community. No responsibility/liability can be held by NJUG for any changes, omissions or other deviations from the advice given.

NJUG 3 advises:

> In the event of congestion of apparatus in the footway/verge (e.g. where less than 2 metres wide) normal distribution mains, pipes, cables and ducts may have to be sited within the carriageway. Transmission and trunk main pipes and cable ducts are invariably of larger dimensions and as a consequence may also need to be located in the carriageway.
>
> Apparatus must be installed below the carriageway construction layers unless special arrangements are made with the relevant authority. Apparatus installed within the carriageway construction is significantly more at risk of damage from traffic loads and excavation and this practice should be avoided whenever possible.

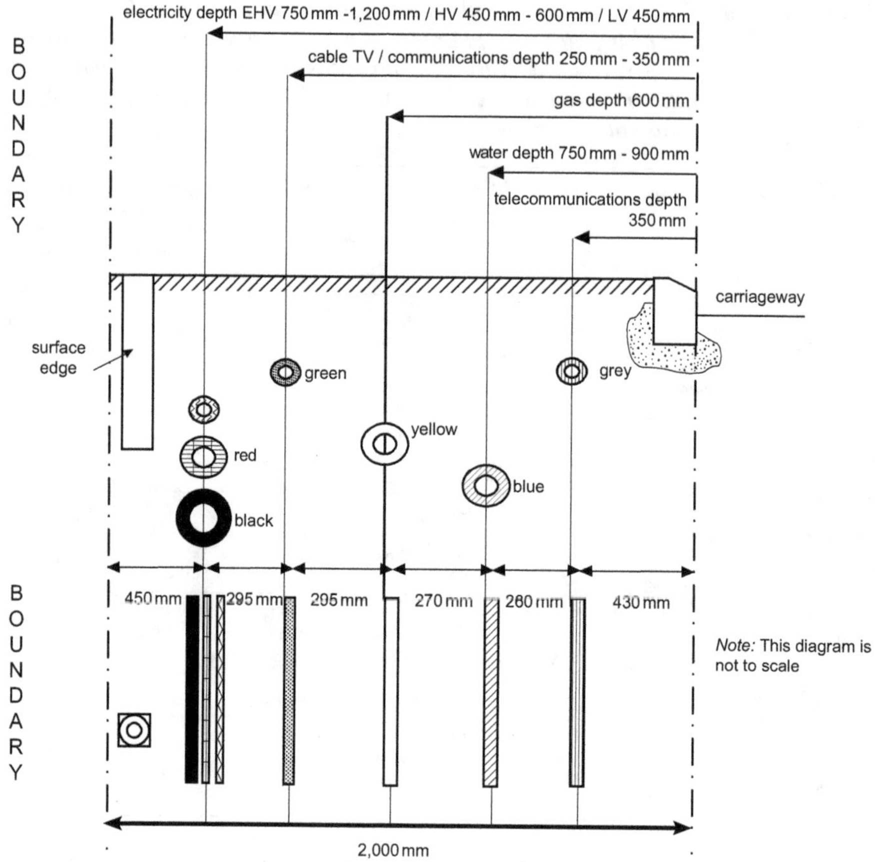

Figure 9.5 Recommended positioning of utility cables, pipes and ducts in a 2 m footway

9.5 Current-carrying capacities of conductors

Section 523, Appendix 4

9.5.1 General

The basic requirement of Regulation 523.1 is that cables shall be selected such that the current, including any harmonic current, to be carried by any conductor for sustained periods during normal operation should not cause a conductor temperature to exceed the maximum operating temperature given in Table 52.1 and that given for each cable type in Appendix 4 of BS 7671. For example, Table 4D1A advises that the conductor operating temperature be 70 °C. These temperatures will not be exceeded if the cable ratings given in the tables and the rating factors (grouping, ambient and thermal insulation) are properly applied. The introduction to Appendix 4 provides comprehensive information on the selection of cables for given currents in specified environments. The maximum conductor operating temperatures from the tables in Appendix 4 are repeated in Table 52.1 of BS 7671.

The use of Appendix 4 is discussed in Chapter 16.

9.5.2 Conductor operating temperature Regulation 523.1, Table 52.1

Table 52.1, note b, has a further requirement that, where the conductor operating temperature exceeds 70 °C, it shall be ascertained that the equipment connected to the conductor is suitable for the conductor operating temperature. This is a very important requirement. While the cables may well be capable of operating at 90 °C, such as multicore armoured cables having thermosetting insulation per Table 4E4A (90 °C), the switchgear connected at either end of this cable is unlikely to have been designed for such a temperature. Switchgear and accessories are usually designed for conductor temperatures not exceeding 70 °C. If it is wished to estimate the current in a thermosetting cable that will result in a conductor operating temperature of 70 °C, the formula in paragraph 9.5.3 may be used, or, more simply, 70 °C cable current ratings. This advice is provided in the notes to the thermosetting cable ratings.

Note that thermosetting-insulated cables to BS 7211 may well be selected for other reasons than their increased current-carrying capacity over thermoplastic-insulated cables. Some have defined characteristics for their performance under fire conditions with respect to the emission of smoke and gases (see paragraph 9.12).

9.5.3 Effect of load on conductor operating temperature

The conductor operating temperature at other than the full load current can be determined from the equation:

$$\frac{I_b}{I_t} = \sqrt{\frac{t_b - t_a}{t_p - t_o}} \text{ hence } t_b = \frac{I_b^2}{I_t^2}(t_p - t_o) + t_a$$

where:

I_b = load current resulting in a conductor temperature t_b at an ambient t_a

I_t = tabulated current rating in Appendix 4 of BS 7671, resulting in a conductor temperature t_p at ambient t_o.

This relationship and many others associated with cable ratings are developed from the concept that the temperature rise of a cable is proportional to the square of the current.

The introduction to Appendix 4 advises on how cables can be selected for particular loads and installation conditions. Appendix 4 is discussed in Chapter 16 of this Commentary.

The current-carrying capacities tabulated in Appendix 4 are maximum thermal ratings. Cables are generally selected by applying current rating factors either to the rating of the overcurrent protective device I_n, or to the design current I_b, and thence selecting a suitable cable.

9.6 Conductors in parallel

Regulation 523.7, Table 4C5, Appendix 10, see also Commentary 6.6

9.6.1 General

Regulation 523.7 requires that conductors in parallel share the load between them, and is considered to be met if they are:

- of the same cross-sectional area;
- of the same material (copper or aluminium);
- of the same length; and
- have no branch circuits.

A further requirement is that the cables be:

- multicore;
- twisted single-core;
- twisted non-sheathed; or
- non-twisted or non-sheathed in particular configurations.

Table 4C5 gives rating factors for certain flat and trefoil arrangements of single-core cables to be applied to the ratings in the rating tables, e.g. 4D1A reference method F. However, the regulation does allow consideration from first principles, i.e. meeting Regulation 523.1.

It is fairly common for cables of dissimilar sizes to be run in parallel, particularly for temporary arrangements. This is not such a disastrous situation as might at first be supposed, as the current is shared between the cables in inverse proportion to the cable impedances. For cables up to 120 mm^2 the impedance is approximately inversely proportional to the cross-sectional area, which itself is also approximately proportional to the current-carrying capacity, so proportionate sharing results. However, the inductance of cables over 120 mm^2 spoils the relationship and they have a proportionately lower current-carrying capacity (I_t/csa) than smaller cables, both of which tend to be self-cancelling. The ratio of heat loss from a small cable to the conductor csa is greater than that of a larger cable. This does mean that if the larger cable of a pair is itself capable of carrying the load current, the addition of another smaller cable of the same construction in parallel will not result in the smaller cable being burnt out. However, it is to be noted that the regulations would require the detailed calculations of Appendix 10 to be carried out.

9.6.2 Single-core cables, flat or trefoil (reference method F)

The Regulations give specific advice with respect to parallel running of single-core cables and the arrangement of the phases so as to ensure optimum balance of load current between them (see Appendix 4 (reference method F of

Table 4A2 and rating factor Table 4C5)). The reactance values given in Appendix 4 of BS 7671, e.g. 4H1B, assume that single-core cables are installed either with an actual spacing of one cable diameter or touching if in trefoil. Larger cable spacing increases reactance. Figure 9.6 gives the increase in loop reactance for unarmoured single-core cables with wider spacing than that specified in the tables of voltage drop.

For example, consider Table 4J3B for single-core cables having thermosetting insulation and non-magnetic armour. The note to Table 4J3B says that spacing larger than one cable diameter will result in larger voltage drops.

The spacing of single-core cables has interesting complications. Reference to Column 5 of Table 4J3B shows that, when in trefoil, the cables have their lowest impedances as a result of reduced reactance. However, putting the cables in close proximity, as in trefoil arrangement, reduces the current-carrying capacity because

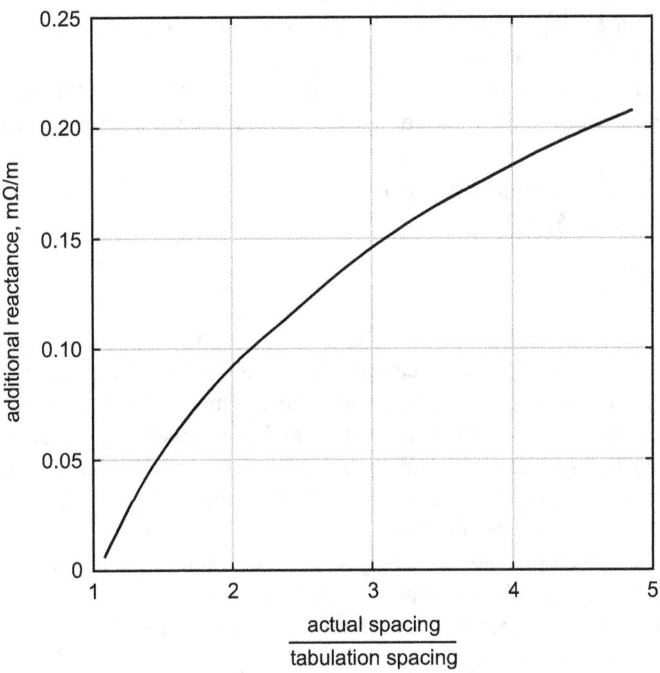

Figure 9.6 Additional reactance for unarmoured single-core spacing cables with wider spacing than specified in the voltage drop tables of BS 7671

the cables are closer together and, in effect, grouped. The effect of spacing is considered below.

Consider a 300 mm^2 aluminium conductor installed as per Reference Method F with the spacing being considered as either one cable diameter or two cable diameters. Column 7 of Table 4J3B tells us that the impedance of the cable at 90 °C in mV/A/m (or mΩ)/m) is:

r	x	$z = \sqrt{r^2+x^2}$
0.29	0.29	0.41

Figure 9.6 shows that, if the actual spacing is twice that tabulated, the additional reactance is 0.1 mΩ/m. The impedance of the cable now becomes:

r	x	$z = \sqrt{r^2+x^2}$
0.29	0.39	0.486

The voltage drop tables make reference to units of mV/A/m for r, x and z. This makes it clear to those using the tables that, if multiplied by the current (A) and length (m), the answer is given in millivolts (mV). The units can also be expressed as mΩ/m (at conductor operating temperature).

Note that cables with uneven spacing will have uneven balance of load between them. There are really only certain cable arrangements that provide reasonable current sharing, examples being given in Figure 9.7.

Other arrangements may provide acceptable current sharing, but consideration will have to be given to the different reactances, and the values must be calculated.

9.7 Cables in thermal insulation
Regulation 523.9

9.7.1 *General*

The basic guidance given in this regulation is self-evident: that cables should preferably not be installed in a location where it is liable to be covered by thermal insulation. Prevention is always better than cure. The best designs avoid the need to derate cables due to thermal insulation, and to a lesser degree grouping. The ratings given in Appendix 4 already provide for a number of the most commonly encountered insulated circumstances such as Reference Method A, and Reference Methods 100 to 103 (see Table 4D5).

The advice given if a cable is totally surrounded by thermal insulation is that, in the absence of more precise information, the rating be taken as 0.5 times the current rating for clipped direct (Reference Method C). Typical thermal conductivities of miscellaneous materials are provided in Table 9.4 (see table on p.208).

9.7.2 *Table 52.2 of BS 7671 – Cable surrounded by thermal insulation*
Regulation 523.9

The derating factors given in Table 52.2 of BS 7671 are intended for insulation having a thermal conductivity greater than 0.04 W/Km.

Regulation 523.9 states that the derating factors in Table 52.2 are applicable for cable sizes up to 10 mm^2. The table may be used for larger cable sizes, but will give conservative results. The larger the cable, the less it need be derated.

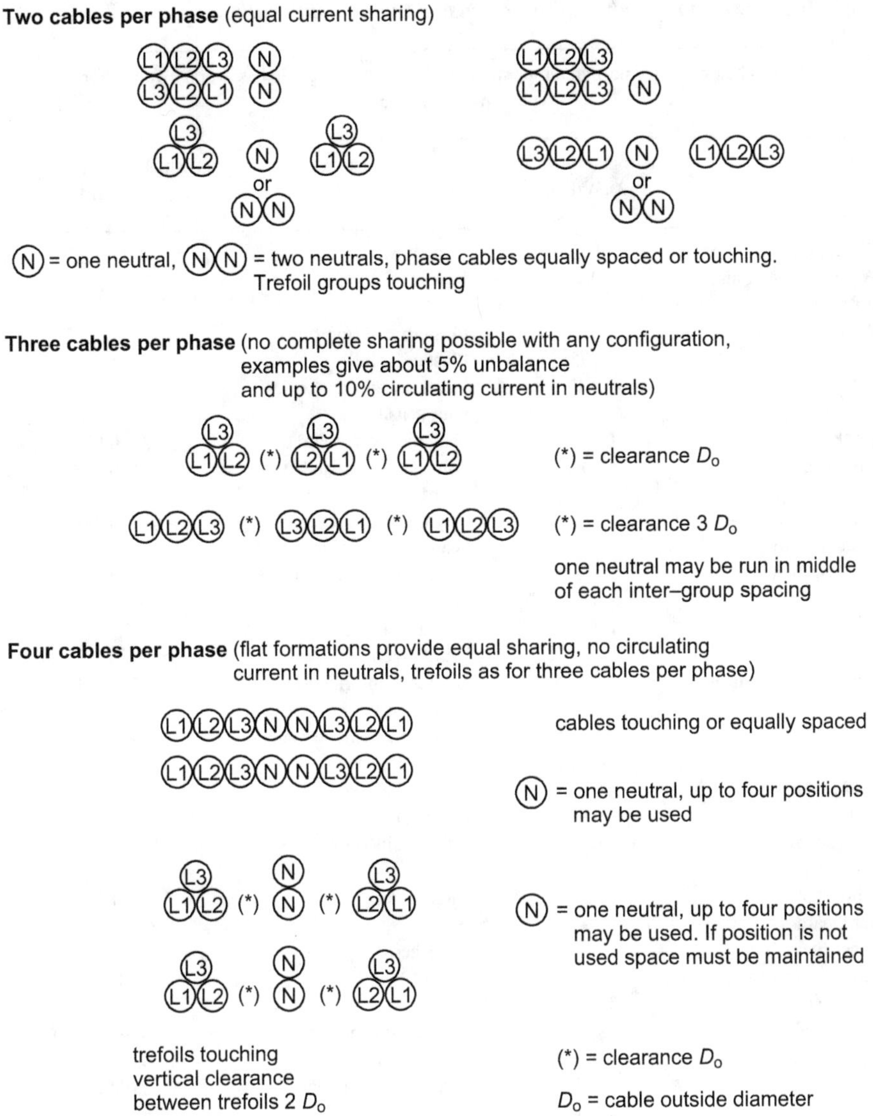

Figure 9.7 Cable configurations for single-core cables in parallel

9.8 Metallic sheaths and/or non-magnetic armour of single-core cables

Regulation 523.201

Single-core cables with ferrous armouring are not allowed by Regulation 521.5.2. However, non-magnetic armouring and metallic sheaths are allowed, and

Table 9.4 Thermal conductivities of miscellaneous materials

The values below are for normal temperature, and should be regarded as average values for the type of material specified.

Substance	Thermal conductivity, λ (W/Km)	Substance	Thermal conductivity, λ (W/Km)
Asbestos cloth	0.125	Mineral wool	0.04
Asbestos insulating board	0.11	Paper	0.06
Asbestos paper	0.104	Plasterboard (gypsum)	0.16
Asbestos wool	0.055	Plastics, solid	
Bitumen	0.17	Plastics, cellular: (varies with density)	
Brisk	0.8–1.2	phenolic foam board	0.031–0.037
Cardboard	0.21	polystyrene, expanded board	0.031–0.038
Concrete (conductivity		polystyrene, expanded bead	0.035–0.055
increases with density)		polyurethane, gas-filled board (fresh)	0.017–0.020
cellular	0.1–0.2	polyurethane, gas-filled board (aged)	0.027
lightweight aggregate	0.2–0.6	polyvinyl chloride, rigid foam board	0.035–0.041
dense	0.6–1.8	urea formaldehyde foam	0.030–0.032
Cork, baked slab	0.038–0.046	Plywood	0.125
Cork, granular	0.04	Rubber, cellular	0.045
Felt	0.04	Rubber, natural	0.15
Fibreboard, insulating	0.055	Rubber, silicone	0.25–0.4
Fibreboard, hardboard	0.125	Sand, silver	0.3–0.4
Glass fibre: (conductivity varies		Silica aerogel powder	0.024
with density)		Soil, clay	1.1
wool blanket	0.035–0.07	Timber, ordinary	0.14–0.17
rigid board	0.030–0.036	Timber, balsa	0.055
Kapok	0.035	Vermiculite granules	0.065
Mica	0.6–0.7	Wool	0.05

are commonly used, particularly for larger single-core cable tails for transformers. Normally, bonding at both ends is required; however, single point bonding is allowed provided that there is a non-conducting outer sheath, suitable insulation at the unbonded ends and that the length of cables is such that, at full load, the voltages from sheath or armour to earth:

(i) do not exceed 25 V;
(ii) do not cause corrosion; and
(iii) do not cause damage with short-circuit conditions.

9.8.1 Calculation of sheath voltages

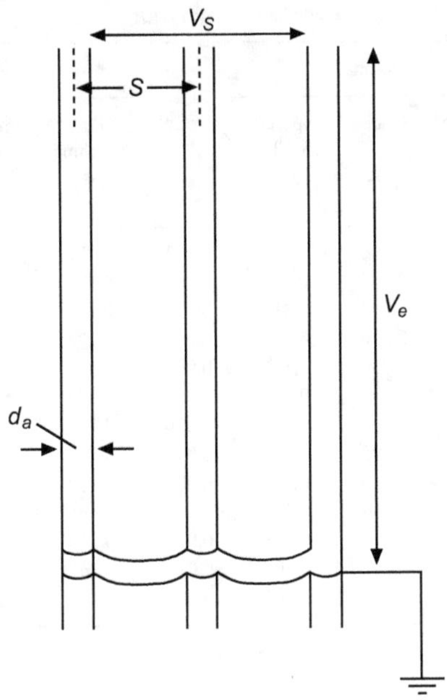

Figure 9.8 Voltages on single-core sheathed cables

Sheath voltages in single-point bonded systems operating at 50 Hz may be calculated using the following equations (see also Figure 9.8):

Single-phase

$$V_s = 2XI \text{ V/km}$$
$$V_e = XI \text{ V/km}$$

Three-phase, trefoil

$$V_s = \sqrt{3}XI \text{ V/km}$$
$$V_e = XI \text{ V/km}$$

Three-phase, flat

$$V_s = \sqrt{3}I\,(X + X_m) \text{ V/km}$$
$$V_e = I\sqrt{X^2 + XX_m + X_m^2} \text{ V/km}$$

where:

$X_m = 0.0436\ \Omega$/km.
$X = 0.0628 \ln(2S/d_a)\ \Omega$/km.
S = spacing between cable centres in mm.
d_a = mean sheath or armour diameter in mm.
V_s = voltage between sheaths, between outers for three flat arrangements.
V_e = voltage sheath to earth, outer to earth for three flat arrangements.
I = line current.

Table 9.5 Sheath to sheath voltage for cables to BS 5467

Single point bonding of sheath or armour, voltages between sheaths or armouring of outer conductors (mV/A/m)

Nominal size	Single-phase				Three-phase*		
	Touching	Spaced 2 × D_e	Spaced 5 × D_e	Trefoil	Flat touching	Spaced 2 × D_e	Spaced 5 × D_e
	(V)	(V)	(V)	(V)	(V)	(V)	(V)
50	0.119	0.206	0.322	0.103	0.179	0.254	0.354
70	0.117	0.204	0.320	0.101	0.177	0.253	0.353
95	0.115	0.203	0.318	0.100	0.175	0.251	0.351
120	0.113	0.200	0.316	0.098	0.173	0.249	0.349
150	0.113	0.200	0.316	0.098	0.173	0.249	0.349
185	0.111	0.199	0.314	0.096	0.172	0.248	0.348
240	0.109	0.196	0.312	0.094	0.170	0.246	0.346
300	0.108	0.195	0.311	0.094	0.169	0.245	0.345
400	0.108	0.195	0.310	0.093	0.169	0.244	0.344
500	0.106	0.194	0.309	0.092	0.168	0.243	0.343
630	0.105	0.192	0.308	0.091	0.166	0.242	0.342
800	0.105	0.192	0.308	0.091	0.167	0.242	0.342
1 000	0.104	0.191	0.307	0.090	0.166	0.241	0.341

* Voltages given for three-phase cables refer to the outer cables.
D_e is the overall cable diameter.

Table 9.6 Sheath or armour voltage to earth for cables to BS 5467 and BS 6346

Single point bonding of sheath or armour, voltages between sheaths or armouring of outer conductors to earth (mV/A/m)

Nominal size	Single-phase				Three-phase*		
	Touching	Spaced 2 × D_e	Spaced 5 × D_e	Trefoil	Flat touching	Spaced 2 × D_e	Spaced 5 × D_e
	(V)	(V)	(V)	(V)	(V)	(V)	(V)
50	0.060	0.103	0.161	0.060	0.090	0.131	0.187
70	0.059	0.102	0.160	0.059	0.089	0.130	0.186
95	0.058	0.101	0.159	0.058	0.088	0.129	0.185
120	0.056	0.100	0.158	0.056	0.087	0.128	0.184
150	0.056	0.100	0.158	0.056	0.087	0.128	0.183
185	0.056	0.099	0.157	0.056	0.086	0.127	0.183
240	0.055	0.098	0.156	0.055	0.085	0.126	0.182
300	0.054	0.098	0.155	0.054	0.085	0.125	0.181
400	0.054	0.097	0.155	0.054	0.085	0.125	0.181
500	0.053	0.097	0.155	0.053	0.084	0.125	0.180
630	0.053	0.096	0.154	0.053	0.083	0.124	0.180
800	0.053	0.096	0.154	0.053	0.083	0.124	0.180
1 000	0.052	0.096	0.153	0.052	0.083	0.123	0.179

* Voltages given for three-phase cables refer to the outer cables.
D_e is the overall cable diameter.

Figure 9.8 shows sheath to sheath and sheath to earth voltages. Tables 9.5 and 9.6 tabulate voltages calculated using the formulae above for cables to BS 5467 and BS 6346.

As an example of their use, consider 2 × 630 mm^2 aluminium conductor cables per phase supplying the load from a 1 MVA transformer. The full load line current per cable is of the order of 700 A. Assuming a two-diameter spacing, the sheath- or armour-to-earth voltage given by Table 9.6 is 0.124 × 700/1 000 V/m = 0.087 V/m. Cable length of the order of 250 m would be necessary to generate 25 V to earth under full load conditions. Such transformer cables are rarely longer than 15 m.

9.8.2 Corrosion

The voltage at which corrosion is likely to be significant is dependent on the conductivity of the soil around the cables. ERA Technology advises that laboratory tests and experience indicate that a 12 V sheath or armour potential to Earth is a safe limit below which there is negligible corrosion due to a.c. electrolysis, although there are installations that have been in service for some years designed to 25 V. A limit of 12 V would reduce the maximum cable length in the above example to about 140 m. This is a much greater length than would normally be required for transformer tails.

9.8.3 Hazard

The voltages generated under fault conditions can be determined using the same equations and tables, but using the maximum prospective fault current. Assuming that the maximum prospective three-phase fault level is of the order of 20 000 A, the sheath-to-sheath voltage between cables given by Table 9.5, for cables spaced at five times their diameter, is of the order of 7 V/m (20 000 × 0.342/1 000). For cable lengths of 20 m, the voltage is 140 V for the duration of the fault.

9.9 Cross-sectional areas of conductors
Section 524

Table 52.3 of BS 7671 provides minimum nominal values of conductor cross-sectional areas. These are specified to provide a certain minimum mechanical strength. Reduced-section neutral conductors are allowed as a generality, provided that the designer has assessed that, in the three-phase circuit being considered, there will be a reduction in the neutral current as compared with the line conductors. In discharge lighting circuits this is not allowed, and the practice is increasingly less common because of the harmonic content of modern commercial and industrial building loads. Where a piece of equipment such as the power supply of a computer generates third or multiples of third harmonics, the current in the neutral does not cancel out, as is the case for the fundamental. In fact, if the current is third harmonic, or a multiple thereof, the neutral current could be as high as three times the line current. In installations supplying computer and other equipment with solid-state power supplies, the designer may well be advised to consult with the client to obtain some details of the equipment likely to be installed.

9.10 Voltage drop
Section 525, Appendix 4 Section 6.4

9.10.1 General and origin of the supply

BS 7671 requires voltage drops in consumers' installations to be limited, such that the voltage at the terminals of any fixed current-using equipment shall be greater than the lower limit corresponding to the product standard relevant to the equipment. This information is often not readily available to the designer, and the voltage drop between the origin of the installation (usually the supply terminals) and a socket-outlet or the terminals of fixed current-using equipment is then often designed to meet the 'deemed to comply' requirement given in Section 6.4 of Appendix 4: which, for low voltage installations supplied directly from a public distribution system, is 3 per cent of the nominal voltage for lighting and 5 per cent for other uses.

For domestic and similar small installations, supplied from a distributor's network subject to the ESQCR, the origin of the installation is clearly the supply terminals at the service head. For larger installations this is not necessarily the case.

9.10.2 Private L V supply
Appendix 4 Section 6.4, Table 4Ab

The circumstances of the additional voltage drop allowed for 'low voltage installations supplied from a private LV supply' are not entirely clear. Presumably, a private LV supply is one supplied from a private generator or from a private transformer. The additional voltage drop allowed compensates for there being no voltage variation (within permitted range) experienced at the origin of an installation supplied from a public distribution system. However, a more practical approach is to design installations such that the voltage range at any point in the installation does not exceed or fall below the limits permitted by the ESQCR (see section 9.10.3).

9.10.3 Large installations

Where the voltage at the origin of the installation is under the designer's control, e.g. when the supply is from an 11 000 V/433 V transformer, use can be made of normal voltage ranges to compensate for voltage drop. The nominal supply voltage in the UK is 230 V + 10% – 6% (see Regulation 27 of the ESQCR).

The upper limit set by statute with respect to electricity distributors is 230 V + 10%, that is 253 V. This can be the no-load voltage on the low voltage side of the transformer. The distributor is allowed by statute to drop his voltage to 230V–6%, that is 216.2 V. However, if a conservative lower limit of 220 V is set, the total allowable voltage drop is 253 – 220 V = 33 V. This is 13 per cent of 253. If a maximum voltage drop of 8 per cent in the distributor's cables plus 5 per cent or

3 per cent in final circuits is adopted, the voltage range will always lie within the 253–220 V range. A typical example is shown in Figure 9.9.

With such an arrangement, the voltage at the distribution point close to the transformer (A) will remain within range, as will those at the extremity (B), both under no-load and full-load conditions, and allow the standard design limit of 5 per cent and 3 per cent for final circuits.

Provided certain conditions are complied with, Regulation 525.203 does allow a higher voltage drop during the starting of motors and for other equipment with high inrush currents. Such starting surges will cause a flicker on the supply, which may be seen

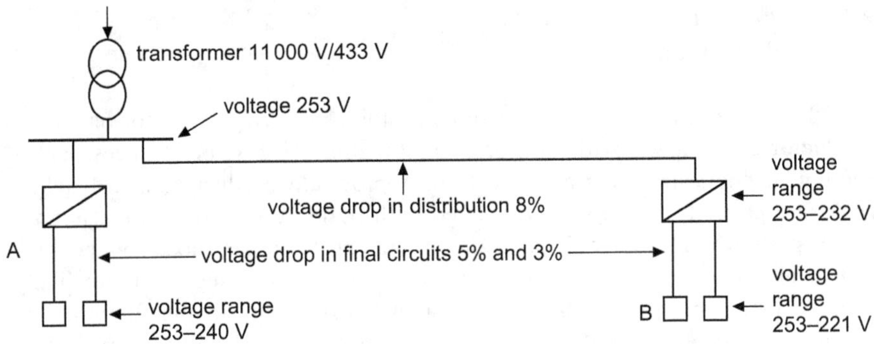

Figure 9.9 Distribution voltage drop

by the dipping of lights. Care has also to be taken that this is not a problem with electronic equipment. Large mainframe computer consumers' installations used to have undervoltage protection, which could drop out on starting surges. This problem is not so common now that personal computers are generally used. Such voltage dips are also quite liable to trip out motor starters, and the designer will probably wish to give consideration to starting sequences for large motors. It is generally preferable to start the larger motors before the smaller motors, as the starting of a smaller motor is less likely to affect the running of the large motor than the reverse.

9.11 Electrical connections
Section 526

9.11.1 Accessibility of connections
Regulation 526.3, Appendix 15

Connections are generally required to be accessible for inspection, testing and maintenance purposes. The exceptions to this requirement include joints:

- designed for use with underground cables;
- made by welding, soldering, brazing or appropriate compression tool within equipment not intended to be accessible;

- cold tails to ceiling heating or floor warming elements or trace heating systems;
- joints or connections made in equipment by the manufacturer of the product and not intended to be inspected or maintained; and
- equipment complying with BS 5733 for a maintenance-free accessory and marked (with the letters MF inside a circle) to be inaccessible if installed in accordance with the manufacturer's instructions.

Note that crimped joints are not included, but compression joints are. *Crimping* is often the term used by manufacturers for cable sizes up to 10 mm^2 and *compression* for larger sizes.

9.11.2 Enclosures for connections
Regulation 526.5

BS 7671 requires every termination and joint in a live conductor or a PEN conductor to be made with one or a combination of a suitable accessory or equipment enclosure complying with an appropriate product standard; or an enclosure partially formed or completed with building material which is non-combustible when tested to BS 476-4 to provide protection to those connections, and to provide some protection against the risk of fire, which is more likely to arise in a connection than perhaps anywhere else in an installation. The requirement for enclosing connections is applicable at all voltages, including extra-low voltage.

9.12 Selection and erection to minimize the spread of fire
Section 527

9.12.1 Precautions within a fire-segregated compartment

The requirements in this section are general, to minimize the spread of fire. The particular requirement is that the wiring system shall be installed so that the building's fire performance and fire safety are not reduced. It deals with precautions within the fire-segregated compartment so as to identify where the sealing of penetrations of the compartment is necessary (see 9.12.2).

Cables are generally required to comply with BS EN 60332-1-2 – *Tests on electric and optical fibre cables under fire conditions*. Most cables meet the requirements of this standard, other than polyethylene cables and some rubber cables.

Cable systems are required to comply with the flame-propagation requirements of the relevant system standard; otherwise, covering with non-combustible building materials is necessary.

For products meeting the resistance to flame propagation of their system standards – such as BS EN 61386 series for *Conduit*, BS EN 50085 series for *Trunking systems and ducting systems*, BS EN 61439-6 for *Busbar trunking*, BS EN 61534-1 for *Powertrack systems*, and BS EN 61537 for *Cable tray systems and cable ladder systems* – no special precautions are required.

9.12.2 Sealing of wiring system penetrations
Regulation group 527.2

The Regulations require that, where wiring systems pass through elements of the building construction such as walls and floors, the openings after passage of the wiring shall be sealed according to the degree of fire resistance required, if any. A further requirement is that cable trunking, conduit etc. shall be internally sealed so as to maintain the degree of fire resistance. Intumescent seals are readily available; other sealing systems such as sand traps are not precluded.

Regulation 527.2.3 exempts a conduit system, cable trunking system, or cable ducting system of internal cross-sectional area not exceeding 710 mm^2 (up to 32 mm diameter conduit) from the requirement to be internally sealed provided they the system is classified as non-flame propagating according to the relevant product standard and the system complies with the two conditions, relating to IP ratings, given in the regulation are complied with, as applicable.

9.13 Proximity to other services
Section 528

9.13.1 Proximity to electrical services
Regulation 528.1

General

Regulation 528.1 imposes requirements for the segregation of Band I and Band II circuits from one another (see Figure 9.10) and from high voltage circuits.

Band I voltages are basically ELV, but the wording of the definition of Band I deliberately includes telecommunication and other data circuits that may have ringing voltages exceeding 50 V.

Band II voltages are generally LV and are limited in BS 7671 to 1000 V a.c. and 1500 V d.c.

Band I (ELV) circuits shall not be contained within the same wiring system (e.g. trunking) as Band II (low voltage) circuits unless:

(i) every cable or conductor is insulated for the highest voltage present; or
(ii) each conductor of a multicore cable is insulated for the highest voltage present in the cable; or
(iii) the cables are insulated for their system voltage and installed in a separate compartment of a cable ducting or trunking system; or
(iv) the cables are installed on a cable tray system where physical separation is provided by a partition; or
(v) a separate conduit, trunking or ducting system is employed; or
(vi) for a multicore cable, the cores of the Band I circuit are separated from the cores of the Band II circuit by an earthed metal screen of equivalent current-carrying capacity to that of the largest core of a Band II circuit.

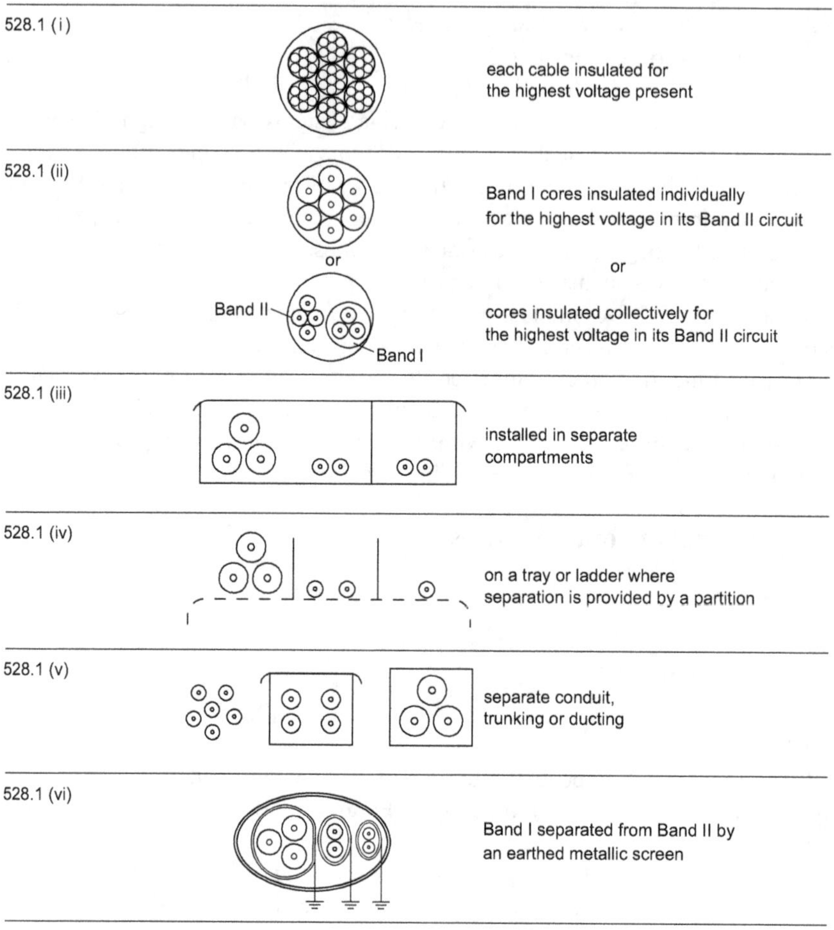

Figure 9.10 Segregation of electrical services

Regulation 528.1 reminds the reader that for SELV and/or PELV systems the requirements of Regulation group 414.4 apply.

9.13.2 Lightning protection
Regulation 528.1 Note 1

BS EN 62305 *Protection against lightning* series, Parts 1, 2, 3 and 4, replaced BS 6651:1999, which was withdrawn on 31 August 2008.

BS EN 62305-1:2011 provides the general principles to be followed in the protection against lightning.

BS EN 62305-2:2012 is applicable to risk assessment for a structure or for a service due to lightning flashes to Earth.

BS EN 62305-3:2011 deals with all the requirements needed to protect buildings and structures against physical damage, and looks at how to protect humans and animals against injury.

BS EN 62305-4:2006 provides information for the design, installation, inspection, maintenance and testing of a lightning electromagnetic impulse (LEMP) protection measures system (LPMS) for electrical and electronic systems within a structure, able to reduce the risk of permanent failures due to lightning electromagnetic impulse.

The basic LEMP protective measures to reduce failure of electrical and electronic systems given in Part 1 are (a) for structures: earthing and bonding measures, magnetic shielding, line routing and SPD protection; (b) for services: surge protective devices (SPDs) at different locations along the length of the line and at the line termination (see Section 443 of BS 7671 and Chapter 7), magnetic shields of cables.

BS EN 62305-1 states that an ideal protection for structures and services is to enclose the object to be protected within an earthed and perfectly conducting continuous shield of adequate thickness, and by providing adequate bonding, at the entrance point into the shield, of the services connected to the structure.

It is to be noted that Section 443 of BS 7671 specifies where overvoltage control is required; that is, it is generally not required where an installation is supplied by a low voltage system containing no overhead lines.

9.13.3 Fire alarms and emergency lighting
Regulation 528.1 Note 2

BS 7671 has requirements for circuits in close proximity to electrical services in general in Regulation 528.1 and safety services in general in Chapter 56. The requirements for fire detection and alarm systems and for emergency lighting systems are given in Regulations 560.9 and 560.10 respectively. However, to avoid duplication and, more importantly, conflict with the relevant parts of BS 5266 series, BS EN 1838, and BS 5839 series, the Joint IET/BSI Technical Committee JPEL/64 decided to direct readers to the respective standards, as appropriate. (See also Chapter 13 for safety systems).

9.13.4 Proximity of communications cables
Regulation 528.2

An adequate separation between telecommunication wiring (Band I) and electric power and lighting (Band II) circuits must be maintained. This is to prevent mains voltage appearing in telecommunication circuits with consequent danger to personnel, and to reduce electromagnetic interference. When cables cross, BS 7671 requires a minimum clearance of 100 m or specified precautions must be taken. BS 6701:2010 *Telecommunications equipment and telecommunications cabling* recommends that the minimum separation distances given in Tables 9.7 and 9.8 should be maintained.

Table 9.7 Cable classification from table 3 of BS EN 50174-2

Cable classification	
a	"a" is used if 1) the mix of applications is unrestricted; 2) the type of cabling to be installed is unrestricted; 3) the relevant cable performance is unknown
b	cables meeting EN 50288-3-1 (EN 50173-1:2011, Category 5) and EN 50288-6-1 (EN 50173-1:2011, Category 6)
c	cables meeting EN 50288-2-1 (EN 50173-1:2011, Category 5) and EN 50288-5-1 (EN 50173-1:2011, Category 6). These cables may deliver performance of Segregation Classification "d" provided that the relevant coupling attenuation requirements are also met.
d	Cables meeting EN 50288-4-1 (EN 50173-1:2011, Category 7)

Table 9.8 Minimum separation S from table 4 of BS EN 50174-2

		Containment applied to information technology or power supply cabling		
Segregation classification (from Table 3)	Separation without electromagnetic barrier mm	Open metallic containment note a mm	Perforated metallic containment notes b, c mm	Solid metallic containment note d mm
d	10	8	5	0
c	50	35	25	0
b	100	75	50	0
a	300	225	150	0

Notes:

a. Screening performance (0 MHz to 100 MHz) equivalent to welded mesh steel basket of mesh size 50 mm × 100 mm (excluding ladders). This screening performance is also achieved with steel tray (trunking without cover) of less than 1.0 mm wall thickness and/or more than 20% equally distributed perforated area.

b. Screening performance (0 MHz to 100 MHz) equivalent to steel tray (trunking without cover) of at least 1.0 mm wall thickness and no more than 20% equally distributed perforated area. This screening performance is also achieved with screened power cables that do not meet the performance defined in footnote d.

c. The upper surface of installed cables shall be at least 10 mm below the top of the barrier.

d. Screening performance (0 MHz to 100 MHz) equivalent to a steel conduit of 1.5 mm wall thickness. Separation specified is in addition to that provided by any divider/barrier.

Table 9.9 Power cabling factor P from table 5 of BS EN 50174-2

Electrical circuit type a, b, c	Quantity of circuits	Power cabling factor P
20 A 230 B 1-phase	1 to 3	0.2
	4 to 6	0.4
	7 to 9	0.6
	10 to 12	0.8
	13 to 15	1.0
	16 to 30	2
	31 to 45	3
	46 to 60	4
	61 to 75	5
	> 75	6

Notes:

(a) 3-phase cables shall be treated as 3 off 1-phase cables.
(b) More than 20 A shall be treated as multiple of 20 A.
(c) Lower voltage AC or DC cable = 5 of 20 A cables ($P = 0.4$)

For example consider two 200 A three-phase cables running close to some IT cabling what separation is required by BS EN 50174-2. The IT cable types are various and there is no electromagnetic barrier.

The IT cables are of many cable standards, so the IT cable classification is "a"; see Table 9.7

The minimum separation "S" from Table 9.8 is 300 mm

The power cabling factor from Table 9.9 is determined as follows:

Note a says a 3-phase cable is treated as 3 single-phase circuits, there fire the equivalent number of circuit is

$2 \times 3 = 6$

From Table 9.9 the power cabling factor is 0.4.

However, note b says cables of loading more than 20 A shall be treated as multiples of 20 A, therefore we have a multiplier of 10.

Hence $P = 10 \times 0.4 = 4$,

and the minimum separation $A = S.\ P = 300 \times 4 = 1200$ mm.

If a solid metal containment system is used for either the power or telecoms cables, the separation distance is 0 mm.

9.13.5 *Proximity to non-electrical services* Regulation group 528.3

General

This group of regulations imposes general requirements to prevent damage or hazard arising from other services, such as those likely to produce heat, smoke or fumes, unless protected from those harmful effects by shielding or similar. There is a specific requirement in Regulation 528.3.5 that no cable shall be run in a lift or hoist shaft unless it forms part of the lift installation as defined in BS EN 81-1 series.

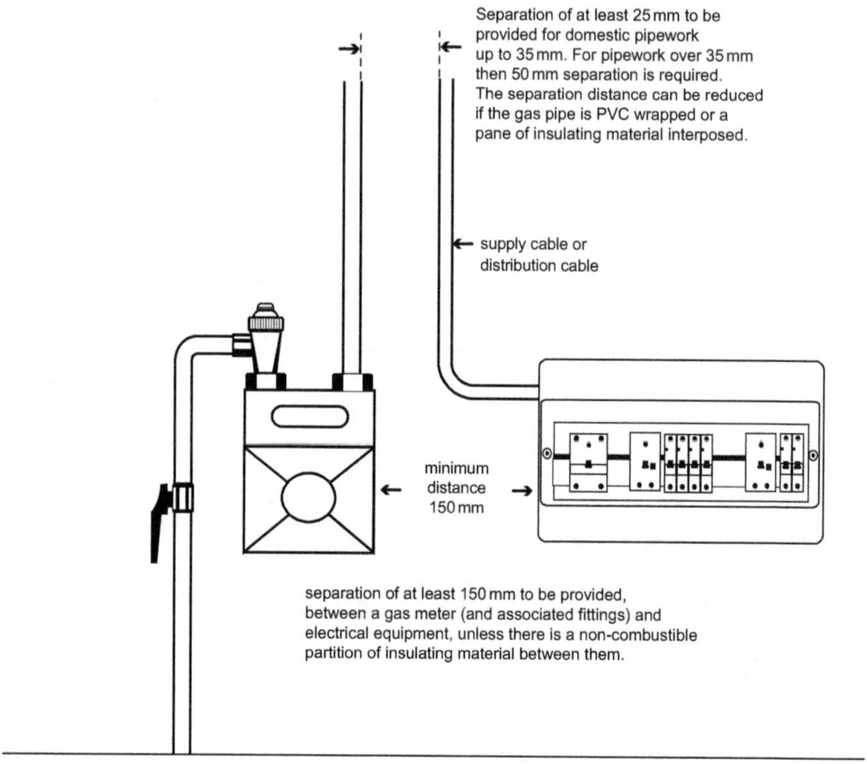

Figure 9.11 Separation from gas pipes and meters

Gas services

BS 6891:2005, as amended, *Installation of low pressure gas pipework of up to 35 mm (R1¼) in domestic premises* (clause 8.16.2) advises that, where gas installation pipes are not separated by electrical insulating material from electrical equipment including cables, they are to be spaced as follows:

(a) at least 150 mm away from electrical equipment (meters, controls, accessories (such as a socket-outlet or switch), distribution boards or consumer units); and
(b) at least 25 mm away from electricity supply and distribution cables.

Chapter 10
Protection, isolation, switching and control

Chapter 53

Adviser: Paul Sayer IEng FIET GCGI, Technical Standards Manager, Hager Engineering Ltd

10.1 Introduction

10.1.1 Scope

Regulation group 530.1

Chapter 53 includes the switchgear requirements for (i) protection against electric shock by automatic disconnection of supply (531), risk of fire (532), overcurrent (533), overvoltage (534), undervoltage (535); (ii) coordination of protective devices (536); (iii) isolation and switching (537); and (iv) monitoring (Sect 538).

As well as including the switchgear requirements for isolation and switching, Chapter 53 also includes in Section 537 the requirements for safety for isolation and switching.

10.1.2 Common requirements

Regulation group 530.3

The common requirements for switchgear are fundamental in that, when all live conductors including the neutral are switched, the construction of the switchgear shall be such that the neutral cannot be disconnected before the line conductors, and is to be reconnected before or at the same time as the connection of the line conductors. Equipment complying with the relevant standards given in Table 53.4 of BS 7671 will meet this requirement.

No fuse, switch or circuit-breaker is to be inserted in the neutral conductor of TN or TT systems, except where linked (Regulation 530.3.2). This ensures that, if a piece of equipment stops operating because of the operation of an overcurrent device, or the operation of a switch, the equipment will not be left in a live condition, i.e. still connected to a line conductor. While competent persons familiar with electrical equipment would never work on it without checking that it is isolated, there is a temptation sometimes to open up equipment on the operation of an overcurrent device, and of course this is particularly dangerous if an overcurrent device has been fitted in the neutral.

Independently operated single-pole switches or protective devices, for whatever purpose, including functional switching, are not allowed in the neutrals of single- or three-phase circuits, be they isolators or functional switches.

10.1.3 Short-circuit rating of low voltage assemblies (switchboards, panel boards, distribution boards, etc.) Regulation 533.3

An assembly will have a fault current (short-circuit) rating specified as the maximum permissible prospective fault current at the point of connection of its supply to the incoming terminals. The assembly manufacturer is responsible for ensuring the capability of the equipment between the incoming and outgoing terminals of the assembly (incoming and outgoing devices, busbars, connections, etc.). The relevant fault current (short-circuit) rating of the assembly should be equal to or exceed the maximum prospective fault current at the point of connection to the system. The terminology to define the short-circuit rating of an assembly is given in the BS EN 61439 series of standards as follows:

- rated short-time withstand current, Icw
- rated peak withstand current, Ipk
- rated conditional short-circuit current, Icc

10.1.4 Conditional rating of switchgear and controlgear assemblies Regulation 530.3.4

Electricity distributors will advise supply applicants that, in accordance with Engineering Recommendation P25 (the short-circuit characteristics of public electricity distributor's low voltage networks and the coordination of protective devices on 230 V single-phase supplies up to 100 A), fault levels at the origin of single-phase supplies may be as high as 16 kA. (Fault levels are rarely as high as this.) This exceeds the fault rating of many domestic consumer units if fitted with circuit-breakers or rewirable fuses. The concept of 'conditionally rated' devices has been developed to make allowances for the 100 A cut-out fuse in practice backing up the consumer unit in domestic premises. Conditionally rated devices can withstand the energy let-through of a 100 A cut-out fuse before the fuse ruptures. In practice one of the consumer unit circuit breakers or fuses will clear the fault, as fault levels very rarely exceed the fault capability (rated short-circuit capacity) of the devices in the consumer unit.

The requirements for conditional rating are found in Annex ZB of BS EN 61439-3 and BS EN 60439-3 Annex A (see below for an excerpt).

Additional test

8.2.3 Verification of the Assembly Capability to withstand a 16 kA fault. The following test is applied to customer distribution boards as covered by the definition in clause 2.1.11.

8.2.3.1 Test arrangements. The Customer Distribution Board shall be set up as in normal use. It will be sufficient to test a single functional use if the remaining functional units are constructed in the same way and cannot affect the test result.

8.2.3.2 Short-circuit test procedure. The following test procedure is intended to verify the performance of the incoming device and its connections, and any other item in the Customer Distribution Board not separately rated in excess of 16 kA, when the complete Customer Distribution Board is protected by a fuse-link complying with BS 1361:1971. This type test shall be deemed to cover the use of any other short-circuit protective device having a Joule integral (I^2t) and cut-off current not exceeding the values given in item (b) below, at the rated voltage, prospective current and power factor.

10.2 Devices for fault protection by automatic disconnection of supply
Sect 531

10.2.1 Overcurrent devices
Regulation group 531.1

This Regulation group brings together the requirements concerning devices for over-current and shock protection. If an overcurrent device is being used additionally for shock protection in the event of a fault, it must ensure that the disconnection is:

(i) in sufficient time to prevent the conductors reaching excessive temperature (Regulation 430.3); and
(ii) within the disconnection time as appropriate for shock protection (Regulation 411.3.2).

10.2.2 Residual current devices
Regulation group 531.2

This section summarizes the requirements for RCDs. They must disconnect all the line conductors of the circuit at substantially the same time; the magnetic circuit (toroid) must encircle all the live conductors (line and neutral); and Regulation 531.2.4 requires that RCDs shall be selected and the electrical circuits so subdivided that any protective conductor current that may be expected to occur during normal operation of the connected load(s) will be unlikely to cause unwanted tripping of the device.

Protective conductor currents arise from two main sources:

(i) functional earthing of load equipment via capacitors to limit electromagnetic effects, and to comply with the EMC requirements in general; and
(ii) capacitance of the circuit conductors to earth.

There will also be leakage currents to earth from tracking across insulation and from capacitance effects of the construction of the equipment.

Reference to British Standards and approval body requirements suggests maximum values of protective conductor current as shown in Table 10.1.

Table 10.1 Typical protective conductor and touch current limits

	Household appliances BS EN 60335-1 Clause 13.2 (mA)	Luminaires BS EN 60598-1 Clause 10.3 (mA)	Hand-held tools BS EN 60745-1:2003, Clause 13 (mA)	Information technology equipment BS EN 60950 Clause 5.1 (mA)
Class 0, Class 0I and Class III appliances	0.50		0.50	
Portable and hand-held Class I appliances	0.75	1.00	0.75	0.75
Stationary Class I motor operated appliances	3.50	N/A	N/A	N/A
Stationary Class I heating appliances	0.75 or 0.75 per kW rated input of the appliance, whichever is the greater with a maximum of 5 mA	N/A	N/A	3.50
Class II appliances	0.25	0.50	0.25	0.25

N/A = Not applicable.

However, these 'leakage' current measurements are taken at a specific point in the test sequence for the appliances and the appliance will have reached normal running temperature and be dry, when leakage currents will be at their minimum. This is particularly true of metal-sheathed elements, which can absorb moisture through the open seals at the end of the element and, as a consequence, on switch-on can be very leaky until they have dried out. As a result, these standard maximum leakage currents are of only minimal assistance in determining the leakage current likely to arise in a circuit when an appliance is first switched on.

Reliability can be increased by limiting the number of sockets (and hence equipment connected at any time) and fixed equipment controlled by each RCD.

BS 7671 used to recommend that protective conductor (leakage) currents do not exceed 25 per cent of the nominal tripping current of an RCD. While this no longer appears in the standard, it's still quite good advice, since the RCD standard allows the tripping current of a 30 mA RCD to be as low as 20 mA.

10.2.2.1 Circuits without protective conductors
Regulation 531.2.5

If an exposed-conductive-part is not connected to a circuit protective conductor, in the event of a fault, even if a 30 mA RCD protects the circuit, the fault will not be cleared until the equipment is touched. This puts a person's life at risk, staked against the reliable operation of the RCD. If the circuit has a protective conductor, in the event of a fault to exposed-conductive-parts the leakage current is much greater than for a person making direct contact with a live part, and the device can operate as soon as the fault arises.

Older installations, i.e. pre-1996, may not have protective conductors in the lighting circuits. A householder may not be aware of the potential danger of fitting a metal Class I luminaire to such an installation. A 30 mA RCD reduces the risk, but such emphasis should not be placed on its reliable operation and an RCD is not permitted as an alternative to a circuit protective conductor where it is reasonably practicable to install such a conductor. The latest guidance on this very subject is covered in item 10 of Electrical Safety First's Best Practice Guide No. 1.

10.2.2.2 Fault withstand of RCDs
Regulation 531.2.8

Regulation 411.3.3 of BS 7671 requires mobile equipment (rating not exceeding 32A) for use outdoors and socket-outlets (of rating not exceeding 20 A) to be protected by 30 mA RCDs.

BS EN 61008-1:2012, as amended *Residual current operated circuit-breakers without integral overcurrent protection for household and similar uses (RCCBs)* specifies rated conditional short-circuit currents for RCDs. Conditional short-circuit current is the short-circuit current withstand when the RCD is protected by a suitable short-circuit protective device. RCDs to BS EN 61008-1 have limited fault withstand, so they must be protected against short-circuit currents by an overcurrent device. The device may be upstream or downstream of the RCD.

Where an RCCB is installed in an assembly, for example, a consumer unit or distribution board, the assembly will have a fault current (short-circuit) rating specified as the maximum permissible prospective fault current at the point of connection of its supply to the incoming terminals. Normally, the RCCB fault current capability will be covered by the assembly rated conditional short-circuit current, Icc; see 10.1.4.

10.2.3 RCD discrimination
Regulation 531.2.9

Regulation 531.2.9 requires discrimination between RCDs where this is necessary to prevent danger. We also have Regulation 314.1, which requires every installation to be divided into circuits as necessary to:

(i) avoid hazards and minimize inconvenience in the event of a fault;
(ii) facilitate safe inspection; testing and maintenance (see also Section 537);

(iii) take account of hazards that may arise from the failure of a single circuit such as a lighting circuit;
(iv) reduce the possibility of unwanted tripping of RCDs due to excessive protective conductor currents not due to a fault;
(v) mitigate the effects of electromagnetic interferences (EMI); and
(vi) prevent the indirect energizing of a circuit intended to be isolated.

RCDs (RCCBs or RCBOs) are required:

(i) where the earth fault loop impedance is too high to provide the required disconnection, for instance where the distributor does not provide an earth, that is in TT systems;
(ii) for socket-outlets with a rated current not exceeding 20 A;
(iii) for LV circuits serving a location containing a bath or shower, and for those passing through zones 1 and/or 2 not serving the location;
(iv) for circuits supplying mobile equipment for use outdoors (not exceeding 32 A);
(v) for cables without earthed metal covering installed in walls or partitions at a depth of less than 50 mm and not protected by earthed steel conduit or similar;
(vi) for cables without earthed metal covering installed in walls or partitions with metal parts (not including screws or nails) and not protected by earthed steel conduit or the like; and
(vii) for circuits in certain special installations and locations in Part 7 of BS 7671:2008.

30 mA RCDs are required for (ii) to (vi) above. There is need for the designer to ensure unwanted tripping is minimized.

Note that RCDs may be omitted for:

(a) where, other than for an installation in a dwelling, a documented risk assessment has determined that RCD protection is not necessary, or
(b) for specific labelled or otherwise suitably identified socket-outlet provided for the connection of a particular item of equipment (such as a freezer or, taking on-board the technical intent of this requirement, something such as a kidney dialysis machine (however, the circuit cable(s) may require RCD protection depending on their construction and how and where they are installed; see (ii) to (vii) above.

10.2.4 RCDs in TN systems
Regulations 411.4.4, 531.3.1

Regulation 411.4.4 allows automatic disconnection in TN systems to be provided by an RCD. This is very useful where circuit lengths are such that disconnection times of overcurrent protective devices are too long. The installation of an RCD having a rated residual operating current of 100 mA will generally provide for automatic disconnection in domestic premises; however, the need for 30 mA RCDs as in 10.2.3 will generally result in all automatic disconnection needs being met. See Figure 10.1.

In other premises a less sensitive RCD would be appropriate. Regulation 531.3.1 allows use to be made of either the TN earth or a separate earth electrode, but in the latter case the circuit concerned has to be treated as part of a TT system.

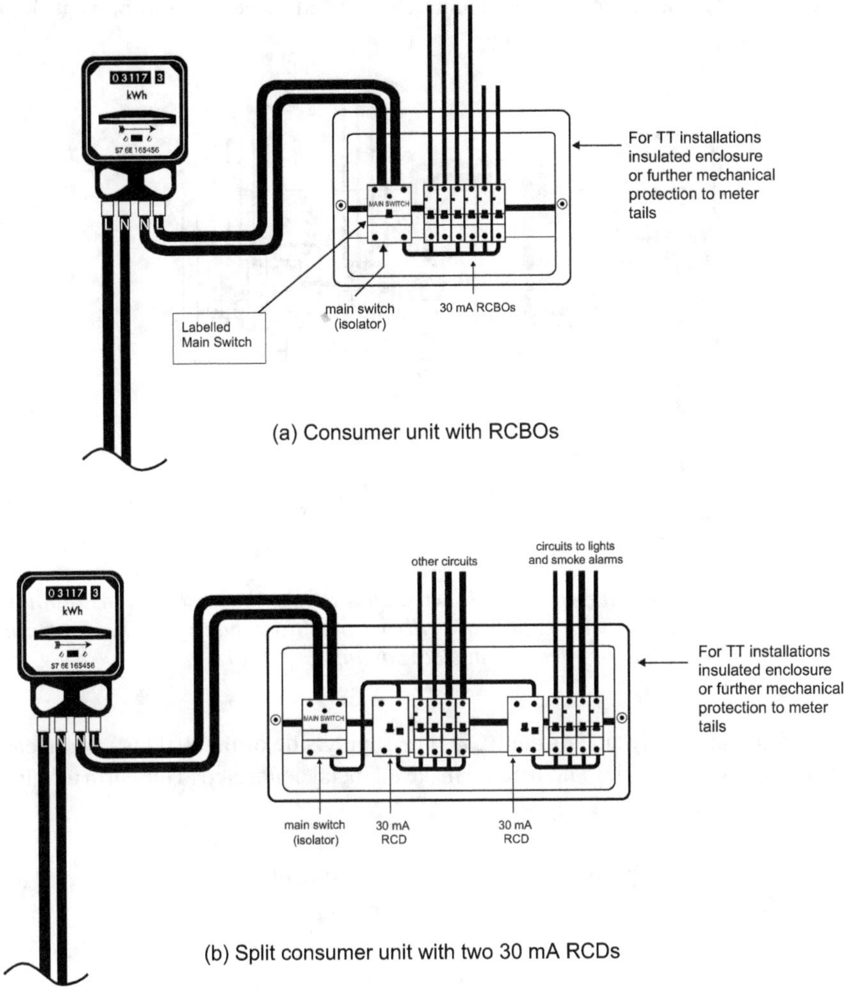

Figure 10.1 Consumer units suitable for TN and TT installations

The installation of a 30 mA RCD at the origin of an installation will result in the disconnection of the complete installation in the event of a fault on one final circuit or one piece of equipment. This is a contravention of Regulation 314.1 (see 10.2.3).

10.2.5 RCDs in TT systems Regulation 531.4.1

The enclosures of RCDs or consumer units incorporating RCDs in TT installations must have an all-insulated or Class II construction, or additional precautions recommended by the manufacturer need to be taken to prevent faults

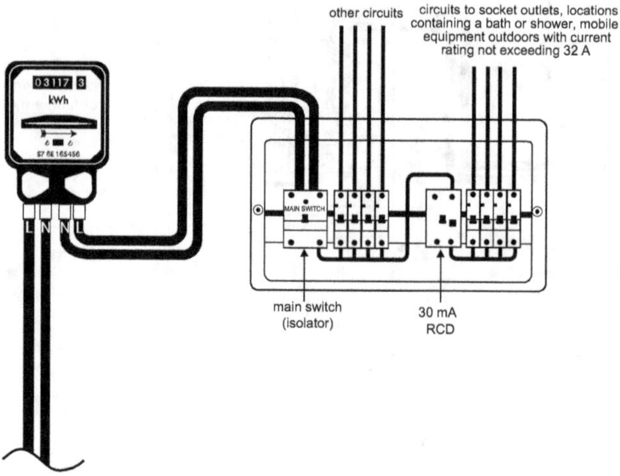

Figure 10.2 Split consumer unit with one 30mA RCD suitable for TN installations only and with cables in walls or partitions having earthed metallic covering or enclosed in steel conduit

to earth on the supply side of the RCD. The supply side of the RCD is unprotected against faults to earth and a fault could make all metal connected to the earth terminal live. See Figure 10.1.

10.3 Devices for protection against overcurrent Section 533

General

Regulation 533.1 includes a list of British Standards and states that devices for protection against overcurrent 'shall comply with' one or more of these standards.

This requirement is, of course, overridden by Regulation 511.1, which allows British or Harmonized Standards but also allows, as an alternative, equipment complying with a foreign national standard based on an IEC standard to be used if the designer or other person responsible verifies that any differences between the standard and the corresponding British or Harmonized Standard will not result in a lesser degree of safety them to that afforded by compliance with the British or Harmonized Standard.

10.3.1 Semi-enclosed fuses
Table 53.1

It is important to note that the sizes of copper wire for use in semi-enclosed fuses are retained. This helps to dispel the widely held misunderstanding that semi-enclosed (rewirable) fuses are not allowed by the current Wiring Regulations. They are.

10.4 Devices for protection against overvoltage
Sections 443, 534

10.4.1 Insulation coordination

The insulation, which may be solid or air gaps, within an electrical installation must be graded or coordinated. Distributor's company equipment at the origin of an installation will have to withstand higher impulse voltages/overvoltages than installation equipment, and this equipment will have to withstand higher impulse voltages that say, appliances. This coordination is specified in BS EN 60664-1. *Insulation coordination for equipment within low-voltage systems. Principles, requirements and tests.* It is a system standard and standards such as the installation standard (BS 7671), equipment standards, e.g. BS EN 60439 and appliance standards, e.g. BS EN 60335 must comply with it to achieve the necessary coordination. Table 44.3 of BS 7671, reproduced in chapter 7 as Table 7.1, is taken from this standard.

BS EN 60664-1 specifies the requirements for clearances, creepage distances and solid insulation for equipment based upon their performance criteria. It includes methods of electrical testing with respect to insulation coordination.

10.4.2 Sources of overvoltage

Overvoltages will arise from:

(i) faults on the supply network, particularly the HV network;
(ii) switching on the supply network or perhaps of large inductive loads. These are not usually severe and protection is not generally required;
(iii) lighting strikes to the supply network or to the lightning protection system of the building.

The overvoltages must be controlled so that they do not exceed the values given in Table 7.1.

Section 7.2 of this Commentary explains the earthing required to limit overvoltages due to HV faults.

The following sub section considers precautions to be taken to protect installations from overvoltages due to lightning strikes.

10.4.3 Overvoltages due to lightning (and switching)

Section 443 of BS 7671 and section 7.4 consider the circumstances where BS 7671 requires protection against overvoltages of atmospheric origin (lightning). In the UK, the frequency of lightning strikes (AQ level) is generally such that even if an installation is supplied by an overhead line the installation of surge protective devices is not required. However, there is a note to Regulation 443.2.2:

Note 1: Irrespective of the AQ value, protection against overvoltages may be necessary in applications where a higher reliability or higher risks (e.g. fire) are expected. See BS EN 62305.

BS EN 62305 *Protection against lightning* requires the installation of surge protective devices where lightning protection is installed. Presumably, the installation of lightning protection, whilst protecting the structure against lightning strikes, increases the probability of the building being struck by lightning. For large tall buildings this must be very likely and surge protection is necessary.

10.4.4 Selection and erection of surge protective devices
Section (534), Appendix 16

Because most industrial and commercial buildings are required by BS EN 62305 to have lightning protection, surge protection in such properties is often commonplace.

BS 7671 unusually provides much specific requirements and guidance on the installation of surge protective devices, concluding in the informative Appendix 16. Consequently, the writer will simply refer the reader to BS 7671 and surge protective device suppliers.

10.5 Devices for protection against undervoltage
Sections 445, 535, Regulation 552.1.3

The requirements for protection against undervoltage call for precautions to be taken where a reduction in voltage or loss of voltage and subsequent restoration could cause danger. Reference is made to the requirements of Regulation 552.1.3, which would require motor starters to have low-/no-voltage relays to prevent automatic restarting.

These are normally necessary to prevent damage to the motor, and there may also be a risk of danger arising from unexpected restarting.

10.6 Co-ordination between device, design current and diversity in assemblies

10.6.1 Rated current of a circuit (I_{nc})

Typical terminology used to describe an LV assembly includes switchboard, panel board distribution board and consumer unit. An assembly comprises a range of electrical components and devices, e.g. busbars, circuit breakers, fuses, switching devices, installed in an enclosure. The rating of a device (I_{nc}) when installed in an assembly might be less than its marked "free air" rated current (I_n) due to the cooling being reduced by the enclosure and heat gain from other equipment in the enclosure.

BS EN 61439-1 defines the rated current of a circuit (I_{nc}) as:

The value of the current that can be carried by this circuit loaded alone, under normal service conditions. This current shall be carried without the temperature rise of the various parts of the ASSEMBLY exceeding the limits specified in 9.2.

Note 1: *The rated current of a circuit may be lower than the rated currents of the devices (according to the respective device standard) installed in this circuit.*

Note 2: *Due to the complex factors determining the rated currents, no standard values can be given.*

For assemblies conforming to the relevant part of the BS EN 61439 series, I_{nc} will be identified in the manufacturer's documentation supplied with the assembly.

The electrical installation designer must ensure as a minimum requirement that for circuits:

$$I_{nc} \geq I_b$$

where I_b is the design current of the circuit.

Note 3: *For assemblies with circuits that are continuously and simultaneously loaded, reductions in the I_{nc} rating of the devices in the assembly could be required, see section 10.6.2 below.*

10.6.2 Rated diversity factor (RDF) for continuously and simultaneously loaded outgoing circuits

Devices, e.g. circuit-breakers installed in an assembly are individually capable of carrying continuously their rated current I_{nc} under conditions defined in BS EN 61439. However, if a group of adjacent circuit devices are to operate continuously and simultaneously, de-rating will probably be necessary to ensure there is no overheating. This de-rating requires application of a rated diversity factor (RDF), sometimes called a loading factor. RDF is applied to outgoing circuit devices of an assembly.

BS EN 61439-1 defines rated diversity factor RDF as:

The per unit value of the rated current, assigned by the ASSEMBLY manufacturer, to which outgoing circuits of an ASSEMBLY can be continuously and simultaneously loaded taking into account the mutual thermal influences.

The assumed diversity factors from two product standards are identified in Table 10.2. The RDF can be stated for groups of circuits or for the whole assembly. These tabulated values may be used unless otherwise specified by the manufacturer.

Table 10.2 Typical values of RDF

Values of rated diversity (RDF)/assumed loading			
BS EN 61439-2 (typically switchboards, panel boards and motor control centres)		BS EN 61439-3 (typically consumer units and distribution boards)	
Number of circuits	RDF (assumed loading)	Number of circuits	RDF (assumed loading)
2 and 3	0.9	2 and 3	0.8
4 and 5	0.8	4 and 5	0.7
6 to 9 inclusive	0.7	6 to 9 inclusive	0.6
10 (and above)	0.6	10 (and above)	0.5

Circuit devices can be rated on the basis of inrush currents, intermittent and/or short duration loads. RDF recognizes that multiple functional units or devices may not be continually loaded simultaneously. Guidance as to what does **not** constitute "continually loaded" can be conservatively given for BS EN 61439 assemblies as; *"a load not exceeding 30 minutes with a maximum duty cycle of 50%, unless otherwise specified by the manufacturer"*. This means that if the load "off" time is equal to or greater than its "on" time, with a maximum "on" time of 30 minutes, diversity need not be applied.

For a circuit supplying a cyclic load i.e. a variable load which is regularly repeated with definable cycles e.g. machine process, punching, welding etc. the installation designer could apply Regulation 533.2.1 of BS 7671. This permits the current rating of the overload protective device and associated cable to be selected on the basis of the design current (I_b) being the thermally equivalent constant load current instead of the maximum peak current of the load cycle; see subsection 6.2.3 for further guidance. It should be noted that, BS EN 61439-1 highlights that care should be taken with ON-times greater than 30 min, since small devices could already reach their thermal equilibrium.

For adjacent circuits that are continuously and simultaneously loaded, the required device/circuit rated current is given by:

$$I_{nc} \geq \frac{I_b}{RDF}$$

Note as described in subsection 10.6.1 I_{nc} may be less than the rated current I_n marked on the device. The assembly manufacturer should provide the I_{nc} rating.

Example

Consider five storage heater circuits supplied from a 5 way consumer unit to BS EN 61439-3, each heater has a load current of 13 A. It is judged, that the circuits will be continuously and simultaneously loaded. The minimum required circuit rated current I_{nc} for each circuit breaker is given by:

$$I_{nc} \geq \frac{1.3}{0.7}$$

$$I_{nc} \geq 18.6\,A$$

10.6.3 Rated current of the assembly and incoming device/circuit (I_{nA})

The rated current of the assembly (I_{nA}) is the maximum load current the assembly is designed to manage and distribute, e.g. the total current which the main busbar(s) is capable of distributing in the particular assembly arrangement. Different incoming device/circuit options could be available for an assembly e.g. main switch, circuit-breaker, contactor that connect to the main busbar(s). Each will have a rated current I_{nc}. The rated current of a device/incoming circuit may be lower than the rated current of the incoming device (according to the respective device standard) when installed in the assembly. Diversity is not applied to the incoming device/circuit I_{nc}, i.e. the incoming circuit can operate at its rated current with all outgoing functional units collectively to their rated current multiplied by the diversity factor.

Therefore, the rated current of the assembly is the smaller of:

- The total rated current of the incoming circuit(s) which is either the rated current of the single incoming circuit or the sum of the rated currents of the incoming circuits within the assembly operated in parallel;
- The total current which the main busbar is capable of distributing in the particular ASSEMBLY arrangement.

The electrical installation designer must specify an appropriate rated current for the assembly or its incoming device/circuit. For example, consider five storage heater circuits supplied from a Consumer Unit, each heater has a load current of 13 A. The minimum required rated current of the Consumer Unit/incoming device/circuit is given by:

$$I_{nA} \geq 5 \times 13 = 65 \text{ A}$$

If a five way Consumer Unit is available with an I_{nA} of 100 A (main busbar), with options of a main switch I_{nc} (not I_n) of 80 or 100 A, the installation designer could specify the 80 A main switch. The rated current of the assembly is the smaller of 100 A and 80 A, thus it is 80 A.

Example of rated current of the assembly and application of diversity

A lighting distribution board is required to supply ten circuits, each circuit having a load current of 3.5 A. A contactor is required as the incoming device. It is judged, that the circuits will be continuously and simultaneously loaded.

(1) From Table 10.2, the RDF is 0.5

(2) The minimum required circuit rated current I_{nc} for each circuit-breaker is given by:

$$I_{nc} \geq 3.5/0.5 = 7 \text{ A}$$

(3) The minimum required rated current of the distribution board incoming device/circuit is given by:

$$I_{nA}/I_{nc} \geq 10 \times 3.5 = 35 \text{ A}$$

(4) The selected distribution board main busbars have a rated current I_{nA} of 80 A and the incoming contactor when installed in the assembly an I_{nc} of 63 A (de-rated from its marked rated current of 100 A). The rated current of the distribution board is the smaller of 80 A and 63 A, thus it is 63 A.

10.7 Coordination of protective devices

10.7.1 Overcurrent devices
Regulation 536.1

Regulation 536.1 requires that, where required (necessary to prevent danger, required for proper functioning), overcurrent devices shall be selected such that intended discrimination is achieved. 'Where required' can be presumed to mean as necessary to prevent danger and as necessary for proper working of the installation. It is appropriate to refer to Regulation 314.1, requiring the division of circuits to minimize inconvenience in the event of a fault. Inconvenience can be minimized in the event of a fault only if there is discrimination between overcurrent devices placed in series.

There are two basic requirements for discrimination. Devices B, C and D in Figure 10.3 need to operate before device A under:

(i) overload conditions; and
(ii) fault conditions.

10.7.2 Fuse-to-fuse discrimination

Discrimination between devices is required for any prospective current ranging from an overload to the maximum prospective fault current at the point of installation.

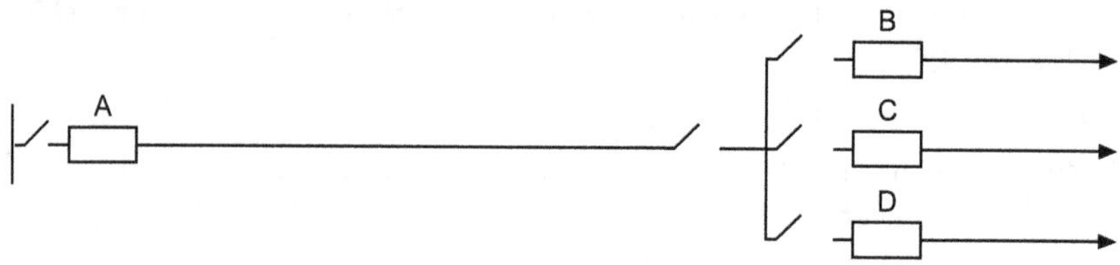

Figure 10.3 Overcurrent devices in series

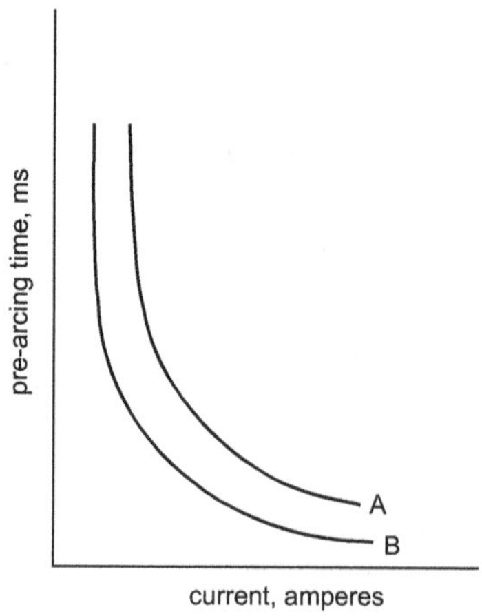

Figure 10.4 Time/current characteristics of cartridge fuses

(Cartridge-fuse rated short-circuit capabilities will almost certainly exceed the maximum prospective fault current; see Table 6.7). It is necessary for the time/current characteristic of the upstream device (A) at all currents to be above that of the downstream device (B) (see Figure 10.4). This will generally be sufficient to achieve discrimination for medium or low overcurrent (e.g. overload).

However, at high overcurrents discrimination is not so easily achieved.

Figure 10.5 indicates the performance of fuses under circumstances of a high prospective fault current (a) and a moderate prospective fault current (b).

For consideration of discrimination, the time for an overcurrent device to interrupt the current can be considered in two parts: the pre-arcing time (a to b) and the arcing time (b to c). The pre-arcing time is the time that elapses before the fuse

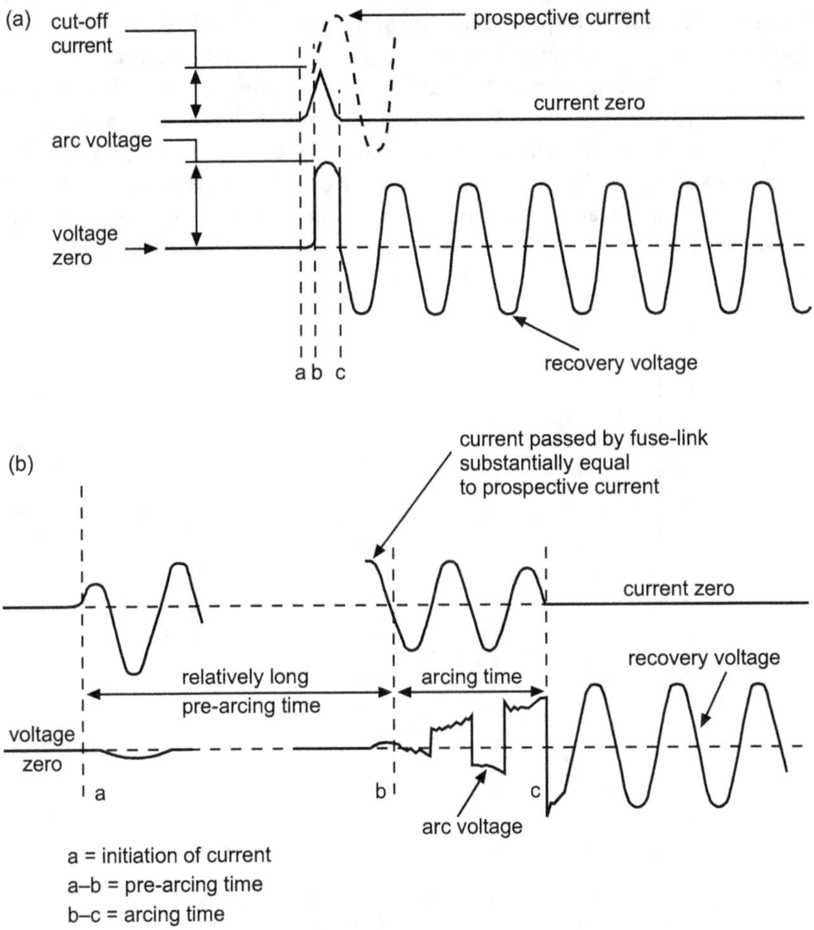

Figure 10.5 Arcing and pre-arcing times for fuses. (a) Typical oscillogram for large a.c. prospective current, e.g. fault. (b) Typical oscillogram for moderate a.c. prospective current, e.g. overload

link has melted. During the arcing time, the fuse link has melted, but there is an arc between the contacts or between the terminals of the fuse. Clearly, if a downstream fuse is to discriminate with another, its total pre-arcing and arcing time must be less than the pre-arcing time of the upstream fuse. Once a fuse has started to arc, there is no going back to the pre-arcing stage; the fuse will be ruptured. However, if the downstream fuse does operate before the end of the pre-arcing time, the upstream fuse can cool down and will not rupture.

Clearly, it is necessary for discrimination between fuses (Figure 10.3) for the characteristics of any fuse B, C or D nearer the fault to lie throughout its length below that of fuse A closer to the supply point. Additionally, the total operating time of fuse B, C or D – that is pre-arcing plus arcing – must be less than the pre-arcing time of fuse A. Consequently, it is not sufficient that simply the one fuse characteristic should always lie below the other, but for high fault levels that the total arcing time of the downstream device be compared with the pre-arcing time of the upstream device.

Quite often the characteristics under high fault conditions are presented showing the cut-off characteristics. Reference to Figure 10.6 will show that, for high prospective fault currents, the fuse will interrupt the current before it reaches its peak prospective value. The cut-off value will depend on the rating of the fuse: the smaller the fuse, the less the maximum value of cut-off current that is reached.

For practical application, discrimination can be achieved between fuses when the rating of the upstream device is twice that of the downstream device.

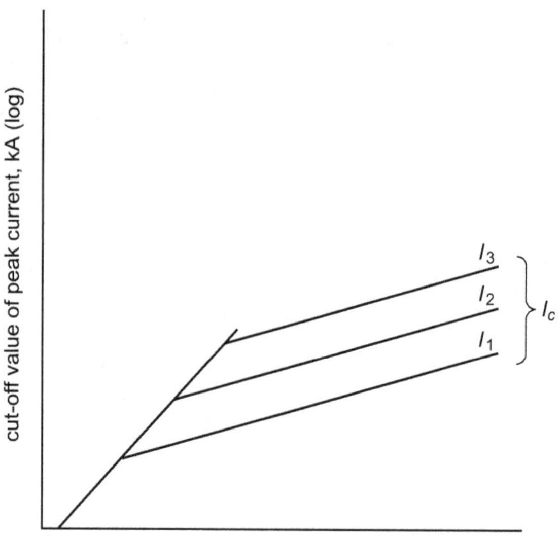

I_1, I_2, I_3 are the rated currents of fuse links (I_1 is smallest)
I_c is the maximum value of cut-off current

Figure 10.6 General presentation of the cut-off characteristics of a series of a.c. fuse link

10.7.3 Circuit-breaker to circuit-breaker discrimination

A circuit-breaker's tripping characteristic(s) is generally in two parts, see Figure 10.7. For low currents (overloads) the circuit-breaker will have a typical inverse time/current (thermal) characteristic, and for higher currents instantaneous (magnetic) characteristics. For high overcurrents such as short-circuit conditions, the operating time (magnetically initiated) is typically 10 ms, or 0.01 s.

Referring to Figure 10.8, if the fault current I_{F2} exceeds that for magnetic operation of the downstream (B) and upstream (A) devices, there will be no discrimination and both devices will operate. If the fault current (I_{F1}) is between that for magnetic operation of the downstream and upstream devices, then discrimination is practicable. This does mean that, if discrimination between circuit-breakers is required under fault conditions, the upstream devices have to be selected on the basis of the fault current and not the load. As shown in Figure 10.8, if device A is to discriminate with device B under fault conditions, the prospective fault current I_{F1} must exceed the magnetic operating current of device B and be in the thermal range of device A.

Electronic circuit-breakers are available with minimum or 'short-time' as opposed to instantaneous characteristics, see Figure 10.9. It is possible to discriminate between such devices over a relatively wide prospective fault current range even if the prospective fault current exceeds that for short-time (instantaneous) operation.

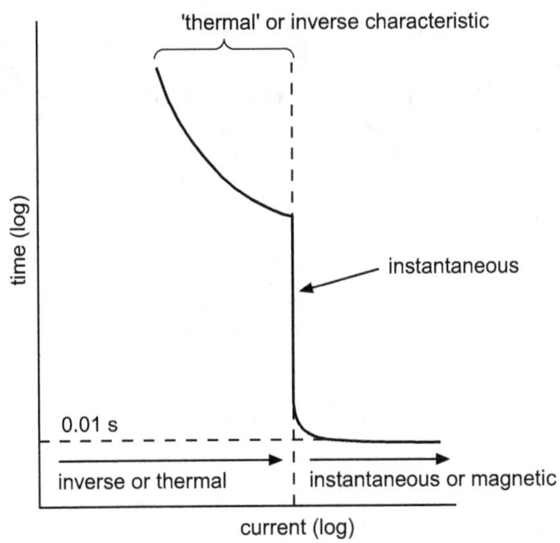

Figure 10.7 Circuit-breaker characteristics

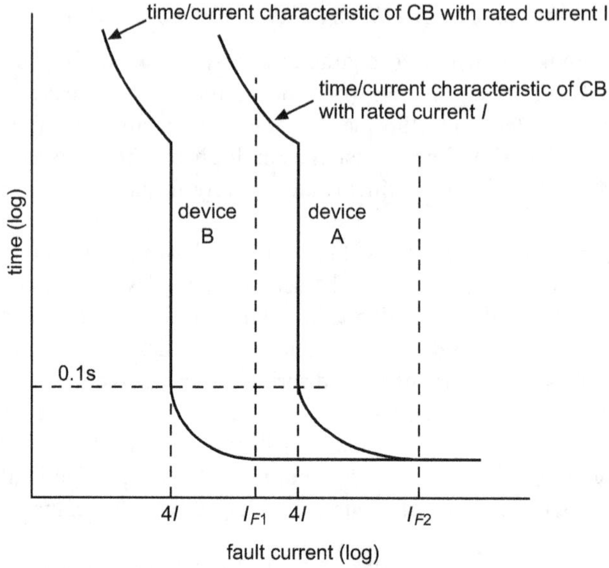

Figure 10.8 Discrimination between circuit-breakers

These devices will discriminate not only at fault current I_{F1}, but also at I_{F2}. However, there remains a fault current above which they will not discriminate, but this approaches the maximum prospective fault current of the devices, which may be above the prospective fault level at that point in the system.

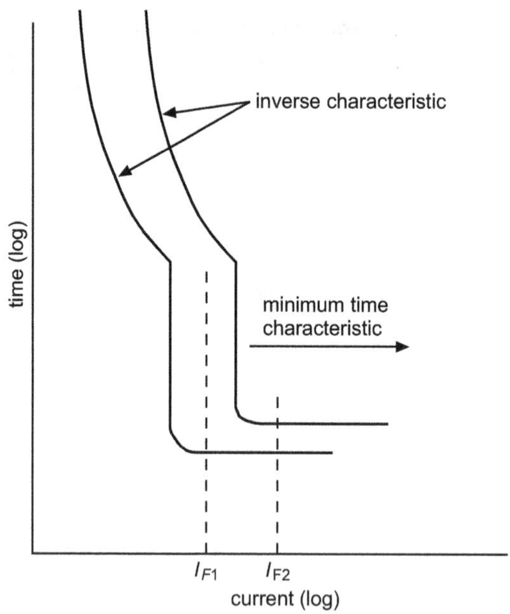

Figure 10.9 Circuit-breakers with minimum time characteristics

10.7.4 Fuse to circuit-breaker discrimination

The fault rating of a circuit-breaker is usually less than that of a fuse. A fuse is often used to back up circuit-breakers, as allowed by Regulation 434.5.1 and as shown in Figure 10.10. This requires the upstream device A to limit the energy let-through such that the downstream device B is not damaged. In the case of fuse/circuit-breaker combinations, this can be achieved. However, it must be remembered that, if the prospective fault current exceeds the rating of the circuit-breaker, and the fuse is correctly installed so that it operates before the circuit-breaker suffers damage, it will of course disconnect circuits (B, C and D of Figure 10.3).

Circuit-breaker time/current characteristics generally level off, for example at approximately 10 ms for miniature circuit-breakers (half a cycle). The intersection of this part of the characteristic curve of the circuit-breaker with the time/current characteristic of the fuse defines the degree of discrimination permissible. For currents less than the crossover current, the circuit-breaker will operate, but for greater currents the fuse will operate (as well as the circuit-breaker). The operation of the fuse is, however, necessary since the fault currents now are such that, while the circuit-breaker would draw an arc, it would probably be unable to extinguish it and the breaker would be damaged. It is a requirement of BS EN 60898 for circuit-breakers that the manufacturer provide advice on the appropriate backup fuse for the circuit-breaker. As discussed in subsection 6.3.4, the takeover current I_0 must not exceed the rated short-circuit capability I_{cn} of the circuit-breaker.

Summary

Discrimination in fault current conditions (short-circuit) between circuit-breakers can be achieved using various methods including upstream intentional time-delay, selective interlocking or a downstream current-limiting circuit-breaker. The manufacturer will provide advice on the appropriate co-ordination.

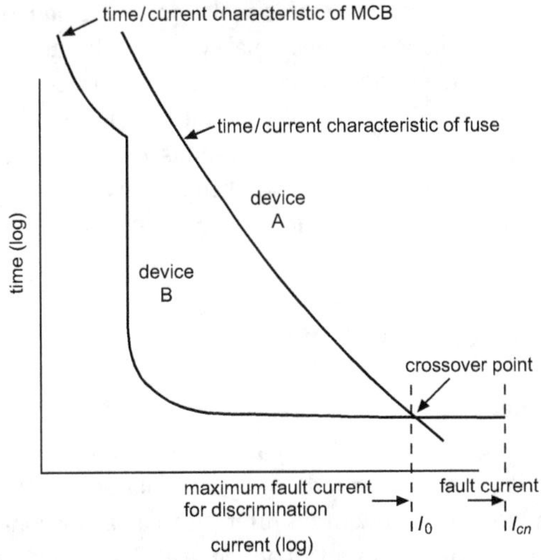

Figure 10.10 Discrimination between miniature circuit-breakers and fuses

10.7.5 Selection of device for protection against overload
Regulation group 533.2

(This subject is discussed in section 6.2.)

10.8 Isolation and switching
Section 537

10.8.1 Definitions

Four types of isolation and switching are considered by the Regulations:

- isolation;
- switching off for mechanical maintenance;
- emergency switching; and
- functional switching or control.

Isolation is a function that cuts off all or part of an installation from all sources of electrical energy, and allows electrically skilled persons to carry out work on, or adjacent to, parts that would otherwise be live.

Switching off for mechanical maintenance deactivates equipment by switching off the electricity supply, to enable non-electrical work to be carried out by persons who may not be electrically skilled.

Emergency switching is an action intended to remove, as quickly as possible, danger, both electrical and mechanical, that may have occurred unexpectedly. Emergency switching does not always include switching off the electricity: there are circumstances where electrical power is applied to remove the danger, such as braking of machinery.

Functional switching is a control mechanism. It may be as simple as the switching on or off of a luminaire, or it may be a complex process-control system.

Irrespective of the perceived need for isolation, switching off for mechanical maintenance, emergency switching or functional switching, Regulation 537.1.3 requires every installation to be provided with a means of disconnection (isolation and switching), and Regulation 537.2.1.1 requires that every circuit be capable of being isolated from each of the live conductors of the supply (the neutral is a live conductor), except that the neutral need not necessarily be isolated or switched in a TN-S or TN-C-S system where it is regarded as being reliably connected to Earth by a suitably low impedance.

10.8.2 Isolation
Regulation groups 537.1 and 537.2

10.8.2.1 Requirements

Isolation is the separation of all or part of an installation from all forms of electrical energy. It is normally carried out by an isolator, which may be an on- or off-load device. An isolator is now often referred to as a 'disconnector'. Similarly, a disconnector may be an on- or off-load device. There are three fundamental requirements for an isolator:

(i) The device must be able to withstand, without breakdown, overvoltages that may occur at the point of installation. Not all switches are suitable for use as isolators in any particular location, as the contact gap may not be sufficient to prevent supply surges bridging the gap and then the arc being maintained with dangerous consequences for any persons working on the installation (Regulation 537.2.2.1).

(ii) The state of the contacts, open or closed, shall be either externally visible or clearly and reliably indicated. The open indication shall be made only when the specified isolating distance has been obtained for each pole (Regulation 537.2.2.2).

(iii) Devices for isolation and switching shall be selected or installed such as to prevent unintentional or inadvertent closure, caused, for example, by mechanical impact or vibration (Regulation 537.2.2.3).

Equipment of overvoltage Categories I and II of BS EN 60664-1 *Insulation Coordination for Low Voltage Systems* is not allowed to be used as isolators. See Table 44.4 of BS 7671 for examples of categories of equipment.

Equipment adjacent to the supply meter should be Category IV, other equipment part of the fixed installation Category III. Table 44.3 of BS 7671 reproduces the requirements of Table F1 of BS EN 60664-1 (see Table 10.3).

Table 10.3a Rated impulse voltage for equipment energized directly from the low voltage mains [Source: Table F1 of BS EN 60664-1]

Nominal voltage of the supply system (1) based on IEC 60038 (3) (V)		Voltage line-to-neutral derived from nominal voltages a.c. or d.c. up to and including	Rated impulse voltage (1), see 443.3 (V)			
Three-phase	Single-phase	(V)	Overvoltage Category I	Overvoltage Category II	Overvoltage Category III	Overvoltage Category IV
		50	330	500	800	1 500
		100	500	800	1 500	2 500
	120–240	150	800	1 500	2 500	4 000
230/400,		300	1 500	2 500	4 000	6 000
277/480		690				
400/690		1 000	2 500	4 000	6 000	8 000
1 000			4 000	6 000	8 000	12 000

Notes:

1. See Annex B of BS EN 60664-1 (Table 10.3b below) for application to existing different low-voltage mains and their nominal voltages.
2. Equipment with these rated voltages can be used in installations in accordance with IEC 60364-4-44.
3. The/mark indicates a four-wire, three-phase distribution system. The lower value is the voltage line-to-neutral, while the higher value is the voltage line-to-line. Where only one value is indicated, it refers to three-wire, three-phase systems and specifies the value line-to-line.
4. See 4.3.3.2. of EN 60664-1 for an explanation of the overvoltage categories.

Table 10.3b Inherent control or equivalent protective control [Source: Table B1 of BS EN 60664-1]

Nominal voltages presently used in the world (V)					Rated impulse voltage for equipment[†] (V)			
Voltage line-to-neutral derived from nominal voltages a.c. or d.c. up to and including	Three-phase, four-wire systems with earthed neutral 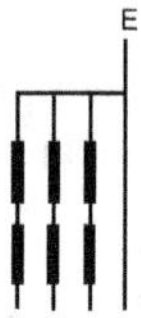 Star secondary	Three-phase, three-wire systems unearthed 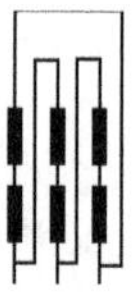 Delta secondary	Single-phase, two-wire systems a.c. or d.c. 	Single-phase, three-wire systems a.c. or d.c. 	Overvoltage category I	Overvoltage category II	Overvoltage category III	Overvoltage category IV
50			12.5, 24, 25, 30, 42, 48	30–60	330	500	800	1 500
100	66/115	66	60		500	800	1 500	2 500
150	120/208*, 127/220	115, 120, 127	100**, 110, 120	100–200** 110–220 120–240	800	1 500	2 500	4 000
300	220/380, 230/400, 240/415, 260/440, 477/480	200**, 220, 230, 240, 260, 277, 380, 400, 415, 440, 480	220	220–440	1 500	2 500	4 000	6 000
600	347/600, 380/660, 400/690, 417/720, 480/830	500, 577, 600	480	480–960	2 500	4 000	6 000	8 000
1 000		660, 690, 720, 830, 1 000	1 000		4 000	6 000	8 000	12 000

* Practice in the United States of America and in Canada.
[†] These columns are taken from Table 10.3a in which the rated impulse voltage values are specified.
** Practice in Japan.

Devices which are suitable for isolation if they have the required voltage withstand include:

(a) isolators (otherwise known as disconnectors)
(b) fuse-switch disconnectors (isolating switches) and switch disconnectors (to BS EN 60947-3)
(c) fuse links
(d) links

(e) circuit-breakers
(f) plugs and socket-outlets.

Further precautions, e.g. locking-off of switches, circuit-breakers etc., and the safekeeping of links or fuse links, once removed, may be necessary to secure the isolation; see section 10.8.2.2.

A disconnector, i.e. isolator, by definition, has to be capable of opening and closing a circuit when either negligible current is interrupted or made, or when no significant change in the voltage across the terminals of each of the poles of the isolator occurs. It is also capable of carrying rated currents under normal conditions and, for specified times, overcurrent. In the open position it must meet the requirements for isolation.

A non-automatic device for isolating and switching is required as near as practicable to the origin of every installation, and as a means of switching (on and off) the supply on load (Regulations 537.1.3 and 537.1.4).

10.8.2.2 Securing the isolation
Regulations 537.2.1.2, 537.2.1.5

Attention is drawn to Regulation 537.2.1.2 requiring suitable provisions to be taken to prevent any equipment from being inadvertently or unintentionally energized. These provisions might include the facility to fit a lock to the isolator or lock the switchgear enclosure or cupboard containing the switchgear. Regulation 537.2.1.5 requires that, where an isolating device for a circuit is placed remotely from the equipment to be isolated, provision shall be made so that the means of isolation can be secured in the open position. As the means of circuit isolation allows work on the circuit, and as the isolator would not be visible from all points on the circuit, means must be provided for securing of circuit isolation. Such securing can be made with a lock. Regulation 537.2.1.3 requires that, where an item of equipment or an enclosure contains live parts connected to more than one supply and is not capable of being isolated by a single device, a suitable and durable warning notice shall be placed in such a position that any persons, before gaining access to live parts, will be warned of the need to isolate further sources of supply (unless an interlock is provided).

10.8.3 Switching PEN and neutral conductors
Regulations 537.1.2, 537.2.1.1

In TN-C systems the PEN conductors must not be isolated or switched, as, not only would the connection with earth be lost (Regulation 537.1.2), but also the potential of exposed-conductive-parts to true Earth will rise, perhaps to the supply voltage. In TN-S and TN-C-S systems, the neutral conductor need not be isolated or switched where the neutral conductor is reliably connected to earth by a suitably low resistance (Regulation 537.2.1.1). For TT systems, both the line and neutral conductors must be isolated and switched.

In all systems the neutral may be switched if the designer wishes.

Regulation 114.1 of BS 7671 states that, for a low voltage supply given in accordance with the ESQCR, it shall be deemed that the connection with Earth of the neutral of the supply is permanent. Outside Great Britain, confirmation shall be sought from the distributor that the supply conforms to the requirements of the ESQCR.

The safety requirement must be that a person can work safely on equipment in which the line conductors have been disconnected, but the neutral remains connected.

Regulations 8(3) and 9 of the ESQC Regulations require:

8. General requirements for connection with earth

(7) A generator or distributor shall, in respect of any low voltage network which he owns or operates, ensure that –

(a) the outer conductor of any electric line which has concentric conductors is connected with earth;

(b) every supply neutral conductor is connected with earth at, or as near as is reasonably practicable to, the source of voltage except that where there is only one point in a network at which consumer's installations are connected to a single source of voltage, that connection may be made at that point, or at another point nearer to the source of voltage; and

(c) no impedance is inserted in any connection with earth of a low voltage network other than that required for the operation of switching devices or of instruments or equipment for control, telemetry or metering.

9. Protective multiple earthing

(1) This regulation applies to distributors' low voltage networks in which the neutral and protective functions are combined.

(2) In addition to the neutral with earth connection required under regulation 8(3)(b) a distributor shall ensure that the supply neutral conductor is connected with earth at –

(a) a point no closer to the distributor's source of voltage (as measured along the distributing main) than the junction between that distributing main and the service line which is most remote from the source; and

(b) such other points as may be necessary to prevent, so far as is reasonably practicable, the risk of danger arising from the supply neutral conductor becoming open circuit.

(3) Paragraph (2)(a) shall only apply where the supply neutral conductor of the service line referred to in paragraph (2)(a) is connected to the protective conductor of a consumer's installation.

(4) The distributor shall not connect his combined neutral and protective conductor to any metalwork in a caravan or boat.

There is a difference between the statement in BS 7671 Regulation 114.1, that the connection with Earth of the neutral of a supply from an electricity company is deemed permanent, and the requirement of Regulation 537.1.2, that the neutral need not be switched where it can be regarded as being reliably connected with Earth by a suitably low impedance.

The DTI guidance to distributors on the ESQCR Regulations (publication reference URN 02/1544) advises:

(i) the supply neutral conductor should be connected with earth at the neutral point of the local transformer, i.e. as close as possible to the source, to ensure that all consumers connected to multiple low voltage cables benefit from a secure common earth at the electrical centre of the network
(ii) The final paragraph of regulation 8(3) places restrictions on duty holders from inserting impedances in any connections with earth, in order to prevent dangerous transient voltages on neutral or protective conductors in the event of an earth fault
(iii) Essentially, distributors' PME networks must be equipped with sufficient multiple earthing connections in order to minimize the risk of the combined neutral and protective conductor becoming disconnected from earth (particularly on overhead line networks), possibly resulting in danger on consumers' premises (see comments for regulation 7 Continuity of the supply neutral conductor and earthing connections). As well as the supply neutral with earth connection at the local transformer (see paragraph 8(3) (b)), other connections with earth are required. Such additional earth connections may include, for example, separate earth electrodes at the service joint on the main farthest from the supplying transformer or alternatively a connection with the supply neutral conductor of another distributing main (which will be connected with earth at the star point of its supplying transformers.

Consequently, the neutral of TN-S or TN-C-S supplies, provided in accordance with the ESQCR can be considered to be reliably connected with Earth by a suitably low resistance.

For private distribution systems the designer must ascertain if the neutral is both reliably connected with earth and at Earth potential, and generally meet all the requirements of the ESQCR. For TN-C distribution systems he or she would need to confirm that the installation met the PME requirements of Regulations 6, 7, 8 and 9 of the ESQCR, as the law requires!

10.8.4 Provision for disconnecting the neutral Regulation 537.2.1.7

Regulation 537.2.1.7 requires that provision be made for disconnecting the neutral. This could be by a linked switch, but a removable link is allowed (removable with a tool). This is a general requirement and the facility should be provided at each isolator. This allows the skilled person who is to work on an installation to disconnect the neutral if they think that the nature of the work requires it, since a potential from neutral to true Earth is likely to exist in all circumstances. Disconnection may also be required when testing is being carried out.

Main switches intended for use by ordinary persons, e.g. in domestic situations, must interrupt both the line and neutral conductors of a single-phase supply (Regulation 537.1.4).

While the Regulations have been constructed on the understanding that all work will be carried out by skilled persons, it has to be recognized that, on domestic and similar installations, unskilled persons are likely to carry out work on the installation and they will not be aware of the potential that might exist between a neutral and earth (see Figure 10.11), and might not be aware of the need for or even be capable of disconnecting the neutral. For this reason this requirement has been included.

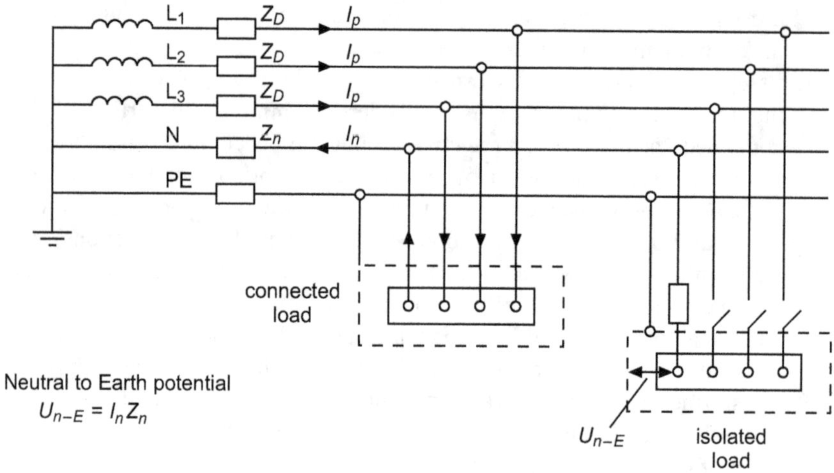

Figure 10.11 Neutral-to-earth voltages on TN-S systems

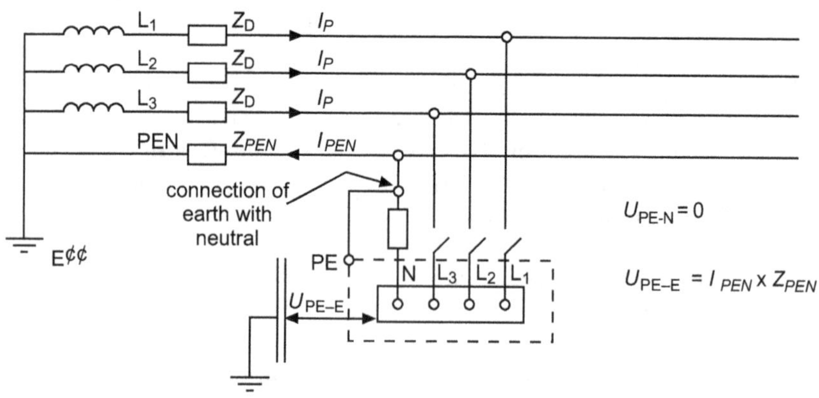

Figure 10.12 Neutral-to-earth voltages on TN-C systems

For TN-C supplies, it is often argued that there can be little advantage in switching the neutral, as all exposed-conductive-parts of a TN-C-S system are connected to the PEN conductor at the incoming supply position, and hence of course to the neutral conductor (see Figure 10.12). Switching the neutral serves no useful purpose, as all parts connected to the earth terminal are themselves connected to the PEN conductor, and so there can be no potential between the neutral and earth conductors.

Figure 10.12 shows how there can be no potential between the neutral and earth of a PME system close to the origin of a TN-C-S installation, where the neutral and earth conductors are connected to the PEN conductor of the supply.

However, it must be remembered that there may be a potential $U_{PE\text{-}E}$ between the PME earth terminal and true Earth. In a large installation, when the neutral is loaded, a potential difference can arise between neutral and earth in a similar way to TN-S systems.

The general requirements for isolation and switching are applicable both at the origin of the installation and for the isolation of circuits or items of equipment (see Figure 10.13). This applies equally to the isolation and switching of the neutral and to the switching of other live conductors. The only exception is the switching of the neutral for single-phase supplies in household or similar premises. The requirement to interrupt both live conductors of a single-phase supply is applicable only to the main switch. This can cause some conflict with the requirement of some appliance standards, which require installation instructions to be provided, advising that local isolation of all live conductors is required.

Single-phase

The same rules apply for single-phase as for three-phase, except that main switches intended for use by persons, e.g. domestic and similar, shall interrupt the line *and* neutral of the supply.

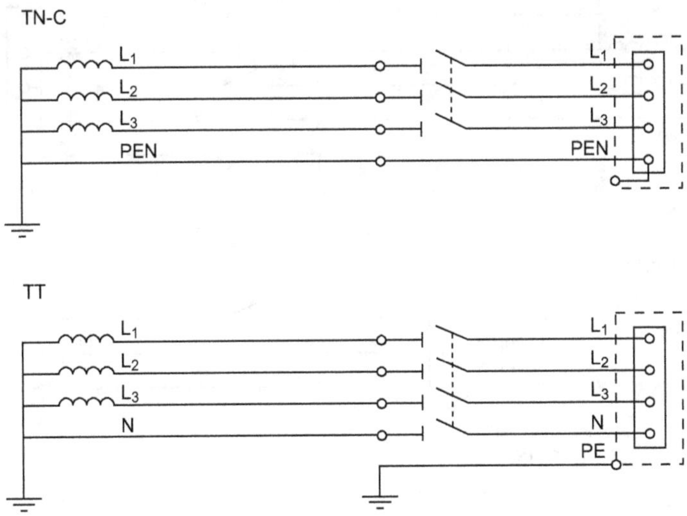

Figure 10.13 Isolation requirements

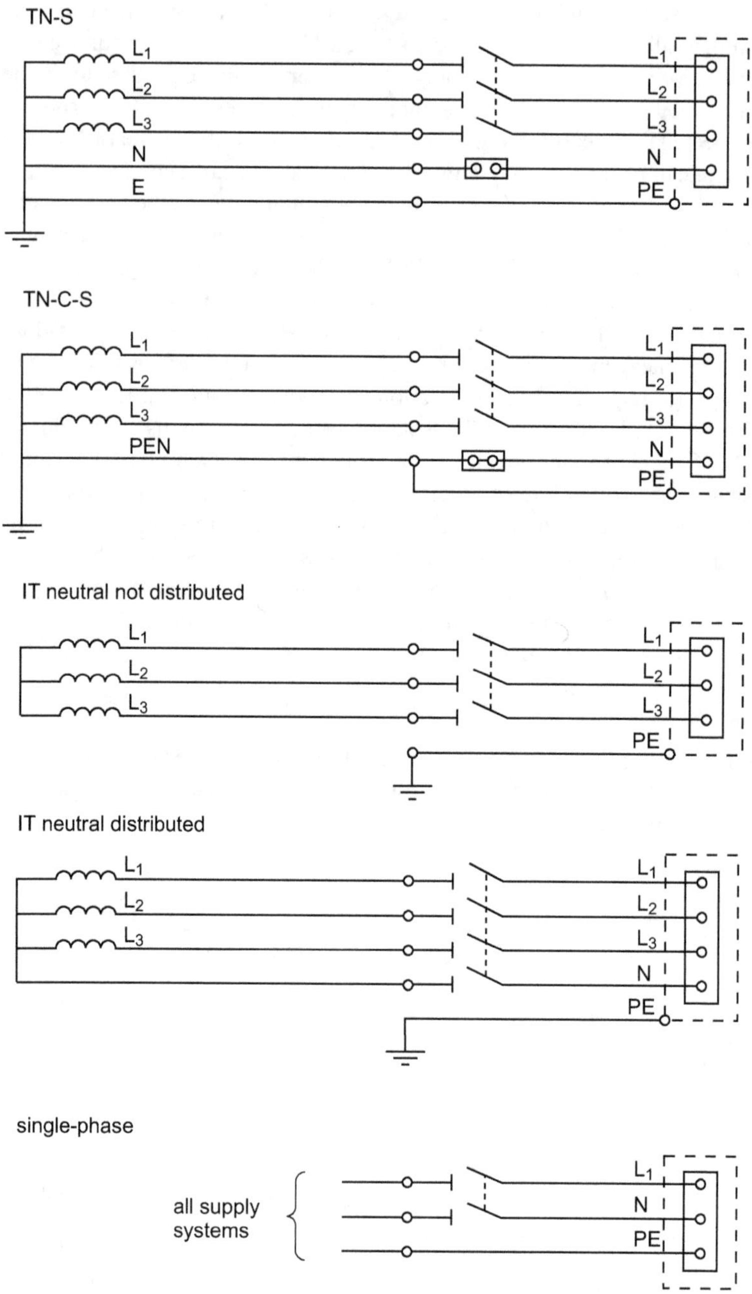

Figure 10.13 Continued

10.8.5 Devices for isolation
Regulation group 537.2.2, Table 53.2

The particular requirements for the devices for isolation are provided in this group. The impulse (overvoltage withstand) requirements are described in Regulation 537.2.2.1 and have been discussed in 10.6.2; see also Table 10.3. Guidance is provided in Table 53.4 of BS 7671 on the functions provided by the devices identified for isolation and switching, together with an indication of the relevant British Standards.

(a) Devices suitable for isolation are marked as follows: 

Semiconductor devices are not permitted as isolators because of the risk of breakdown (Regulation 537.2.2.1). They may be used for functional switching, or for switching-off for mechanical maintenance. The state of an isolator, open or closed, is required to be clearly and reliably indicated for the avoidance of accidents (Regulation 537.2.2.2). The requirement for reliable indication of the state of the isolator (open or closed) is understood to require direct linking to the supply conductors that are to be isolated, and not initiated by any control circuits. There must be no possibility of a control circuit indicating that an isolator is open when the main contacts are closed. A device that gives no indication as to whether the supply contacts are open or closed may not be used as an isolator.

Isolators are not required to be on-load devices, but if they are off-load they must be prevented from operating on-load, for example by interlocks (Regulation 537.2.2.4).

A plug and socket-outlet make a very effective isolator but they should be treated as offload devices for load currents above 16 A (see Table 53.4).

10.9 Switching-off for mechanical maintenance
Regulation group 537.3

10.9.1 General
Regulation 537.3.1

One would expect mechanical maintenance to be the repair or replacement of equipment not requiring access to live parts. However, by definition, the replacement, refurbishment or cleaning of lamps is included as well as 'non-electrical parts of equipment, plant and machinery'. The requirements for mechanical maintenance assume that this work is being carried out without access to live parts. If maintenance involves access to live parts, then isolation of the equipment is necessary and, by definition, is not mechanical maintenance. However, by definition, replacement of lamps does not require isolation, simply switching off.

Although removing lamps may provide access to live parts, it is an activity that is recognized as being capable of being carried out by unskilled persons with adequate job instruction. However, lamp replacement should be carried out with the circuit live conductor switched functionally off. If the light switch is not in the same room, it should be lockable, or there should be an overriding key switch. The requirement for manual operation (Regulation 537.3.2.2) even of computer-controlled luminaires will avoid hazardous surprise switch-ons when changing lamps at the top of a ladder.

The requirement for mechanical maintenance (that does not provide any access to live parts) is that the person carrying out the maintenance shall be made safe from mechanical hazards. This means that the supply does not need to be isolated, but the switching off must be secure. Switching off for mechanical maintenance is required where the mechanical maintenance may involve a risk of physical injury, including burns (Regulation 537.3.1.1). Unless the means of switching off is continuously under the control of the person performing the maintenance, provision has to be made to prevent the equipment from being unintentionally reactivated; this in practice means there must be a facility to lock-off the switch (Regulation 537.3.1.2).

10.9.2 Devices for mechanical maintenance
Regulation group 537.3.2

Switches for mechanical maintenance require:

- manual operation;
- clear and reliable indication of the open position of the contacts;
- installation so as to prevent inadvertent or unintentional operation; and
- to be readily identifiable.

A device is required for switching off for mechanical maintenance only if there is a need for mechanical maintenance. For example, a domestic shower with, say, a functional switch operated by water pressure would not necessarily need a further switch for mechanical maintenance, which would involve, say, cleaning of the shower head.

But, for example, a switch is required for a compressor motor. An operator must not change the belts by allowing the compressor to raise the system pressure so that the pressure switch stops the motor and then closes the outlet valve.

10.10 Emergency switching

10.10.1 General
Regulation group 537.4

Regulation 537.4.1.1 states that emergency switching shall be provided for any part of an installation from where it may be necessary to control the supply to remove an unexpected danger, which, as mentioned at the beginning of this chapter, may be emergency switching on or emergency switching off. The danger may be mechanical, thermal or electrical. A significant aspect to consider is that while it is often necessary to switch off the supply rapidly to remove the danger, in the case of emergency stopping, where the hazard is mechanical, prevention of the hazard may require the supply of electricity to be maintained.

The switching requirements for emergency switching where the risk is electric shock are essentially the same as for isolation in terms of switching line and neutral conductors.

Devices for emergency switching should switch the circuit conductors. This is not always practical, but the intent is clear: the emergency switching device should, as directly as practicable, switch off or stop the supply to the equipment (Regulation 537.4.1.3).

10.10.2 Safety of machinery BS EN 60204 Regulation 537.4.1

Regulation 537.4.1 advises that, for equipment within the scope of BS EN 60204 *Safety of machinery, Electrical equipment of machines,* the requirements of that standard (for emergency switching) apply.

BS EN 60204-1 is a European standard and it has been prepared under a mandate given to CEN/CENELEC by the European Commission and the European Free Trade Association. It covers the essential requirements of two EC Directives:

- Low Voltage Directive (2014/35/EU); and
- Machinery Directive (2006/42/EC as amended).

The standard has the status of a horizontal standard and may be used as a reference standard by technical committees in the preparation of product standards and can be applied by the supplier of a machine for which no product family or dedicated product standard exists. The standard provides requirements for the electrical equipment of machines so as to provide for:

- safety of persons and property;
- consistency of control response; and
- ease of maintenance.

The requirements for emergency switching are a very small part of the standard. Requirements for 'Emergency operations' include:

- emergency stop;
- emergency start;
- emergency switching off; and
- emergency switching on.

There are three categories of emergency stop:

- Category 0: stopping by immediate removal of power to the machine actuators (i.e. an uncontrolled stop);
- Category 1: a controlled stop with power available to the machine actuators to achieve the stop and then removal of power when the stop is achieved; and
- Category 2: a controlled stop with power left available to the machine actuators.

Emergency stop in BS EN 60204 is 'An emergency operation intended to stop a process or a movement that has become hazardous'.

Emergency switching off in BS EN 60204 is 'An emergency operation intended to switch off the supply of electrical energy to all or a part of an installation where a risk of electric shock or another risk of electrical origin is involved'.

These are similar to the definitions in BS 7671, though somewhat more detailed.

10.10.3 Emergency switching off

Emergency switching off, as defined above, is required by BS EN 60204-1 to be provided where:

(i) protection against direct contact with live parts (e.g. with collector wires, collector bars, slip-ring assemblies, controlgear in electrical operating areas) is achieved only by placing out of reach or by obstacles; or
(ii) there is the possibility of other hazards or damage caused by electricity.

Emergency switching off is preferably achieved by disconnecting the incoming supply of the machine (category 0 stop). When this is not practicable – because, say, the machine cannot be safely stopped in this way – it may be necessary to provide other protection, for example against direct contact, so that emergency switching off is not necessary, but perhaps emergency stopping is appropriate.

10.10.4 Devices for emergency switching off
Regulation group 537.4.2

Regulation 537.4.1.2 requires that, where a risk of electric shock is involved, the emergency switching device shall be an isolating device. BS EN 60204-1 *Safety of machinery, Electrical equipment of machines* requires the functional requirements of IEC 60364-4-46 to be met, which can be understood to mean BS 7671 Regulation group 537.4.

10.10.5 Devices for emergency switching
Regulation group 537.4.2

Regulation 537.4.2.6 requires the operating means to be of the latching type, or capable of being restrained in the 'off' or 'stop' position, unless both the means of operation or emergency switching and for re-energizing are under the control of the same person. The intent is fairly clear. If you have caught your clothing in a lathe and just have time to depress the emergency stop button before being pulled into the machine, and your hand pulled from the stop button, the lathe must not restart after the removal of your hand from the stop button. Additionally, it must not be possible to restart the machine from a remote switch not under the control of the operator, again to prevent restarting other than by the operator.

The Supply of Machinery (Safety) Regulations 2008 (SI 2008 No. 1597) as amended have requirements in this respect. These Regulations implement the Machinery Directive in the UK. Annex I to the Machinery Directive and Schedule 3 to the Machinery Regulations state, with respect to emergency stop buttons, the following:

Emergency stop

Each machine must be fitted with one or more emergency stop devices to enable actual or impending danger to be averted. The following exceptions apply:

- machines in which an emergency stop device would not lessen the risk, either because it would not reduce the stopping time or because it would not enable the special measures required to deal with the risk to be taken;
- hand-held portable machines and hand-guided machines.

This device must:

- have clearly identifiable, clearly visible and quickly accessible controls;
- stop the dangerous process as quickly as possible, without creating additional hazards;
- where necessary, trigger or permit the triggering of certain safeguard movements.

The emergency stop control must remain engaged; it must be possible to disengage it only by an appropriate operation; disengaging the control must not restart the machinery, but only permit restarting; the stop control must not trigger the stopping function before being in the engaged position.

Complex installations

In the case of machinery or parts of machinery designed to work together, the manufacturer must so design and construct the machinery that the stop controls, including the emergency stop, can stop not only the machinery itself but also all equipment upstream and/or downstream if its continued operation can be dangerous.

The last requirement under Emergency stop is perhaps clearer than Regulation 537.4.2.6: 'disengaging the (emergency stop) control must not restart the machinery, but only permit restarting'.

10.10.6 Push-buttons

BS EN 60204-1:2006, as amended, *Safety of machinery, Electrical equipment of machines* provides information on operator interface and machine-mounted control devices. Tables 10.4 and 10.5 are taken from clause 10.2 Push-buttons and Table 10.2 of the standard.

Table 10.4 Colour-coding for push-button actuators

Colour	Meaning	Explanation	Examples of application
RED	Emergency	Actuate in the event of a hazardous condition or emergency	Emergency stop Initiation of emergency function (see also 10.2.1)
YELLOW	Abnormal	Actuate in the event of an abnormal condition	Intervention to suppress abnormal condition Intervention to restart an interrupted automatic cycle
GREEN	Normal	Actuate to initiate normal conditions	(See 10.2.1)
BLUE	Mandatory	Actuate for a condition requiring mandatory action	Reset function
WHITE	No specific meaning assigned	For general initiation of functions except for emergency stop (see note)	START/ON (preferred) STOP/OFF
GREY	No specific meaning assigned	For general initiation of functions except for emergency stop (see note)	START/ON STOP/OFF
BLACK	No specific meaning assigned	For general initiation of functions except for emergency stop (see note)	START/ON STOP/OFF (preferred)

Note: Where a supplemental means of coding (e.g. shape, position, texture) is used for the identification of push-button actuators, the same colour, white, grey or black, may be used for various functions (e.g. white for start/on and for stop/off actuators).

10.11 Functional switching
Regulation group 537.5

Functional switching is, by definition, an operation intended to switch on or off or vary the supply of electrical energy to all or part of an installation for normal operating purposes.

Regulation 537.5.1.1 requires functional switching for each part of a circuit that may require to be controlled independently of other parts of the installation. There is a requirement (Regulation 537.5.1.3), in general, that all current-using equipment requiring control is provided with this function.

Control circuits are required to be designed, arranged and protected so as to limit danger that might arise from faults within them. When sophisticated control systems are being installed, a designer must assess the hazards that might arise from malfunction and take precautions as necessary (Regulation 557.3.201). Circuit-breakers and

Table 10.5 Function colours

Function	Preferred colour	Allowed	Not allowed
START/ON actuators	WHITE, GREY or BLACK with a preference for WHITE.	GREEN is also permitted	RED shall not be used.
Emergency stop and emergency switching-off actuators.	RED		
STOP/OFF actuators	BLACK, GREY, or WHITE with a preference for BLACK.	RED is also permitted, but it is recommended that RED is not used near an emergency operation device.	GREEN shall not be used

Notes:

- The colours for start/on actuators should be white, grey or black with a preference for white. Green is also permitted. Red shall not be used. The colour red shall be used for emergency stop and emergency switching off actuators.
- The colours for stop/off actuators should be: black, grey or white with a preference for black. Green shall not be used. Red is also permitted, but it is recommended that red be not used near an emergency operation device.
- White, grey or black are the preferred colours for push-button actuators that alternately act as start/on and stop/off push-buttons. The colour red, yellow or green shall not be used (see also clause 9.2.6 of BS EN 60204-1).
- White, grey or black are the preferred colours for push-button actuators that cause operation while they are actuated and cease the operation when they are released (e.g. hold-to-run). The colour red, yellow or green shall not be used.
- Reset push-buttons shall be blue, white, grey or black. Where they also act as a stop/off button, the colour white, grey or black is preferred with the main preference being for black. Green shall not be used.

RCBOs are primarily circuit-protective devices and, as such, they are not intended for frequent load switching. Infrequent switching of circuit-breakers on-load is admissible for the purposes of isolation or emergency switching.

For a more frequent duty the number of operations and load characteristics according to the manufacturer's instructions should be taken into account. Preferably an alternative device should be selected (See Table 53.4 of BS 7671) e.g. a contactor to BS EN 60947-4 or BS EN 61095 or a switch to the BS EN 60669 series.

10.12 Plugs and sockets
Regulations 537.2.2.1, 537.3.2.6, 537.4.2.8 and Table 53.4

Plugs and sockets are recognized by the Regulations for use as switches and isolators in certain circumstances in Table 53.2 of BS 7671 and there are further references in:

- Regulation 537.3.2.6: Allowed for mechanical maintenance up to 16 A; and
- Regulation 537.4.2.8: Not allowed for emergency switching.

However, the requirements are not entirely clear as Regulation 537.3.2.6 limits the use of a plug and socket for mechanical maintenance (isolation) to 13A whilst Table 53.4 allows industrial plugs and sockets to be used for functional switching up to 32 A. Table 53.4 does not allow functional switching of plugs and sockets for ratings greater than 32A but does allow on-load isolation.

Table 10.6, may provide the reader with helpful guidance, but of course reference must be made to Table 53.4 of BS 7671 for the actual requirements of the standard.

Plugs and sockets (both switched and unswitched) are not permitted to be selected as devices for emergency switching. This recognizes that a plug and socket may very well be used as a device for emergency switching, but the need for an emergency switching device is not met by installing a plug and socket.

Table 10.6 Use of plugs and sockets for switching and isolation

Plug and socket	Standard	Isolation on-load Table 53.4	Isolation off-load Table 53.4	Emergency switching 537.4.2.8	Functional switching Table 53.2	Mechanical maintenance 537.3.2.6
13 A plug and unswitched socket	BS 1363	Yes (by withdrawing Plug)	Yes (by withdrawing Plug)	No	Yes	Yes
13 A plug and switched socket	BS 1363	Yes (by withdrawing Plug)	Yes (by withdrawing Plug)	No	Yes	Yes
Industrial plug and socket ≤32 A	BS EN 60309 IEC 60384 IEC 60906	No, unless switched	Yes	No	No unless ≤16 A or switched	No (unless ≤16 A)
Industrial plug and socket ≥32 A	BS EN 60309	No, unless switched	Yes	No	No unless switched	No (Unless ≤16 A)
Plug and socket	BS 5733	Yes up to 13 A	Yes	No	No (Unless ≤16 A)	No (Unless ≤16 A)

Plugs and socket-outlets are recognised by the Regulations for:

(a) use as an isolator – Regulation 537.2.2.1 and Table 53.4 (no limit as to the rating provided that, where the rating exceeds 16 A, the isolation must be recognized as being off-load);

(b) switching-off for mechanical maintenance (Regulation 537.3.2.6) for ratings and loads not exceeding 16 A

Chapter 11
Earthing arrangements and protective conductors

Chapter 54

11.1 Earthing arrangements
Section 542

11.1.1 General

The requirements in Chapter 54 facilitate compliance with the requirements of other parts of the Regulations, for example Section 411, 'Automatic disconnection of supply'. However, the requirements of Chapter 54 are specific to earthing arrangements and protective conductors. Section 542 reminds the user of the Regulations that the purposes of the earthing arrangements are:

(i) to provide a sufficiently low earth fault loop impedance such that the shock protection requirements of Chapter 41 can be met;
(ii) to provide paths to earth of adequate current-carrying capacity so that protection of conductors and equipment against faults to earth can be provided (protective earthing); and
(iii) to provide a functional conductor for devices needing a functional earth, for example for EMC requirements.

The earthing facilities themselves have to be sufficiently robust, or have sufficient mechanical protection to enable them to continue functioning in normal use, including during a fault, or such misuse as might reasonably be expected.

11.1.2 Earthing arrangements, TN systems
Regulation 542.1

The ESQC Regulations impose only general requirements for connection with Earth (Regulations 8 and 9). Reference also needs to be made to Electricity Association publication 'Guidance for the design, installation, testing and maintenance of main earthing systems in substations' – ENATS 41-24. This recommends that the potential rise be limited to 430 V in the event of an HV fault. Reference can also be made to Engineering Recommendation G12, 'Requirements for the application of PME to LV networks'.

Where the HV equipment earth and the LV neutral earth are connected, the Electricity Supply Regulations 1988 required the combined resistance to earth not to exceed 1 Ω (see Figure 11.1), and this provides general guidance on how low the resistance to earth must be if a common LV neutral/HV earth is adopted. However, the Electricity Safety, Quality and Continuity Regulations have no specific requirements and designers must now meet the requirements of Chapter 44 of BS 7671 (see Chapter 7 of this Commentary). This limits the voltage imposed on the LV system by any HV fault.

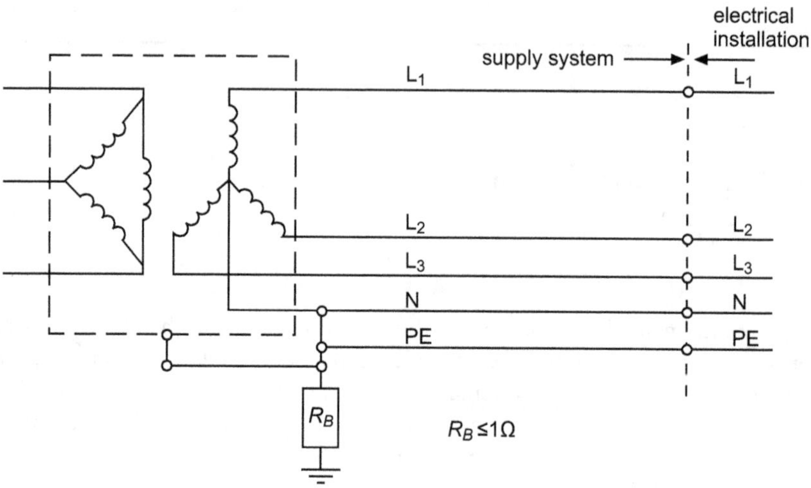

Figure 11.1 Combined HV and LV earths in a TN-S system

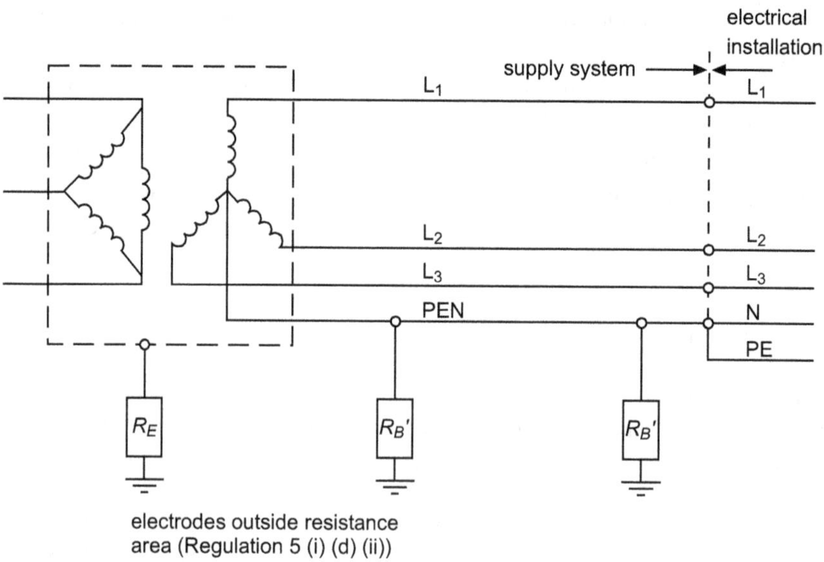

Figure 11.2 Separated HV and LV earths in a TN-C-S system

However, where there is no common HV/LV Earth, a value is not specified except for protective multiple earthed systems (TN-C-S in the UK), where the resistance to earth of the supply neutral must not exceed 20 Ω anywhere on the system (see Figure 11.2). The bulk of public supplies will have a neutral-to-earth impedance of less than 20 Ω and be suitable for PME.

This may at first sight be surprising. However, the electrical installation in TN systems does not rely on the resistance to true Earth (R_E) in any way for faults to earth. The return path in TN systems is that either the PE conductor or the PE + PEN conductor runs all the way back to the neutral point of the transformer (or generator). The earth acts as a reference earth to limit voltage disturbances.

11.1.3 Earthing arrangements, TT systems

In TT systems the earth resistance R_B is part of any earth fault loop impedance, and, if RCDs are not to be used for automatic disconnection, R_B will need to be maintained at a sufficiently low value to allow overcurrent devices to operate in the event of a fault (see 11.1.2).

As general guidance to meet the requirement of the Electricity Safety, Quality and Continuity Regulations 2002 for a connection with earth, the resistance to earth should not exceed 20 Ω.

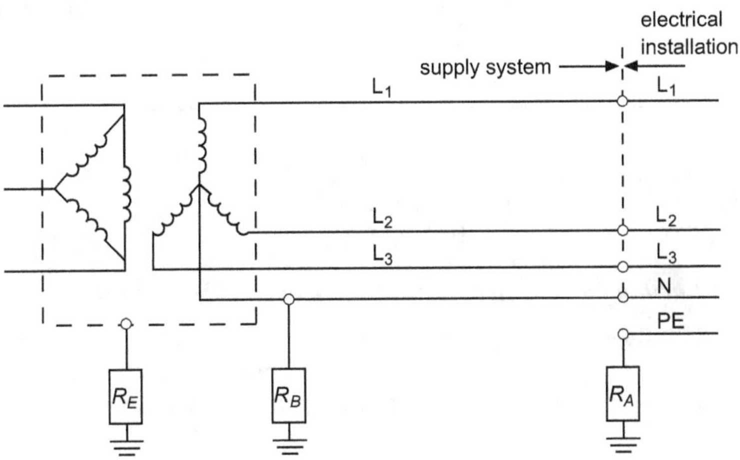

Figure 11.3 TT system

In a TT installation the earth fault current is limited by the supply transformer earth impedance R_B and that of the installation R_A (see Figure 11.3). As is recommended in Regulation 411.5.2, fault protection is usually provided by an RCD. RCDs to BS EN 61008-1 have maximum disconnection times of:

- 0.5 s for S type;
- 0.3 s for general-type.

If an earth fault loop impedance as low as 1 Ω is assumed, and consequently a fault current of 230 A and a disconnection time of the maximum of 0.5 s, then the equation

$$S = \frac{\sqrt{I^2 t}}{k}$$

produces an earthing conductor minimum size of 1.41 mm² copper.

Earthing conductor sizes are limited by Table 54.1 and Regulation 543.1.1. Table 11.1 has been prepared on this basis.

If earth fault loop impedances are less than 1 Ω it is possible that RCDs will not be used and use will be made of overcurrent devices.

Table 11.1 Copper earthing conductor cross-sectional areas (csa) for TT installations for earth fault loop impedances not less than 1 Ω (Note 2), with automatic disconnection provided by an RCD (Note 3)

Buried			Not buried		
Unprotected (mm²)	Protected against corrosion, Note 1 (mm²)	Protected against corrosion and mechanical damage (mm²)	Unprotected (mm²)	Protected against corrosion, Note 1 (mm²)	Protected against corrosion and mechanical damage (mm²)
25	16	2.5	4	4	2.5

Notes:

1. Protected against corrosion by a sheath.
2. For impedances less than 1 Ω determine as per Regulation 543.1.2.
3. Disconnection time not to exceed 0.5 s.

11.1.4 Earthing arrangements, protective neutral bonding (PNB)

The ESQC Regulation 8(4)(b) allows the connection with Earth of the source of supply at some other point, if there are no consumers on the transformer side of the connection with earth often used for overhead supplies to small groups of dwellings, (see Figure 11.4).

11.2 Earth electrodes
Regulation group 542.2

The value of earth electrode resistance is discussed in section 11.1 above and in section 4.4.7.

The designer has to ensure that the connections of the installation with Earth will be of a permanent nature. The range of means of connecting with earth is listed in Regulation 542.2.3, including earth rods or pipe, earth tapes, plates, underground metal structures, reinforcement of concrete, lead sheath of other cables, and other

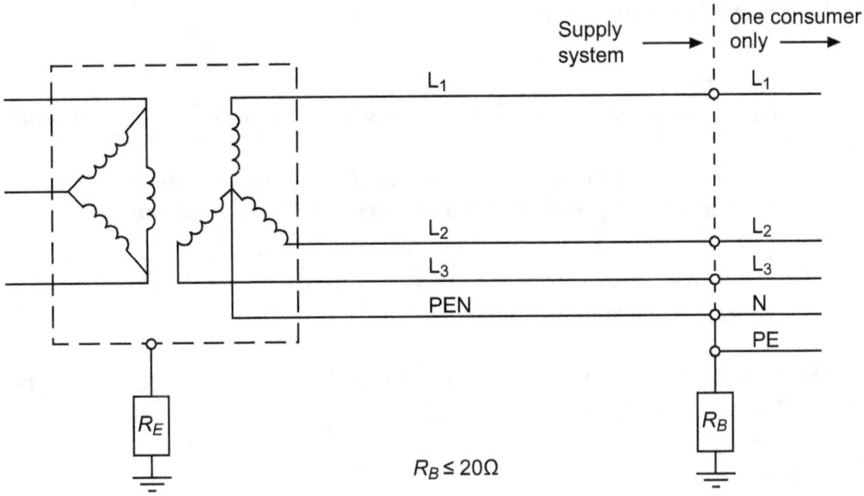

Figure 11.4 *Protective neutral bonding*

suitable underground metalwork. Note that Regulation 542.2.6 does not allow the metalwork of gas, water utility or other services to be used as the protective earth electrode as such, although of course they must be bonded as required by Regulation 411.3. This is because in many circumstances the designer cannot be confident that gas and water pipes will continue to provide a good connection with earth throughout the life of the installation, as they may be replaced by plastics pipes, or at least have plastics inserts placed in them during the lifetime of the electrical installation. For an installation completely under the control of one person, including cables and pipes, it may seem an onerous requirement, but there is always a considerable risk even in a privately owned water or gas distribution system that someone unbeknown to the operators will insert plastics into the system. This could, of course, considerably affect the connection of the electrical installation with earth.

11.3 Earthing conductors
Regulation group 542.3

The minimum cross-sectional area requirements of buried earthing conductors given in Table 54.1 are necessary to provide for protection against corrosion and mechanical damage, which are obviously particularly relevant to earthing conductors.

The connection of an earthing conductor to an earth electrode, or to any other means of earthing, must be soundly made, be electrically and mechanically satisfactory, and be labelled. The labelling is most important, particularly where the connections with earth are remote or separate from the main switchgear (Regulation 514.13.1).

Earthing conductors are protective conductors and are sized as discussed in sections 11.4.1 and 11.5.1 below.

11.4 Protective conductors
Section 543

Protective conductors are required to carry leakage currents and earth fault currents.

The requirement to carry leakage currents generally imposes no constraints on the size of the conductors, other than that necessary for mechanical strength.

11.4.1 Protective conductor cross-sectional area
Regulation group 543.1

The requirement of Regulation 543.1.3 is that the cross-sectional area shall be not less than that determined by the formula:

$$S = \frac{\sqrt{I^2 t}}{k}$$

where:

- S is the nominal cross-sectional area of the conductor in mm^2
- I is the value in amperes (rms for a.c.) of fault current for a fault of negligible impedance, which can flow through the associated protective device, due account being taken of the current limiting effect of the circuit impedances and the limiting capability (I^2t) of that protective device
- t is the operating time of the disconnecting device in seconds corresponding to the fault current I amperes
- k is a factor taking account of the resistivity, temperature coefficient and heat capacity of the conductor material, and the appropriate initial and final temperatures

(or by reference to BS 7454 concerning the non-adiabatic calculation described in section 6.5).

The above equation is the same as that in Regulation 434.5.2, except that the presumption is now made that the overcurrent device has been selected, the disconnection time is known, and consequently the minimum size of protective conductor can be determined. (The emphasis in Regulation 434.5.2 is on selecting a protective device with characteristics such that it will disconnect, in the event of a fault, sufficiently quickly to prevent damage to the live conductors already selected.) The values for k given in Table 43.1 are for conductors incorporated in a cable or bunched together, and this is the same as Table 54.3 of Chapter 54. However, protective conductors are often installed in different circumstances to the live conductors, and Chapter 54 needs to provide k factors for:

(i) conductors run separately and not incorporated in a cable, or not bunched with cables – Table 54.2;
(ii) conductors incorporated in a cable or bunched with cables – Table 54.3;
(iii) the sheath or armour of a cable – Table 54.4;

(iv) steel conduit, ducting or trunking – Table 54.5; and
(v) bare conductors located where there is no risk of damage to neighbouring materials – Table 54.6.

The minimum cross-sectional area for all of these conditions can be calculated using the k values from Tables 54.2 to 54.6. Most commonly, protective conductors are:

(i) incorporated with a cable and of reduced cross-sectional area;
(ii) the armour of a cable; or
(iii) steel conduit or trunking.

The assessment of these types of protective conductors is considered below.

11.4.2 Protective conductors of reduced cross-sectional area Regulation 543.1.3

The requirement that $S \geq \{\sqrt{(I^2{}_t)}\}/k$ (the adiabatic equation) is to be met not only by live conductors (line and neutral) but also by protective conductors. If the protective conductor has the same current-carrying capability as the line conductors, and the overcurrent device is providing protection against overload and fault currents, no further checks need be carried out. However, if the equivalent csa of the protective conductor is less than that of the live conductors, compliance with the adiabatic equation must be checked. This is a common situation in the UK.

Within the fault rating of the device the most onerous condition may occur when the fault current is low and disconnection time long. The suitability of a protective conductor can be determined graphically (see Figure 11.5). The thermal characteristics of the protective conductor are plotted on the device time/current characteristics. (For a given S, values of I are assumed and values of t then determined from the equation above. The plot will be a straight line on the log/log graph paper of the device time/current characteristics.) The point of intersection is the minimum fault current (or maximum loop impedance) at which the protective conductor is protected by the particular device.

11.4.2.1 Fuses

Figure 11.5 shows the adiabatic characteristics of copper 70 °C thermoplastic-insulated conductors incorporated in a cable or bunched (e.g. 70 °C thermoplastic-insulated twin with earth cables) plotted on BS 88-3 fuse characteristics.

The intersection of the conductor characteristic with that of the fuse gives the minimum adiabatic prospective earth fault current, I_a, that the conductor can withstand. From this minimum adiabatic current (I_a), the maximum loop impedance, Z_a, which can be allowed, can be calculated, i.e.

$$Z_a = \frac{C_{min} 230}{I_a}$$

where C_{min} is a voltage factor to allow for the voltage being less than nominal voltage, say under full load conditions. CENELEC technical report CLC/TR 50480: *Determination of cross-sectional area of conductors and selection of protective devices*, recommends a value of 0.95 (see section 4.4.6). This is not a particularly conservative value.

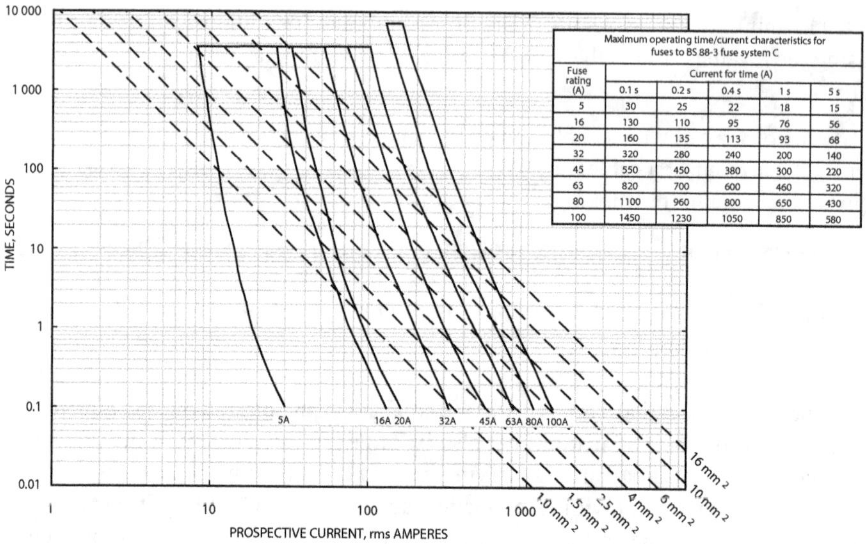

Figure 11.5 Conductor adiabatics plotted on BS 88-3 fuse characteristics from Figure 3A1 of Appendix 3 of BS 7671

If we consider in Figure 11.5 the intersection of the 1.0 mm^2 conductor with the fuse characteristics, we can estimate the minimum adiabatic fault current (I_a) and the maximum adiabatic loop impedance (Z_a) as below:

Fuse rating (A)	5	16	20	32	45
Minimum adiabatic fault current, I_a (A)	11	60	80	290	NP
Maximum adiabatic loop impedance, Z_a (Ω)	20	3.6	2.7	0.75	NP

NP = Not permitted.

These maximum adiabatic loop impedances (Z_a) can be used to calculate the maximum circuit length allowed, in the same way as Z_s is used to determine the maximum circuit length for shock protection. The corrections for temperature are identical. The maximum adiabatic loop impedances for the fuses in BS 7671 are found in Tables 11.2 to 11.4.

Table 11.2 Cartridge fuses to BS 88-3: Maximum adiabatic loop impedance Z_a for copper conductors with 70 °C thermoplastic insulation incorporated in a cable or bunched with cables

Conductor csa (mm²)	*5 A		16 A		20 A		32 A		45 A		63 A		80 A		100 A	
	I_a A	Z_a Ω	I_a A	Z_a Ω	I_a A	Z_a Ω	I_a A	Z_a Ω	I_a A	Z_a Ω	I_a A	Z_a Ω	I_a A	Z_a Ω	I_a A	Z_a Ω
1.0	11	20‡	60	3.6*	80	2.7*	300	0.8*		NP		NP		NP		NP
1.5	10	22‡	50	4.2*‡	63	3.1*	210	1.1*		NP		NP		NP		NP
2.5		‡			51	4.4‡	140	1.82‡	300	0.73*	700	0.27		NP		NP
4.0		‡		‡		‡	110	2.1‡	210	0.99	420	0.49	800	0.27	1100	0.20
6.0		‡		‡		‡		‡		‡	300	0.70	550	0.40	1000	0.22
10.0		‡		‡		‡		‡		‡	200	‡	330	0.66‡	600	0.36
16		‡		‡		‡		‡			170	‡	270		440	‡
$Z^{\dagger}_s$		15.33		4.11		3.38		1.64		1.04		0.72		0.53		0.40

Notes:

* Loop impedance is less than that required for 5 *s* disconnection – see Table 41.4.

† Maximum loop impedance to meet the 5 *s* disconnection limit of Table 41.4.

‡ Loop impedance well exceeds that required for 5 s disconnection.

NP = Not permitted at any loop impedance.

Table 11.3 Semi-enclosed fuses to BS 3036: Maximum adiabatic loop impedance Z_a for copper conductors with 70 °C thermoplastic insulation incorporated in a cable or bunched with cables

Device rating	5 A		15 A		20 A		30 A		45 A		60 A		100 A	
Conductor csa (mm²)	I_a A	Z_a Ω	I_a A	Z_a Ω	I_a A	Z_a Ω	I_a A	Z_a Ω	I_a A	Z_a Ω	I_a A	Z_a Ω	I_a A	Z_a Ω
1.0	10	22‡	42	5.20‡	68	3.21*		NP		NP		NP		NP
1.5		‡	36	6.10‡	54	4.05‡	92	2.4*		NP		NP		NP
2.5		‡		‡	48	‡	72	3.03‡	160	1.37*		NP		NP
4.0		‡		‡		‡		‡	120	1.82‡	220	1.10‡		NP
6.0		‡		‡		‡		‡	110	‡	160	‡		NP
10.0		‡		‡		‡		‡		‡		‡	360	0.70‡
$Z^{\dagger}_s$		16.8		5.08		3.64		2.51		1.51		1.07		0.51

Notes:

* Loop impedance is less than that required for 5 *s* disconnection – see Table 41.4.

† Maximum loop impedance to meet the 5 *s* disconnection limit of Table 41.4.

‡ Loop impedance well exceeds that required for 5 s disconnection.

NP = Not permitted at any loop impedance.

Table 11.4 Cartridge fuses to BS 88-2: Maximum adiabatic loop impedance Z_a (Ω) for copper conductors with 70 °C thermoplastic insulation incorporated in a cable or bunched with cables

Protective conductor csa (mm^2)	Device rating								
	2 A	4 A	6 A	10 A	16 A	20 A	25 A	32 A	40 A
1.0	87‡	34‡	19‡	8.7	3.8*	2.2*	1.4*	0.95*	NP
1.5	95‡	35‡	20‡	9.9‡	4.4‡	2.4‡	1.7*	1.04*	0.68*
2.5			‡	10.9‡	5.5‡	3.5‡	2.4*‡	1.82‡	1.04*
4.0			‡	‡	6.8‡	4.4‡	3.1‡	2.30‡	1.56‡
6.0			‡	‡	‡	‡	‡	‡	‡
$Z^{\dagger}_s$	44	21	12	6.8	4.0	2.8	2.2	1.7	1.3

Protective conductor csa (mm^2)	Device rating						
	50 A	63 A	80 A	100 A	125 A	160 A	200 A
1.5	NP	NP	NP	NP	NP	NP	NP
2.5	0.70*	0.36*	NP	NP	NP	NP	NP
4.0	1.04‡	0.66*	0.29*	0.17*	NP	NP	NP
6.0	1.36‡	0.91‡	0.45*	0.26*	0.16*	NP	NP
10.0	‡	1.15‡	0.72‡	0.44‡	0.27*	0.18*	0.1*
16.0	‡	‡	‡	‡	0.36‡	0.27*	0.13*
25.0	‡	‡	‡	‡	‡	0.34‡	0.20‡
$Z^{\dagger}_s$	0.99	0.78	0.55	0.42	0.32	0.27	0.18

Notes:

* Loop impedance is less than that required for 5 *s* disconnection – see Table 41.4.

† Maximum loop impedance to meet the 5 *s* disconnection limit of Table 41.4.

‡ Loop impedance well exceeds that required for 5 s disconnection.

NP = Not permitted at any loop impedance.

11.4.2.2 Types B, C and D circuit-breakers

Provided a circuit-breaker is operating instantaneously, either using the adiabatic formula $S \geq \frac{\sqrt{I^2 t}}{k}$ (see Table 11.5) or by plotting the protective conductor adiabatics on the circuit-breaker characteristics (see Figure 11.6), it can be shown that relatively small protective conductors can handle the energy let-through (see 3 of Table 11.5).

However for currents less than that for instantaneous operation, for example the 5 second disconnection recognised for type D circuit breakers, larger protective conductors are required, see column 6 of table 11.5.

Table 11.5 Type D circuit breakers, Minimum protective conductor cross-sectional area S_a for copper conductors with 70 °C thermoplastic insulation incorporated in a cable or bunched for type D circuit-breakers operating instantaneously and within 5 seconds

1	2	3	4	5	6	7
c.b. rating In A	Current instantaneous operation (t = 0.1 sec) (A)	S_a calculated csa for instantaneous operation (note 1) (mm^2)	S_a selected minimum csa for instantaneous (mm^2)	Current for operation in 5 sec (A)	S_a calculated csa for 5 sec operation (mm^2)	S_a selected minimum csa for 5 sec (mm^2)
6	120	0.32	1.0	60	1.16	1.5
10	200	0.55	1.0	100	1.93	2.5
16	320	0.88	1.0	160	3.10	4
20	400	1.09	1.5	200	3.87	4
25	500	1.37	1.5	250	4.84	6
32	640	1.75	2.5	320	6.20	6
40	800	2.20	2.5	400	7.76	10
50	1000	2.75	4	500	9.70	10
63	1260	3.46	4	630	12.22	16
80	1600	4.40	10	800	15.51	16
100	2000	5.50	10	1000	19.39	25

Note:
$k = 115$, t = 0.1 sec

For molded case circuit breakers (to BS EN 60947) the minimum protective conductor csa must be calculated using energy let-through figures provided by the manufacturer as the standard does not specify any limits, see section 6.9.

11.4.2.3 Energy let-through for high fault currents
Regulation 434.5.2, Appendix 3, Figure 3.4

As discussed in 11.4.2.2, for currents sufficient just to cause the circuit-breaker to operate instantaneously, relatively small protective conductors can handle the energy let-through; however, if fault currents exceed that value of instantaneous operation the disconnection time remains much the same, but the energy let-through increases.

Regulation 434.5.2 requires:

For a fault of very short duration (less than 0.1 s), for current limiting devices k^2S^2 shall be greater than the value of let-through energy (I^2t) quoted for the Class of protective device to BS EN 60898-1, BS EN 60898-2 or BS EN 61009-1, or as quoted by the manufacturer.

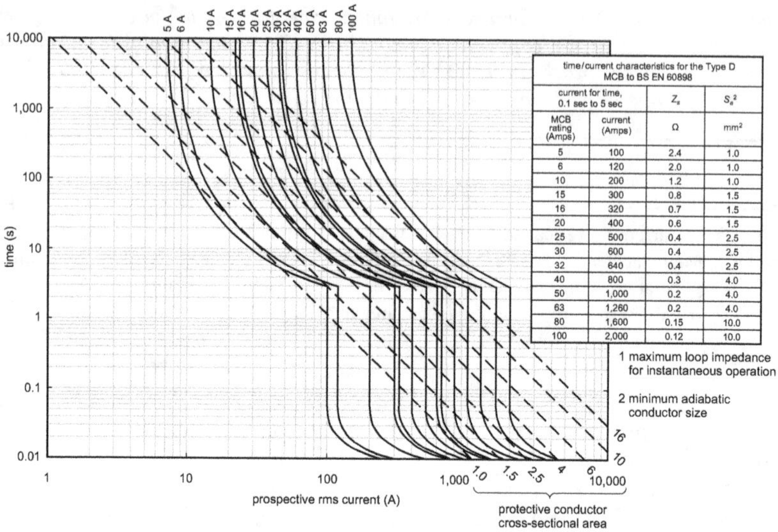

time/current characteristics for the Type D MCB to BS EN 60898			
current for time, 0.1 sec to 5 sec		Z_s	S_a^2
MCB rating (Amps)	current (Amps)	Ω	mm²
5	100	2.4	1.0
6	120	2.0	1.0
10	200	1.2	1.0
15	300	0.8	1.5
16	320	0.7	1.5
20	400	0.6	1.5
25	500	0.4	2.5
30	600	0.4	2.5
32	640	0.4	2.5
40	800	0.3	4.0
50	1,000	0.2	4.0
63	1,260	0.2	4.0
80	1,600	0.15	10.0
100	2,000	0.12	10.0

Figure 11.6 Adiabatic characteristics of 70 °C thermoplastic insulated copper conductors incorporated in a cable or bunched endorsed on the characteristics of a Type D circuit-breaker, for fault currents not exceeding those for instantaneous operation

Clause 4.6 of BS EN 60898-1 states that circuit-breakers may be classified by energy limiting class. Table 6.10 of section 6.3.6 tabulates the energy-limiting classes with the maximum energy let-through as given in Table ZA of BS EN 60898-1 and endorsed with the minimum conductor csa (S) in mm^2 allowed by these energy let-throughs (I^2t) calculated using the formula:

$$S \geq \frac{\sqrt{\text{energy let-through } (I^2t)}}{k}$$

As the prospective fault current increases, so the tripping time is reduced until the device operates instantaneously, e.g. at between 3 and 5 I_n for a Type B circuit-breaker (see Figure 11.7), and energy let-through decreases (see Figure 11.8). For greater fault currents there is no further reduction in disconnection time and energy let-through increases (see again Figures 11.7 and 11.8).

Each of the figures of circuit-breaker device characteristics in Appendix 3 of BS 7671 is endorsed with the statement, 'For prospective fault currents in excess of those providing instantaneous operation refer to the manufacturer's let-through energy data.'

Figure 11.8 shows that, for Hager Type B circuit-breakers, the energy let-through starts to increase at $5I_n$, as would be expected. Often, a manufacturer will provide energy let-through curves only for prospective fault levels above which the energy let-through is increasing (see Figure 11.9 from the MK catalogue).

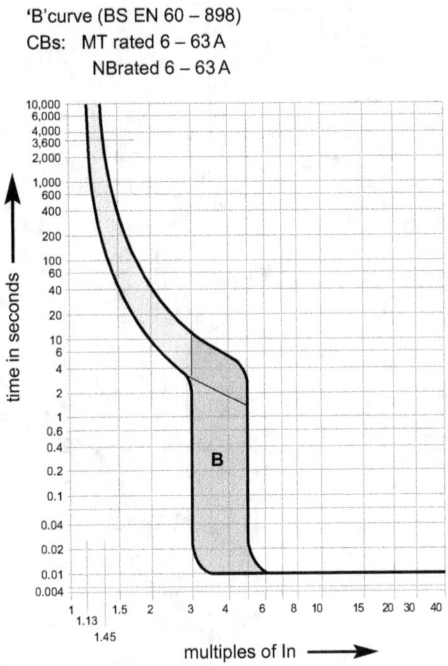

Figure 11.7 Time/current characteristics – Hager Type B circuit-breakers

In Regulation 434.5.2, for circuit-breakers to BS EN 60898, that is type B, C or D, there is a requirement that for a fault of very short duration (less than 0.1 s):

$$k^2 S^2 > I^2 t$$

where:

$S =$ nominal cross-sectional area of the conductor in mm^2
$k =$ factor from Table 43.1
$I^2t =$ energy let-through quoted for the class of device in BS EN 60898, BS EN 61009 or as quoted by the manufacturer.

Minimum conductor csa can be determined from appendix ZA (see Table 11.6) and section 6.3.6. This check must particularly be carried out for reduced-csa protective conductors.

Data is not provided in BS EN 60898 for type D devices, so manufacturer's data must be used.

An alternative approach is to use manufacturers data for energy let-through, these values being less than those quoted in BS EN 60898-1 and BS EN 61009-1; see Tables 11.7a, b.

Not all manufacturers provide data for type D circuit-breakers. In this circumstance a prudent approach for type D circuit-breakers particularly if 5-second disconnection (see Table 11.5) is required, is not use reduced csa protective conductors.

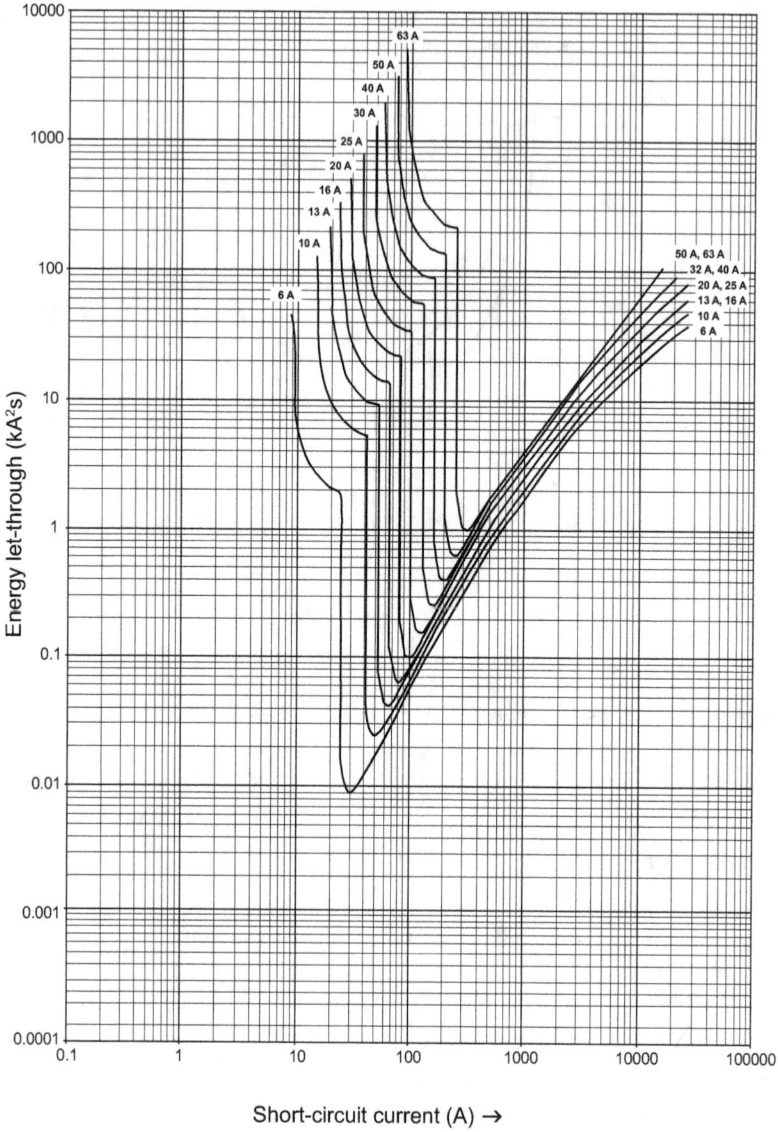

Figure 11.8 Hager Type B circuit-breaker energy let-through

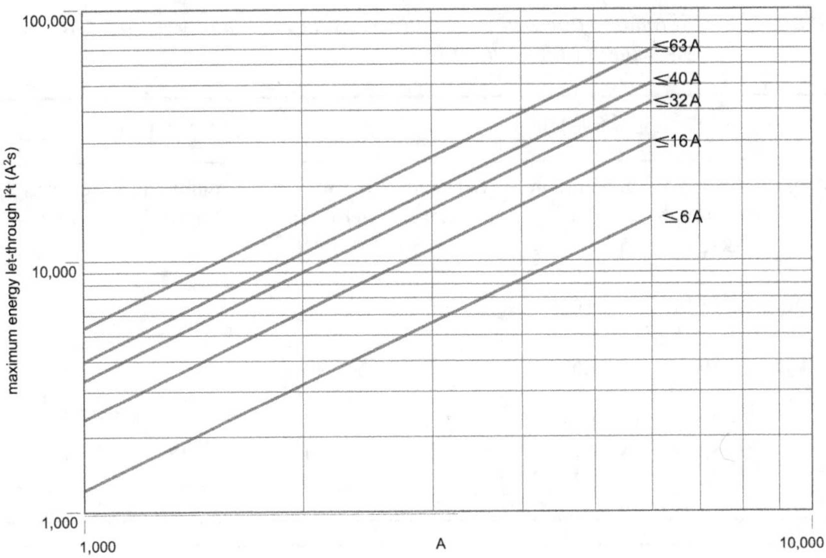

Figure 11.9 MK Type B circuit-breaker energy let-through

Table 11.6 Energy-limiting Class 3 Type B and C circuit-breakers: minimum copper protective conductor sizes determined from maximum energy let-through allowed by Table ZA of BS EN 60898-1 and table ZD1 of BS EN 61009-1, k = 115

		Type B			Type C		
Device rating (A)	Prospective fault current (A)	Energy let-through (A^2s)	Minimum copper protective conductor csa, S (note 1) (mm^2)	Selected copper protective conductor (mm^2)	Energy let-through (A^2s)	Minimum copper protective conductor csa, S note 1 (mm^2)	Selected copper protective conductor csa (mm^2)
3 to 16	3 000	15 000	1.06	1.0	17 000	1.13	1.5
	4 500	25 000	1.4	1.5	28 000	1.45	1.5
	6 000	35 000	1.63	2.5	40 000	1.74	2.5
	10 000	70 000	2.3	2.5	80 000	1.72	2.5
20 to 32	3 000	18 000	1.17	1.5	20 000	1.23	1.5
	4 500	32 000	1.56	1.5	37 000	1.67	1.5
	6 000	45 000	1.84	2.5	52 000	1.98	2.5
	10 000	90 000	2.61	4	100 000	2.75	4
40	3 000	21 600	1.28	1.5	24 000	1.34	1.5
	4 500	38 400	1.7	1.5	45 000	1.84	2.5
	6 000	54 000	2.02	2.5	63 000	2.18	2.5
	10 000	108 000	2.86	4	120 000	3.01	4

Note:

S = energy let-through (12t) $S \geq \frac{\sqrt{energy\ let\text{-}through\,(I^2t)}}{k}$

Table 11.7a Minimum values of S calculated using manufacturers' quoted energy let-through at 6 kA fault level

	MK B type			MK C type		
Rating	Energy let-through kA^2s	S minimum mm^2	Selected S mm^2	Energy let-through kA^2s	S minimum mm^2	Selected S mm^2
6	14	1.02	1	17	1.1	1.5
10			1.5			2.5
16	30	1.51	1.5	40	1.74	2.5
20						2.5
32	42	1.78	2.5	50	1.94	2.5
40	60	2.13	2.5	60	2.13	2.5
63	70	2.3	2.5	70	2.45	2.5

	Hager B and C type			Hager D type		
Rating	Energy let-through kA^2s	S minimum mm^2	Selected S mm^2	Energy let-through kA^2s	S minimum mm^2	Selected S mm^2
6	13	0.99	1	14	1.03	1.5
10	15	1.06	1.5	18	1.17	1.5
16	18	1.17	1.5	23	1.32	1.5
20	23	1.31	1.5	33	1.58	2.5
32	29	1.48	1.5	47	1.89	2.5
40	29	1.48	1.5	47	1.89	2.5

11.4.2.4 Manufacturers energy let through calculations

See section 6.3.9 for the calculation of minimum conductor csa S using manufacturers energy let-through data.

11.4.3 Protective conductors as the sheath or armour of a cable

The armour or metal sheath of a cable may be for use as the circuit protective conductor.

Regulation 543.1.4 allows Table 54.7 to be used where it is not wished to calculate the minimum cross-sectional area of a protective conductor. It will be found that there are many cables, particularly thermosetting cables, that do not meet the simplified requirements of Table 54.7, and consequently the adiabatic calculation of Regulation 543.1.3 needs to be carried out to determine whether the area of the cable armouring is sufficient for the earth fault current. This calculation can effectively be carried out in either of two ways:

Table 11.7b Minimum values of S calculated using manufacturers' quoted energy let-through at 3kA fault level

	MK B type			MK C type		
Rating	Energy let-through kA^2s	S minimum mm^2	Selected S mm^2	Energy let-through kA^2s	S minimum mm^2	Selected S mm^2
6	6	0.67	1	7	0.7	1
10						
16	12	0.95	1	16	1.1	1.5
20						1.5
32	16	1.1	1.5	20	1.2	1.5
40	19	1.2	1.5	24	1.3	1.5
63	26	1.4	1.5	32	1.5	1.5

	Hager B and C type			Hager D type		
Rating	Energy let-through kA^2s	S minimum mm^2	Selected S mm^2	Energy let-through kA^2s	S minimum mm^2	Selected S mm^2
6	6.7	0.7	1	7.1	0.7	1
10	7.7	0.8	1	9	0.8	1
16	9.1	0.8	1	11	0.9	1
20	11.3	0.9	1	16	1.1	1.5
32	14.3	1.0	1	21	1.2	1.5
40	14.3	1.0	1	21	1.2	1.5
63	16	1.1	1.5	26	1.4	1.5

(i) by plotting the adiabatic characteristics of the sheath or armouring on to the overcurrent device characteristics; or

(ii) by individual calculation.

11.4.3.1 Plotting of cable armour adiabatics

Figure 11.10 shows the adiabatic characteristics of the armouring of four-core copper conductor thermosetting cables to BS 5467, plotted over BS 88-2 fuse characteristics. The characteristics are plotted in a similar manner to that described in section 11.4.2, except that the value for k selected is that from Table 54.4, e.g. k = 46 for the steel-wire armouring of a 90 °C thermosetting cable. The cross-sectional area of the armouring is found in Tables 11.8 and 11.9.

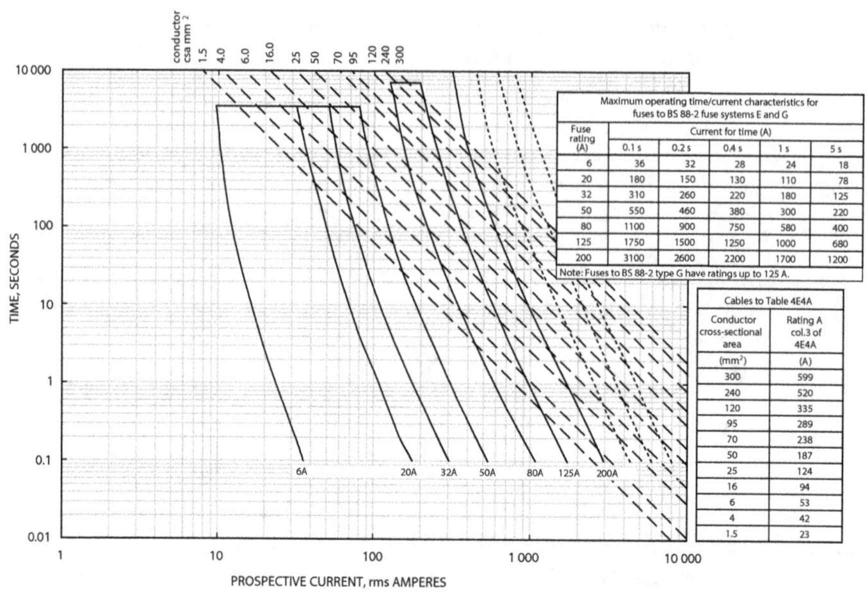

Maximum operating time/current characteristics for fuses to BS 88-2 fuse systems E and G					
Fuse rating (A)	Current for time (A)				
	0.1 s	0.2 s	0.4 s	1 s	5 s
6	36	32	28	24	18
20	180	150	130	110	78
32	310	260	220	180	125
50	550	460	380	300	220
80	1100	900	750	580	400
125	1750	1500	1250	1000	680
200	3100	2600	2200	1700	1200

Note: Fuses to BS 88-2 type G have ratings up to 125 A.

Cables to Table 4E4A	
Conductor cross-sectional area	Rating A col.3 of 4E4A
(mm^2)	(A)
300	599
240	520
120	335
95	289
70	238
50	187
25	124
16	94
6	53
4	42
1.5	23

Figure 11.10 Adiabatic characteristics of armouring of four-core copper conductor cable with thermosetting insulation to BS 5467 plotted on BS 88-2 fuse characteristics from Figure 3A3(a) of BS 7671

The graph indicates that, where the rating of the fuse is less than that of the cable, as is usual for the four-core cables considered, the armouring is of sufficient cross-sectional area, provided loop impedances are such that disconnection occurs within 5 s. Where fuse ratings exceed the rating of the cable, as may be the case for motor fuses, a check is necessary.

It is not practical to plot characteristics for all fuse/protective conductor combinations, so that calculation is often necessary.

11.4.3.2 Calculation of armouring capability

If it is wished to determine by calculation whether the armouring is of adequate size for the particular installation, the following procedure may be followed:

(i) select the overcurrent device;
(ii) calculate earth fault impedance (see Chapter 18);
(iii) determine the fault current I_f from the earth fault loop impedance, and the time for disconnection t from the characteristics of the overcurrent device, e.g. for BS 88-2 fuses from Appendix 3, Fig. 3A3 of BS 7671;
(iv) use the equation:

$$S \geq \frac{\sqrt{I^2 t}}{k}$$

to calculate the minimum cross-sectional area of armouring.

Table 11.8 Gross cross-sectional area of steel-wire armour for two-core, three-core and four-core 600/1000 V cables with thermosetting insulation to BS 5467

Nominal area of conductor	Cross-sectional area of round armour wires (mm^2)						
	Cables with stranded copper conductors				Cables with solid aluminium conductors		
	Two-core	Three-core	Four-core	Four-core (reduced neutral)	Two-core	Three-core	Four-core
1.5*	15	16	17	–	–	–	–
2.5*	17	19	20	–	–	–	–
4*	19	21	23	–	–	–	–
6*	22	23	36	–	–	–	–
10*	26	39	43	–	–	–	–
16*	41	44	49	–	40	42	46
25	42	62	70	70	48	58	66
35	62	70	80	76	54	64	72
50	68	78	90	86	60	72	82
70	80	90	131	128	70	84	122
95	113	128	147	144	100	119	135
120	125	141	206	163	–	131	191
150	138	201	230	220	–	181	211
185	191	220	255	250	–	206	235
240	215	250	289	279	–	230	265
300	235	269	319	304	–	250	289
400	265	304	452	343	–	–	–

* Circular conductors.
Source: BS 5467, Table N2

This can be compared with the data in Table 11.8 for cables with thermosetting insulation and Table 11.9 for thermoplastic insulated cables.

To simplify the selection, minimum areas of armouring for BS 88 fuses are given in Table 11.10. This table assumes disconnection in 5 s. The check is simple. Table 11.10 gives the minimum area of armouring for the particular fuse; this can be compared with the data in Table 11.8 or 11.9.

Table 11.9 shows the maximum BS 88 fuse ratings for armoured cables with thermosetting insulation to BS 5467 aligned with Table 4E4A of Appendix 4 of BS 7671.

For devices other than fuses to BS 88, in particular molded-case circuit-breakers, the energy let-through (I^2t) can be obtained for the particular device in question. Knowing k, and the area of the armouring, a check of the armouring csa is made to determine its suitability for use as the protective conductor (see below).

Table 11.9 Gross cross-sectional area of steel-wire armour for two-core, three-core and four-core 600/1000 V thermoplastic insulated cables to BS 6346

Nominal area of conductor	Cross-sectional area of round armour wires (mm^2)						
	Cables with stranded copper conductors				Cables with solid aluminium conductors		
	Two-core	Three-core	Four-core	Four-core (reduced neutral)	Two-core	Three-core	Four-core
1.5*	15	16	17	–	–	–	–
2.5*	17	19	20	–	–	–	–
4*	21	23	35	–	–	–	–
6*	24	36	40	–	–	–	–
10*	41	44	49	–	–	–	–
16*	46	50	72	–	42	46	66
25	60	66	76	76	54	62	70
35	66	74	84	82	58	68	78
50	74	84	122	94	66	78	113
70	84	119	138	135	74	113	128
95	122	138	160	157	109	128	147
120	131	150	220	215	–	138	201
150	144	211	240	235	–	191	220
185	201	230	265	260	–	215	245
240	225	260	299	289	–	240	274
300	250	289	333	323	–	265	304
400	279	319	467	452	–	–	–

* Circular conductors.
Source: BS 6346, Table N2

11.4.3.3 Energy let-through calculation and circuit-breakers

The energy let through by a circuit-breaker determines the minimum size of the downstream cable. The cable withstand depends on the conductor material, the insulation used and the conductor size. Manufacturers of circuit-breakers provide energy let-through data for their devices, which allow the adequacy of protective conductors to be checked. The particular manufacturer's data must be used for the particular device (frame size and rating), as they differ from one manufacturer to another, for the product standard is not sufficiently specific for standard tables to be prepared. Data will be presented in a similar manner to Table 11.12.

Using the formula:

$$S \geq \frac{\sqrt{I^2 t}}{k}$$

Table 11.10 Minimum area of steel cable armouring when used as a protective conductor in a circuit protected by a BS 88 fuse

Fuse rating (A)	Current for 5 s disconnection (A)	Minimum area of armouring*	
		90° C Thermosetting insulation (mm^2)	70° C *thermoplastic* insulation (mm^2)
6	18	0.9	0.7
10	32	1.6	1.3
16	55	2.7	2.4
20	78	3.8	3.4
25	100	4.9	4.3
32	125	6.1	5.4
40	170	8.3	7.4
50	220	10.7	9.6
63	280	13.6	12
80	400	19.4	17
100	520	25.3	23
125	680	33.0	30
160	820	39	36
200	1 200	58	52
250	1 450	70	63
315	2 000	97	87
400	2 580	125	113
500	3 590	174	157
630	4 420	215	194
800	6 870	333	301
1 000	8 000	388	350
1 250	10 950	532	480

* Determined from the adiabatic equation, $S \geq \{\sqrt{(I^2t)}\}/k$, using $k = 46$, $t = 5$ for 90° C thermosetting insulation and $k = 51$, $t = 5$ for 70° C *thermoplastic insulation*; see Table 54.4 of BS 7671.

and looking up k in Tables 54.2 to 54.6, the minimum cross-sectional area can be calculated. For example, consider a 200 A circuit-breaker (frame type CN or CH) is installed in a location where the fault level is 20 kA protecting a 4-core 95 mm^2 steel-wire armoured thermoplastic cable. Is the armouring of sufficient size? From Table 11.12 the I^2t energy let-through would be 0.7×10^6 ampere^2 seconds.

From Table 54.4, k is 51.

$$\text{Now } S \geq \frac{\sqrt{I^2t}}{k} = \frac{\sqrt{0.7 \times 10^6}}{51}$$

therefore $S \geq 16.4$ mm^2

From Table 11.9, the area of the cable armouring is 160 mm^2, which is clearly more than sufficient.

Table 11.11 Maximum BS 88 fuse ratings when the armouring is to be used as the protective conductor

Size endorsed on Table 4E4A of BS 7671
Multicore armoured cables having thermosetting insulation (copper conductors)
Current-carrying capacity BS 5467, BS 6724 Ambient temperature: 30 °C
Conductor operating temperature: 90 °C

Conductor cross-sectional area	Reference Method C (clipped direct)		Reference Method E (in free air or on a perforated horizontal or vertical cable tray) or Reference 13 (free air)		Maximum BS 88 fuse rating note 4	
	1 two-core cable, single-phase a.c. or d.c.	1 three- or four-core cable, three-phase a.c.	1 two-core cable	1 three- or four-core cable, three-phase a.c.	two-core cable	four-core cable
1	2	3	4	5	6	7
(mm^2)	(A)	(A)	(A)	(A)	(A)	(A)
1.5	27	23	29	25	63	63
2.5	36	31	39	33	63	63
4	49	42	52	44	63	80
6	62	53	66	56	80	125
10	85	73	90	78	80	125
16	110	94	115	99	125	160
25	146	124	152	131	125	200
35	180	154	188	162	200	200
50	219	187	288	197	200	200
70	279	238	291	251	200	200
95	338	289	354	304	200	400
120	392	335	410	353	315	500
150	451	386	472	406	315	500
185	515	441	539	463	400	500
240	607	520	636	546	500	500
300	698	599	732	628	500	630
400	787	673	847	728	500	800

Notes:

1. Where the conductor is to be protected by a semi-enclosed fuse to BS 3036, see item 5.1.1 of Appendix 4, preceding the tables.
2. Where a conductor operates at a temperature exceeding 70 °C it shall be ascertained that the equipment connected to the conductor is suitable for the conductor operating temperature (see Table 52.1 of BS 7671).
3. Where cables in this table are connected to equipment or accessories designed to operate at a temperature not exceeding 70 °C, the current ratings given in the equivalent table for 70 °C *thermoplastic* insulated cables (BS 6004, BS 6346) shall be used (see also Regulation 523.1).
4. The maximum BS 88 fuse size has been calculated assuming the earth fault loop impedance results in disconnection in 5 s.

Table 11.12 Manufacturer's energy let-through data for circuit-breakers

Maximum let-through at various prospective fault currents in amperes^2 seconds × 10^6

Frame	Ratings (A)	Prospective fault current						
		10 kA	20 kA	25 kA	30 kA	36 kA	40 kA	50 kA
CD	16–100	0.28	0.42	0.47				
CN	125–250	0.52	0.70	0.71	0.72	0.73		
CH	16–250	0.52	0.70	0.71	0.72	0.73	0.74	0.75
SMA	300–800		8.6	12	15	21	25	36

For *thermoplastic* insulated cables the thermal withstand in amperes^2 seconds × 10^6 ($k = 115$)

Cable size (mm^2)	4	35	50				
Maximum thermal I_2t	0.212	0.476	1.32	3.4	8.26	16.2	33.1

For the steel wire armouring of 4-core cables to say BS 6346, the thermal withstand in amperes^2 seconds × 10^6 ($k = 51$)

Cable size (mm^2)	1.5	4	10	16	50	120	240	400
Armour csa (mm^2)	17	35	49	72	122	220	299	452
Maximum thermal $I^2t = s^2k^2$	0.75	3.4	6.2	13.5	39	125	232	531

11.5 Sizing protective conductors

11.5.1 Earthing conductors
Regulation groups 542.3 and 543.1

Earthing conductors are protective conductors that connect the main earthing terminal to the means of earthing and they are sized in the same way as for circuit protective conductors, except that certain additional minimum sizes are specified when the conductor is part of a TN-C-S (PME) system or is buried in the ground (see Regulation groups 542.3 and 543.1 and Table 54.1).

The size, subject to the minimum requirements above, can be determined using the formula of Regulation 543.1.3:

$$S = \frac{\sqrt{I^2t}}{k}$$

or by using Table 54.7 of BS 7671. The use of Table 54.7 will provide a conservatively sized (oversized) earthing conductor.

Earthing conductors are subject to the minimum sizes of Regulation 543.1.1 as follows:

- 2.5 mm^2 copper equivalent if protected against mechanical damage is provided; and
- 4.0 mm^2 copper equivalent otherwise.

Table 11.13 has been prepared on the basis of Table 54.7 for line or neutral conductor sizes of up to 50 mm^2. For larger sizes the adiabatic equation has been used assuming a fuse size and assuming that the earth fault loop impedance is such as to give a 5 s disconnection.

Electricity companies will normally require a minimum earthing conductor size of 16 mm^2 for supplies up to 100 A. Electricity distribution systems are excluded from the scope of BS 7671 (Regulation 110.2) and so a disconnection of supply in 5 s for a line-to-neutral earth fault cannot be assumed.

11.5.2 Main protective bonding conductors Regulation group 544.1

Non-PME supplies

The prime purpose of main protective equipotential bonding is to reduce the touch voltage in the event of a fault. (For information on touch voltages, see Chapter 4.) The body current that is likely to flow from contact with an exposed-conductive-part made live by a fault is very low and insignificant in terms of the sizing of bonding conductors. In the event of a fault to an exposed-conductive-part, the fault current will return to the transformer via the earth provided by the distributor and via connections with earth arising from bonding of extraneous-conductive-parts.

The magnitude of the current depends on the loop impedance and the relative impedances to earth of the main earth and the extraneous-conductive-parts (see Figure 11.11).

Without knowledge of the relative impedances of the parallel paths, the current actually transmitted by the bonding conductors cannot be known and the sizing recommendations in BS 7671 are somewhat arbitrary, with a minimum size of 6 mm^2 and a cross-sectional area not less than half the cross-sectional area required for the earthing conductor with a maximum of 25 mm^2 copper equivalent. These requirements again are summarized in Table 11.13.

It is necessary to distinguish between an earthing conductor and main bonding conductor. A connection to, say, a piece of structural steelwork, where the steelwork is part of the main earth, could be an earthing conductor and not a main bonding conductor. Its cross-sectional area would then have to satisfy the requirements for an earthing conductor. The bonding to gas and water utility services would always be considered as main bonding, as Regulation 542.2.4 prohibits the use of such metalwork as earth electrodes.

11.5.3 PME main protective bonding conductors

The Electricity Safety, Quality and Continuity Regulations 2002 (ESQCR) are less specific than the Electricity Supply Regulations 1988 they superseded with respect

Table 11.13 *Earthing conductor and main protective bonding conductor sizes (copper equivalent) for TN-S and TN-C-S supplies*

Line conductor or neutral conductor of PME supplies	mm^2	4	6	10	16	25	35	50	70	95	120	150	185	240	300
Assumed BS 88 fuse size	A				Note 6				200	250	315	400	400	500	630
Earthing conductor not buried or buried and protected against corrosion and mechanical damage Notes 1, 2, 3	mm^2	4	6	10	16	16	16	25	25	35	35	50	50	70	70
Main protective bonding conductor Notes 1, 4	mm^2	6	6	6	10	10	10	16	16	25	25	25	25	25	25
Main protective bonding conductor for PME supplies (TN-C-S) Note 4	mm^2	10	10	10	10	10	10	16	25	25	35	35	50	50	50

Notes:

1. Protective conductors (including earthing and bonding conductors) of 10 mm^2 cross-sectional area or less shall be copper (Regulation 543.2.4).
2. The distributor may require a minimum size of earthing conductor at the origin of the supply of 16 mm^2 copper or greater for TN-S and TN-C-S supplies.
3. Buried earthing conductors must be at least:
   25 mm^2 copper if not protected against mechanical damage or corrosion
   50 mm^2 steel if not protected against mechanical damage or corrosion
   16 mm^2 copper if not protected against mechanical damage but protected against corrosion
   16 mm^2 coated steel if not protected against mechanical damage but protected against corrosion
   (Table 54.1 and Regulation 542.3.1).
4. See Regulation 544.1.1 and Table 54.8.
5. The distributor should be consulted when in doubt.
6. Conductor size determined using Table 54.7 of BS 7671.

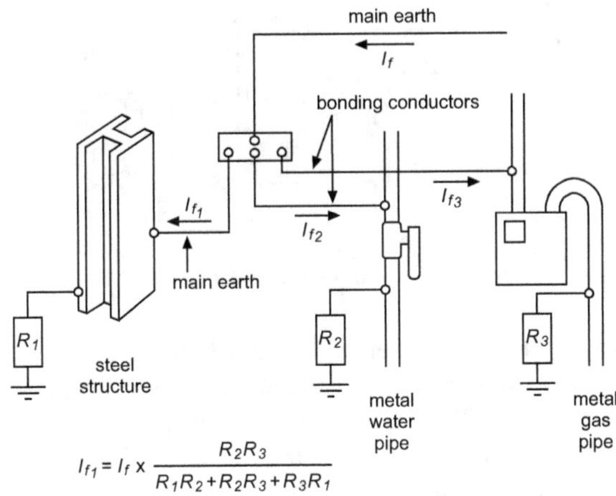

Figure 11.11 *Fault currents in main bonding conductors*

to the requirement for TN-C networks and TN-C-S installations. Regulation group 544.1.1 and Table 54.8 of BS 7671 reproduce the requirement of the earlier Electricity Supply Regulations. Where a consumer's installation is to be connected to a supply the distributor must give his consent if the installation complies with the requirements of BS 7671. The ESQC Regulations impose requirements for

TN-C networks, although they are less specific than the Supply Regulations 1988. The guidance expects compliance with Engineering Recommendation G12/3. (See Table 2.1 of this Commentary.)

The fault condition of concern in a PME supply is an open circuit in the PEN conductor of the distribution system (see Figure 11.12). For such a fault condition, load currents may very well return to the source of supply via the connections with earth of a consumer's installation. The magnitudes of the currents will depend on the load and on the effectiveness of the connection of the consumer's installation bonding with earth and that of other consumers. The risk that will arise in this situation is of shock, as the potential of the main bonding of such installations relative to Earth will rise. The fault voltage U_f is given approximately by

$$U_f \simeq \frac{R_A}{R_A + R_B + R_D} \times 230\ \text{V}$$

where:

R_A = resistance to earth of the consumer's installation main bonding
R_B = resistance to earth of the source of supply
R_D = line resistance of the distribution system.

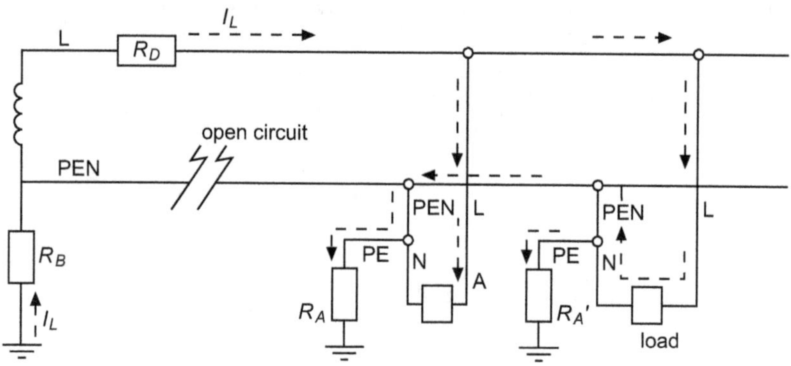

Figure 11.12 Open-circuit PEN conductors in a PME system

R_B may very well be of the order of 1 Ω and R_D even lower, so that the touch voltage U_t can very well approach 230 V. To reduce the risk of a person experiencing this touch voltage, PME installations are required to be bonded and, this being so, the consumer will have little opportunity for direct contact with earth potential within the building. This means that the main bonding must be installed and that it must be capable of carrying any currents that might flow in the event of an open-circuit PEN conductor, hence main bonding conductors need to be substantially sized.

11.5.4 Protective measures for PME supplies

Regulation 9(3) of the ESQC Regulations forbids the distributor from providing a PME earth to a caravan or boat. The notes of guidance provide as examples premises such as construction sites, farms, petrol filling stations and swimming pools, where additional protective measures may be warranted. This guidance is very much in line with that given in Energy Networks publication G12/3.

The further protective measures that can be taken do not include the installation of RCDs, as RCDs will not necessarily operate on loss of the PEN conductor. Load current returning via, say, the connections with earth of a particular premises will flow only in the PEN conductor of the distributing main cable and the earthing conductor of the installation and will not unbalance the current through the RCD. The further precaution that can be taken is the installation of supplementary earth electrodes at the consumer's premises, which will reduce touch voltages in the event of an open circuit PEN conductor (see Figure 11.13).

By the use of a simplified circuit, Table 11.14 can be prepared. It can be seen that the greater the load the lower must be the resistance to earth of the supplementary earth electrode if the touch voltage is to be maintained at a low level. For domestic premises it is suggested that the resistance of the supplementary earth electrode should be of the order of 20 Ω. For commercial premises the designer will have to take note of the actual load and make a decision as to what maximum touch voltage

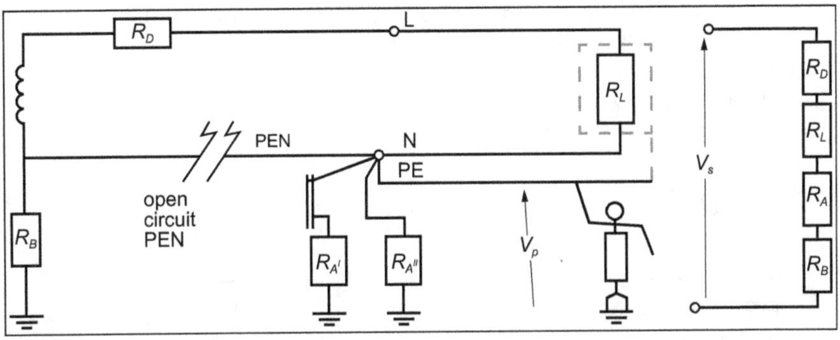

Figure 11.13 Schematic of a PME system with an open-circuit PEN conductor

he or she thinks appropriate, e.g. 50 V. Neglecting R_D, as it is small compared with R_I and R_A, and neglecting R_B as this errs on the safe side, the network can be simplified as in Figure 11.14.

11.6 Supplementary equipotential bonding
Regulation groups 544.2, 415.2

11.6.1 Sizing supplementary bonding conductors

The sizing of supplementary bonding conductors is specified in Regulation group 544.2. These requirements have been summarized in Table 11.15. Regulation 544.2.5 has significance in particular for bathrooms as it allows the use of the protective conductor within a short length of flexible cable supplying a fixed appliance from

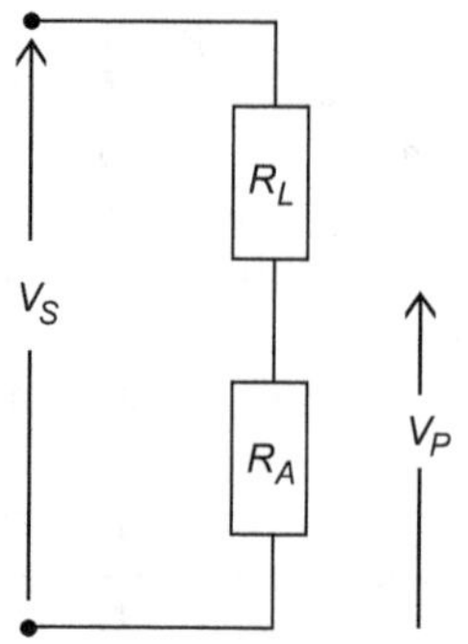

$$\frac{V_s}{V_p} = \frac{R_L + R_A}{R_A}$$

$$R_A = R_L \times \frac{V_p}{(V_s - V_p)}$$

V_s = the nominal supply voltage
V_p = the touch voltage
R_L = the load resistance (V_s^2/wattage)
R_A = the resistance of the earth electrode including parallel earths (e.g. water and gas pipes)

Figure 11.14 Voltage division for supplementary earth electrode

Table 11.14 Supplementary electrode resistance to earth R_A necessary to reduce touch voltages V_p to 50 V and 100 V

Load (kW)	R_L (Ω)	R_A (Ω)	
		For Vp = 50 V	For Vp = 100 V
7	8.2	2.1	5.8
3	19.2	5.1	13.7
2	28.8	7.5	20.5
1	57.6	15.1	41.0

an adjacent connection unit or other accessory, incorporating a flex outlet, to be the supplementary bonding connection. This removes the need for some unsightly bonds to electric panel radiators or fires.

11.6.2 Maximum length of supplementary bonding conductors Regulation 415.2.2

Regulation 415.2.2 requires a further condition to be met:

415.2.2 Where doubt exists regarding the effectiveness of supplementary bonding, it shall be confirmed that the resistance R between simultaneously accessible exposed-conductive-parts and extraneous-conductive-parts fulfils the following condition:

$R \leq 50\ \mathrm{V}/I_a$ in a.c. systems
$R \leq 120\ \mathrm{V}/I_a$ in d.c. systems

where I_a is the operating current in amperes of the protective device

for RCDs, $I_{\Delta n}$.
for overcurrent devices, the current causing automatic operation in 5 s.

Part 7

For Part 7 considerations this condition is almost always complied with, as Regulation 544.2.1 sets a minimum supplementary bonding conductor conductance of that of the circuit protective conductor.

The maximum supplementary bonding conductor resistance R and length L for various supplementary bonding conductor sizes are calculated in Tables 11.16 and 11.17.

Supplementary bonding is normally applied only to final circuits, so larger device ratings have not been considered. Supplementary bonding conductors are unlikely to exceed 10 m in length, so, as can be seen from the tables mentioned above, lengths are unlikely to be limited.

Table 11.15 Supplementary bonding conductors

Size of circuit protective conductor (mm²)	Minimum cross-sectional area of supplementary bonding conductors (mm²)					
	Exposed-conductive-part to extraneous-conductive-part		Exposed-conductive-part to exposed-conductive-part		Extraneous-conductive-part to extraneous-conductive-part*	
	Mechanically protected	Not mechanically protected	Mechanically protected	Not mechanically protected	Mechanically protected	Not mechanically protected
1	2	3	4	5	6	7
1.0	1.0	4.0	1.0	4.0	2.5	4.0
1.5	1.0	4.0	1.5	4.0	2.5	4.0
2.5	1.5	4.0	2.5	4.0	2.5	4.0
4.0	2.5	4.0	4.0	4.0	2.5	4.0
6.0	4.0	4.0	6.0	6.0	2.5	4.0
10.0	6.0	6.0	10.0	10.0	2.5	4.0
16.0	10.0	10.0	16.0	16.0	2.5	4.0

* If one of the extraneous-conductive-parts is connected to an exposed-conductive-part, the bond must be no smaller than that required for bonds between exposed-conductive-parts and extraneous-conductive-parts – Column 2 or 3.

Table 11.16 Type C circuit-breaker, maximum length of supplementary bonding conductors to comply with Regulations 411.4.5 and 415.2.2

Circuit breaker rating, I_n (A)	Current, I_a Note 1 (A)	$R = \frac{50}{I_a}$ (Ω)	S_a Note 2 (mm^2)	Conductor resistance, R_2 Note 3 (mΩ/m)	Maximum length (L) of conductor (area S_a mm^2) $L = R.1000/R_2$ (m)
6	60	0.83	1.0	18.10	46
10	100	0.5	1.0	18.10	27.6
16	160	0.312	1.0	18.10	17.2
20	200	0.25	1.0	18.10	13.8
25	250	0.20	1.0	18.10	11
32	320	0.156	1.5	12.10	12.8
40	400	0.125	1.5	12.10	10.3
50	500	0.10	1.5	12.10	8.26
63	630	0.079	2.5	7.41	10.7
80	800	0.0625	4.0	4.61	13.55
100	1000	0.05	4.0	4.61	10.8

1. From table to Fig. 3A5 of Appendix 3 of BS 7671.
2. From Table 11.5.
3. From Table C.1.

Table 11.17 BS 88-2 fuse, maximum length of supplementary bonding conductors to comply with Regulations 411.4.5 and 415.2.2

BS 88 fuse rating, I_n (A)	Current (0.4 s), I_a Note 2 (A)	$R = \frac{50}{I_a}$ (Ω)	S_a Note 1 (mm^2)	Conductor resistance, R_2 Note 3 (mΩ/m)	Maximum length (L) of conductor (area S_a mm^2) $L = R.1000/R^2$ (m)
10	45	1.11	1.5	12.10	91
16	85	0.59	1.5	12.10	48
32	220	0.227	1.5	12.10	18
40	280	0.1768	2.5	7.41	24

1. From Table 11.4.
2. From Table to Fig. 3A3 of Appendix 3 of BS 7671.
3. From Table C.1.

11.7 Non-conducting service pipes

Plastic gas and water pipes are now used both for the services to properties and for plumbing within the premises. Copper pipework usually continues to be used for the visible parts of such installations. The question arises: should such services be bonded and, if so, how? Some d.c. measurements of water resistance indicate the following resistivities:

- distilled water: $\rho = 200\ \Omega\text{m}$;
- tap water: $\rho = 20\ \Omega\text{m}$;
- water with corrosion inhibitor: $\rho = 2\ \Omega\text{m}$.

Assuming that shock currents are to be limited to 10 mA, the impedance of the pipe and contents must exceed 230 V/10 mA = 23 kΩ if the shock risk is to be negligible. Table 11.18 indicates the minimum length of pipe of various diameters necessary to provide such impedance. In domestic and similar premises, when pipework does not exceed 25 mm^2, there is no need and it is somewhat impractical to main bond plastic service pipes. Where the installation pipes are metal it is usual practice to main bond these unless it is confirmed the pipes are not in contact with earth. It is not necessary

Table 11.18 Minimum length of non-conductive pipe to provide a resistance of 23 kΩ

Pipe diameter, D (mm)	Water resistivity, ρ (Ωm) Note 1	Length, L (m) Note 2
15 (13)	200 (distilled)	0.020
15	20 (tap)	0.20
15	2 (inhibited)	2.000
25	200 (distilled)	0.055
25	20 (tap)	0.550
25	2 (inhibited)	5.500
50	200 (distilled)	0.220
50	20 (tap)	2.250
50	2 (inhibited)	22.2

1. $\rho = 2\ \Omega\text{m}$ for water heavily doped with inhibitor.
   $\rho = 20\ \Omega\text{m}$ for typical tap water.

2. $L = \frac{R_A}{\rho} = \frac{23\pi D^2}{4\rho} \times 10^{-3}\ \text{m}$

   where:
   R = maximum resistance to be allowed of 23 kΩ
   D = diameter of the pipe in mm
   A = internal csa of pipe in mm^2
   ρ = water resistivity in Ωm.

to supplementary bond (in say bathrooms) metal equipment such as metal hot water radiators supplied by plastic pipes.

11.8 Earthing requirements for the installation of equipment having high protective conductor currents

Regulation 543.7

11.8.1 General

The risk associated with equipment having high protective conductor currents is that of electric shock, should the earth connection be lost. For equipment with low earth leakage (say below 3.8 mA), if the earth connection is lost, and if the metal case is touched, there is little if any shock perception. However, equipment that has high protective conductor currents, say exceeding 10 mA, will give a perceptible shock should the earth connection be lost, the shock potential being applied to all exposed-conductive-parts connected to the earthing terminal of the equipment. Reference to Figure 4.3 and Table 4.1 of Chapter 4 of this Commentary shows that, above 10 mA, harmful physiological effects can arise, the risk increasing with magnitude of the current and with time. Three items of IT equipment only, each leaking up to 3.5 mA, could produce such effects if the earth connection was lost.

A protective conductor current of 10 mA equates to a loop impedance of 23 000 ohms (230 V/10 mA). If the connection with earth is lost and a person in contact with earth touches exposed-conductive-parts of the equipment, they put themselves in series and the equivalent loop impedance increases to say 24 000 ohms. This produces a potential body or touch current of 9.6 mA (see Figure 11.15). This could be quite unpleasant (see Figure 4.3/Table 4.1 and zone AC-2, where perception and involuntary muscular contractions are likely but usually no harmful electrical physiological effects).

11.8.2 Equipment with protective conductor current exceeding 3.5 mA

Regulation 543.7.1.201

The regulation requires equipment having a protective conductor current exceeding 3.5 mA to be either permanently connected to the fixed wiring of the installation without the use of a plug and socket-outlet or connected by means of a industrial type plug and socket-outlet complying with BS EN 60309-2.

The standard for household and similar appliances, BS EN 60335-1, generally requires pluggable equipment protective conductor current to be less than 3.5 mA.

The standard for IT equipment, BS EN 60950-1, requires 'pluggable Type A' equipment', that is equipment to be connected with a standard 13 A plug, to have a protective conductor current not exceeding 3.5 mA. Where the protective conductor current exceeds 3.5 mA the standard requires the equipment to be permanently connected or connected via an industrial plug and socket. This is called 'pluggable Type B equipment'. There is also a requirement for a label as shown below (see clause 5.1.7 of BS EN 60950-1).

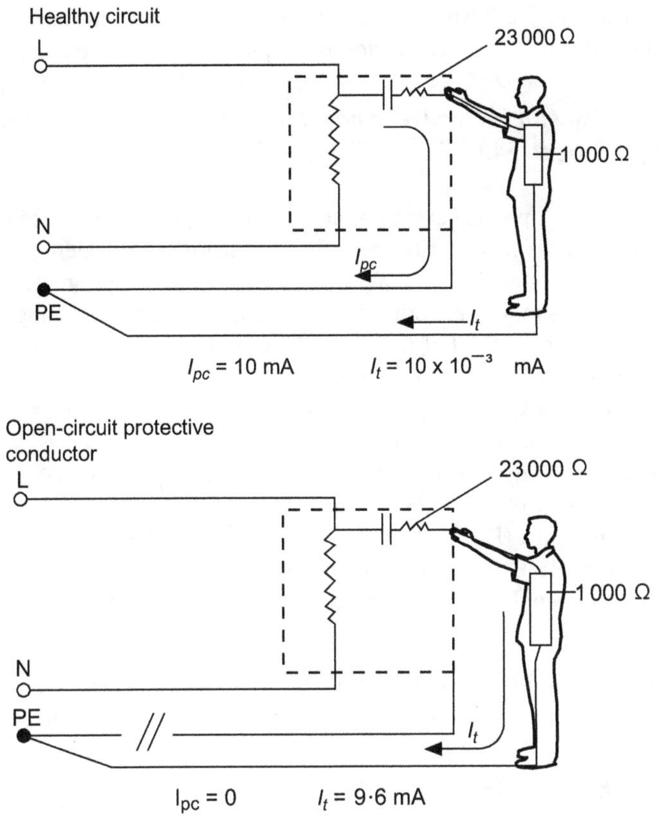

Figure 11.15 *Touch currents I_t with sound and with open-circuit protective conductor*

Equipment with touch current exceeding 3.5 mA – one of the following labels, or a label with similar wording, shall be affixed adjacent to the equipment AC MAINS SUPPLY connection, such that it will be visible to a person...

WARNING
HIGH TOUCH CURRENT
EARTH CONNECTION ESSENTIAL BEFORE CONNECTING SUPPLY

11.8.3 Equipment with protective conductor current exceeding 10 mA Regulation 543.7.1.202

Equipment having protective conductor current exceeding 10 mA must be permanently connected to the fixed wiring (with minimum protective conductor requirements); alternatively, it may be connected by an industrial plug and socket-outlet (with minimum flex protective conductor csa), or the connection be monitored.

11.8.4 Circuits with protective conductor current exceeding 10 mA Regulations 543.7.1.203, 533.2.1, 531.2.4

The circuits to equipment having protective conductor currents exceeding 10 mA are required to have a high-integrity protective conductor or equivalent arrangements made. Circuits supplying IT equipment may have protective conductor currents exceeding 10 mA, as may many other circuits, including certain lighting circuits.

As discussed above, BS EN 60950-1 requires 'pluggable type A' equipment, that is equipment to be connected with a standard 13 A plug, to have a protective conductor current not exceeding 3.5 mA. In practice flat-screen PCs may have a protective conductor current of the order of 1 mA.

Luminaire high-frequency ballasts have RFI filter networks that will include capacitor connections to earth. Typically, the protective conductor current per ballasted luminaire is up to 1 mA (BS EN 60598-1:1989 allows 1 mA for fixed luminaires up 1 kVA rated input; BS EN 60598-1:2008 allows up to 3.5 mA for luminaires with a supply current of 7 A or less (see clause 10.3 reproduced below)).

10.3 Touch current, protective conductor current and electric burn

The touch current or protective conductor current that may occur during normal operation of the luminaire shall not exceed the values given in Table 10.3 when measured in accordance with Annex G.

Table 10.3 Limits of touch current or protective conductor current and electric burn

	Touch current	Max. limit (peak)
All luminaires of Class II and Class I luminaires rated up to and including 16 A fitted with a plug connectable to an unearthed socket-outlet		0.7 mA
Protective conductor current	**Supply current**	**Max. limit (rms)**
Class I luminaires fitted with a single or multiphase plug, rated up to and including 32 A	≤4 A	2 mA
	> 4 A but ≤10 A	0.5 mA/A
	>10 A	5 mA
Class I luminaires intended for permanent connection	≤7 A	3.5 mA
	≥7 A but ≤20 A	0.5 mA/A
	≥20 A	10 mA
Electric burn		Under consideration

When considering such circuits the designer will have to consider limits imposed by:

1. the 10 mA protective conductor current limit (Regulation 543.7.1.203);
2. inrush current limit of the protective device (Regulation 533.2.1), see appendix C of *Electrical Installation Design Guide*; and
3. RCD protective conductor current limits (Regulation 531.2.4).

Rule-of-thumb protective conductor current

As a rule of thumb the following protective conductor currents will assist in preliminary designs:

Table 11.19 Equipment protective conductor currents

Equipment	Protective conductor current (mA)
PC plus flat screen	0.75
PC plus CRT screen	1.25
36 W ballasted luminaire	0.5
60 W ballasted luminaire	1.0

11.8.5 Socket-outlet circuits
Regulation 543.7.1.203

For a final circuit with socket-outlets or connection units, where the protective conductor current in normal service will exceed 10 mA, particular arrangements are allowed, the most common being:

(i) a ring final circuit with a ring protective conductor. Spurs, if provided, require high integrity protective conductor connections (Figure 11.17), or
(ii) a radial final circuit with
  (a) a protective conductor connected as a ring (Figure 11.16) or
  (b) with an additional protective conductor provided by metal conduit or ducting.

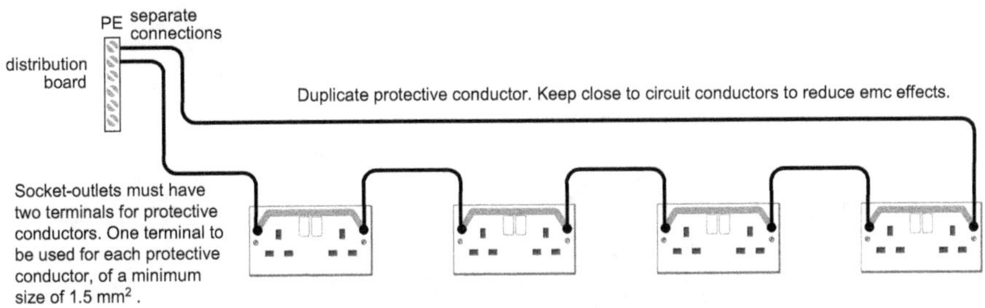

Figure 11.16 Radial circuit supplying twin socket-outlets (total protective conductor current exceeding 10 mA), with duplicate protective conductor

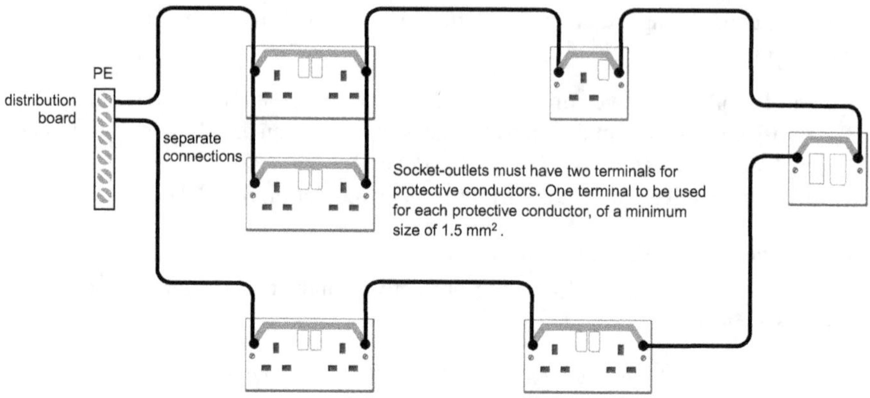

Figure 11.17 Ring final circuit supplying socket-outlets (total protective conductor current exceeding 10 mA)

Chapter 12
Generators and other equipment

Chapter 55

12.1 Low voltage generating sets

Section 551

12.1.1 General

The legal requirements for the installation of generating sets are found in Part IV 'Generation' of the Electricity Safety, Quality and Continuity Regulations 2002 (ESQCR). The guidance on the ESQCR issued by the Department of Trade and Industry (now BIS) draws particular attention to the Energy Networks Association Engineering Recommendation G59/3, *Recommendations for the connection of embedded generating plant to the regional electricity companies* distribution systems (see also Engineering Recommendation G75/1, *Recommendations for the connection of embedded generating plant to the public distribution systems above 20 kV or with outputs over 5 MW*).

The ESQCR require that, where an operation in parallel with a public supply network is intended, the agreement of the distributor has first to be obtained (except for small-scale embedded generators with an output of less than 16 A – see section 12.2).

12.1.2 Switched alternative supplies

Regulations 551.4.3.2, 551.6

Regulation 21 of the ESQCR requires the operator of a switched alternative source to ensure that the source cannot operate in parallel with the supply and to comply with BS 7671.

The switching and earthing arrangements for generators installed as a switched alternative to the public supply for single generators are shown in Figures 12.1 and 12.2 (these figures are taken from Figures 4.1 and 4.3 of withdrawn Technical Report 113).

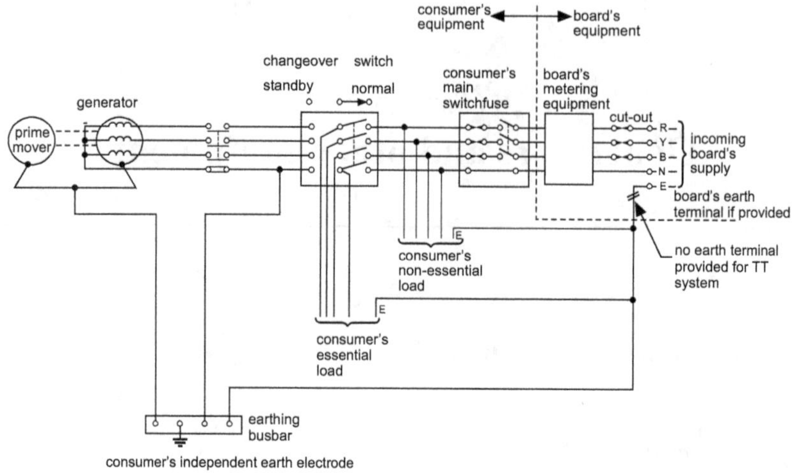

Figure 12.1 Private generator with alternative connections to TN-S or TT low voltage system

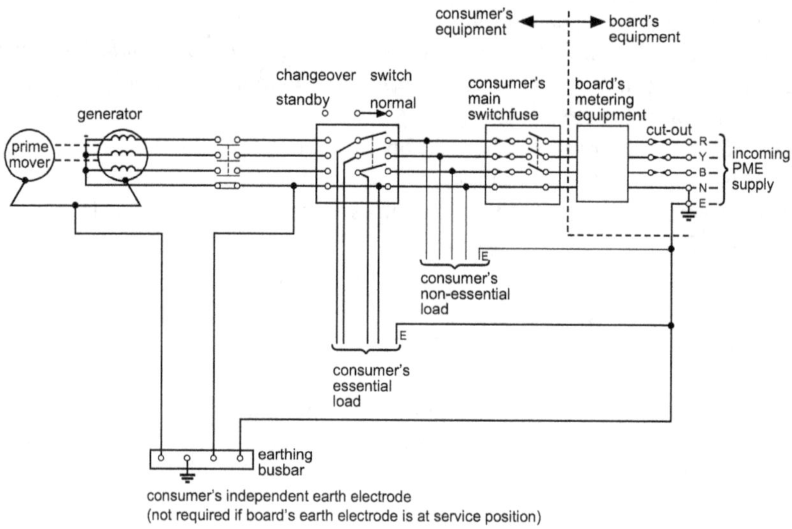

Figure 12.2 Private generator with alternative connections to PME low voltage system

12.1.3 Earthing general
Regulation 551.4.3.2.1

Regulation 551.4.3.2.1 requires that when a generator is operating as a switched alternative source, and where protection is by automatic disconnection of supply,

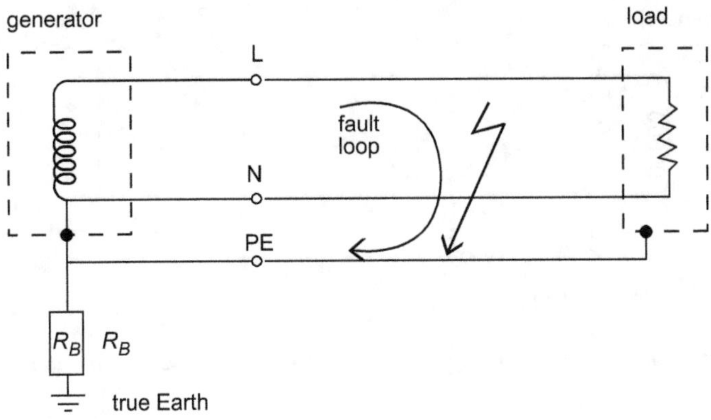

Figure 12.3 Generator fault loop

a suitable means of earthing must be provided. This implies the generator neutral requires earthing only if necessary for automatic disconnection. This distinction is important for small (less than 10 kVA) generators. Regulation 8(3)b of the ESQCR requires that every low voltage network neutral conductor be connected with Earth; again, this would exempt generators supplying a small installation (see subsection 12.1.6 below).

Where a protective conductor connected to the neutral point is distributed throughout the installation (TN-S or TN-C-S system), automatic disconnection is not dependent on a connection with true Earth of the neutral point (see Figure 12.3).

12.1.4 Earth electrode
Regulations 551.4.3.2, 542.2

Regulation 551.4.3.2 requires a suitable means of earthing to be provided for generators operating as a switched alternative source, independent of that provided by the public electricity supply.

For independent earth electrodes associated with the local earthing of the star point of generating plant, BS 7430 recommends that the earth resistance should not exceed 20 Ω.

12.1.5 Low voltage three-phase mobile generators 15 kW and above (clause 7.1.1 of BS 7430)
Connections to generator windings

The generators considered here may be trailer-, skid- or vehicle-mounted with outputs from about 15 kVA upwards.

Neutral/star points of the windings, the generator frame/chassis, all exposed metalwork and the outgoing circuit neutral and outgoing circuit protective conductor should be connected to a common 'earth' bar or reference point and this reference point connected to Earth via an earth electrode (see Regulation 542.2.3).

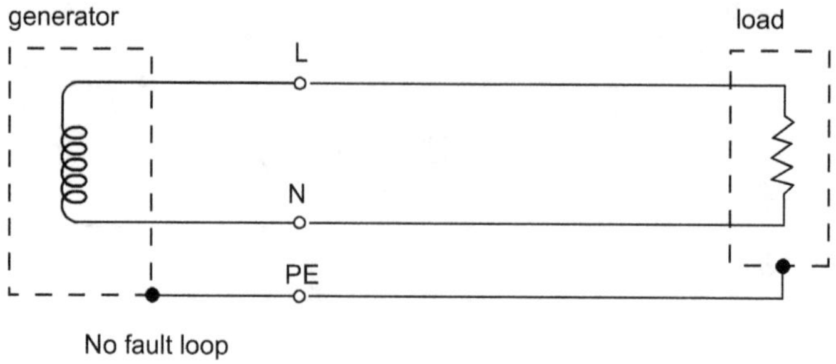

Figure 12.4 Small single-phase generator with floating system (unearthed)

12.1.6 Portable and transportable low voltage single-phase generators (clause 7.1.4) from 0.3 to 10 kVA

The generator winding is often isolated from the frame. It is accepted in BS 7430 that small single-phase generators can be run as floating systems, i.e. without the winding connected to the frame or to earth (see Figure 12.4). The generator frame and enclosure should be connected to the circuit protective conductor and hence to all exposed-conductive-parts of the connected equipment. The use of a 30 mA RCD may reduce the risk of electric shock. The first fault to 'earth' in such systems will not be detected, but in the event of a second fault the RCD may operate.

12.2 Small-scale embedded generators (SSEGs)
Regulations 551.1, 551.7

12.2.1 The law

The ESQCR exempt sources of energy with an electrical output not exceeding 16 amperes per phase at low voltage from Regulations 22(1)(b) and 22(1)(d) of the Regulations. This relaxation allows the use of alternative sources of energy such as renewable and combined heating and power. Such generators are required to meet Regulation 22(1) (a) and (c) as follows:

22. Parallel operation

(1) No person shall install or operate a source of energy which may be connected in parallel with a distributor's network unless he –
- (a) has the necessary and appropriate equipment to prevent danger or interference with that network or with the supply to consumers, so far as is reasonably practicable;
- (b) where the source is part of a low voltage consumer's installation, complies with the provisions of British Standard requirements.

The British Standard requirements referred to are those of BS 7671 and the section specific to generators is 551: 'Low voltage generating sets'.

There are special requirements for small-scale embedded generators (SSEG) in Regulation 22(2) as follows:

> **22.** – (2) (b) the source of energy is configured to disconnect itself electrically from the parallel connection when the distributor's equipment disconnects the supply of electricity to the person's installation; and
>
> (c) the person installing the equipment ensures that the distributor is advised of the intention to use the source of energy in parallel with the network before, or at the time of, commissioning the source.

It is necessary for such equipment to be type-tested and approved by a recognized approval body.

12.2.2 Engineering Recommendation G83 (BS EN 50438)

To assist network operators and installers the Energy Networks Association has prepared Engineering Recommendation G83: 'Recommendations for the connection of type-tested small scale embedded generators (up to 16 amperes per phase) in parallel with public low-voltage distribution networks'. The recommendation is for all small-scale embedded generator installations with an output up to 16 amperes per phase, single- or polyphase, 230/400 V including:

- domestic combined heat and power (DCHP);
- micro hydro;
- micro wind;
- photovoltaic; and
- fuel cells.

Engineering Recommendation G83 includes application forms for the connection of multiple SSEGs and commissioning confirmation for a single SSEG. The supply of information in this form, for a type-tested and approved unit, is intended to satisfy the legal requirements of the ESQC Regulations. The recommendations of G83 are now included in BS EN 50438.

Appendices from G83/2 (2012) have been made available in MS Word format and are free to download from the Energy Networks Association Website site:

(http://www.energynetworks.org/electricity/engineering/distributed-generation/distributed-generation.htm)

- Appendix 2 – Application for Connection
- Appendix 3 – SSEG Installation Commissioning Confirmation
- Appendix 4 – Type Verification Test Report
- Appendix 5 – SSEG Decommissioning Confirmation.

A guidance document on G83 is available as a free download, but the engineering recommendation must be bought.

12.2.3 Engineering Recommendation G59/2-1 (2011)

Engineering Recommendation G 59/2: *Recommendations for the connection of generating plant to the distribution systems of licensed distribution network operators*, provides the general requirements for connecting the generation plant to the distribution network, as distinct from the requirements for small-scale generating plant of G 83.

The Data Registration Code of the Distribution Code sets out the obligations on the generator and DNO to exchange data as part of the design process and lists the data items that may need to be exchanged. The purpose of the application forms is to simplify and clarify this data exchange process.

Appendices from G59/2-1 (2011) – A13.1, A13.2, A13.2 and A13.4 – have been made available in MS word format and are free to download from the ENA website.

- A13.1 Generating Plant Type Verification Test Sheet for type approved Generating Plant (>16 A per phase but < 50 k W 3 phase or 17 k W 1 phase)
- A13.2 Generating Plant Installation and Commissioning Tests for type approved Generating Plant (>16 A per phase but < 50 k W 3 phase or 17 k W 1 phase)
- A13.3 Generating Plant Installation and Commissioning Tests for non type approved Generating Plant and Generating Plant > 50 k W 3 phase or 17 k W 1 phase
- A13.4 Generating Plant Decommissioning Confirmation form.

Completed forms are sent to the appropriate Distribution Network Operator.

A guidance document on G59 is available as a free download, but the engineering recommendation must be bought.

12.2.3.1 Overcurrent protection
Regulations 551.5, 551.7.2

A suitably rated overcurrent protective device is required to protect the wiring between the electricity supply terminals and the embedded generator.

Where a generator is not connected via its own circuit, there are additional requirements as described in Regulation 551.7.2. This may seem appropriate, particularly, say, when a combined-heat-and-power generator replaces a gas boiler.

However, if an SSEG is not supplied from its own circuit (see Figure 12.5) the following additional requirements apply:

(i) Installed cable rating I_z

$I_z \geq I_n + I_g$

where:

I_z is the current-carrying capacity of the final circuit conductors

I_n is the rated current of the protective device of the final circuit
I_g is the rated output current of the generating set.

This is to compensate for the generator reducing the current in the overcurrent device, but not in the cable downstream of the generator (see Figure 12.5);

(ii) the generator must not be connected by a plug and socket;
(iii) RCDs shall switch all poles;
(iv) line and neutral conductors of the final circuit and of the generating set shall not be connected to Earth; and
(v) the device protecting the final circuit shall disconnect all poles.

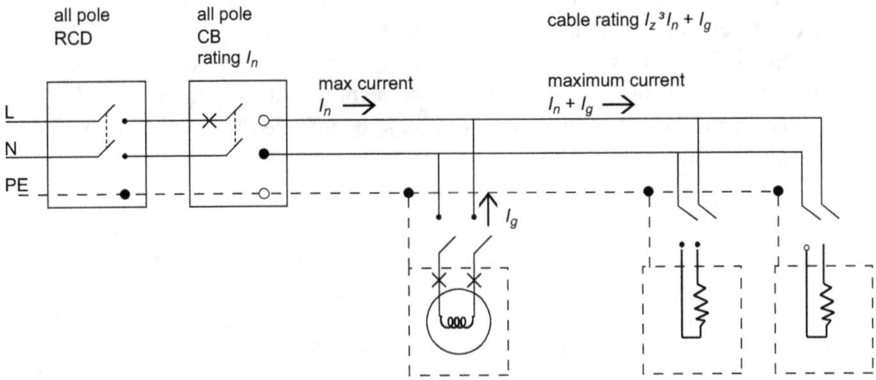

Figure 12.5 SSEG connected via a shared circuit

12.2.3.2 Isolation

Regulation 551.7.6

The SSEG is required to be connected directly to a local isolating switch, see Figure 12.6. For single-phase machines the line and neutral are to be isolated and for three-phase machines all live conductors (includes the neutral) are to be isolated. In all instances the switch, which must be manual, shall be capable of being secured in the off isolating position. The switch is to be located in an accessible position in the customer's installation.

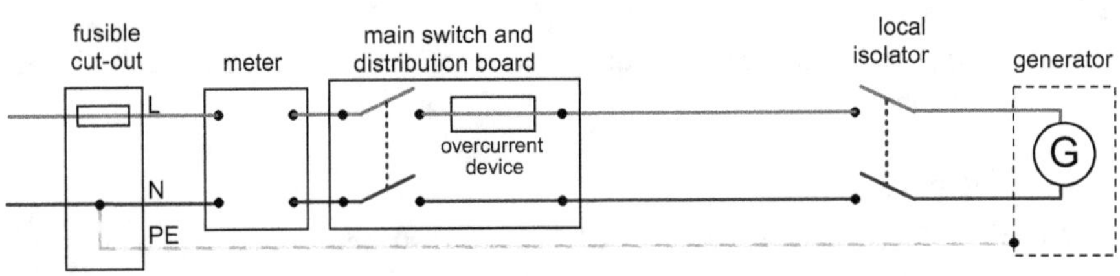

Figure 12.6 Isolation of small-scale embedded generator (SSEG) on its own circuit

Reparagulation 551.7.6 requires that means shall be provided to enable the generating set to be isolated from the public supply and the means of isolation shall, for generating sets with an output exceeding 16 A, be accessible to the supplier (or distributor) at all times. For generators with an output not exceeding 16 A, the accessibility of the means of isolation shall comply with BS EN 50438 (see subsection 12.2.3.3 below).

12.2.3.3 Accessibility of isolation switching devices

BS EN 50438 *Requirements for the connection of micro-generators in parallel with public low voltage distribution networks* advises:

> Under HD 384 series (BS 7671) there is a requirement that means shall be provided to enable a generator set to be isolated from the public supply. Where this means of isolation is not accessible for the DNO at all times it is acceptable to provide two means of automatic disconnection, with a single control. At least one of the means of disconnection must be afforded by the separation of mechanical contacts.

It is understood that the electricity distribution network operators will accept the distribution cut-out (fused unit) as the means of isolation. This is considered to be accessible in the sense that all metering equipment is accessible to the distributor. In practice, when small-scale embedded generators become more common, it will be difficult for a distributor to identify all the locations where embedded generators are installed and equally difficult to isolate all such equipment.

12.2.3.4 Interface protection

The legal requirement for all generators is that the source of energy must be configured to disconnect itself electrically from the parallel connection when the distributor's equipment disconnects the supply of electricity to the consumer's installation. See ESQCR 22(2). This is an essential requirement.

The DTI (BIS) advise that the means of disconnection for interface protection should preferably be by mechanical separation. However, a suitable solid-state switching device is permitted if it is equipped with fail-safe monitoring to ensure the line to neutral voltage on the mains side of the device reduces to less than 50 V within 0.5 s of the device failing to operate when required to do so.

12.2.3.5 Approval

Only type-tested and approved equipment that meets this requirement will be allowed to be connected.

12.2.4 Earthing

When an SSEG is operating in parallel with a distributor's network, there shall be no direct connection between the generator winding (or pole of the primary energy source in the case of a PV array or fuel cell) and the network operator's earthing terminal (see Figure 12.7).

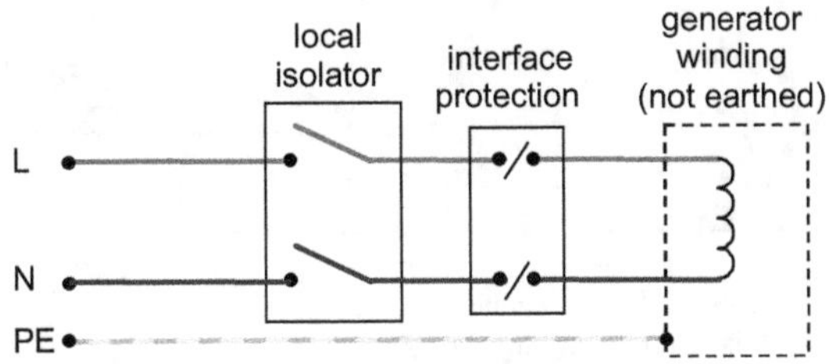

Figure 12.7 Earthing for small-scale embedded generators

12.2.5 Labelling
Regulation 514.15.1

There are specific requirements for labelling in installations that include an alternative or additional sources of supply. Labels are required at:

- the supply terminals (fused cut-out);
- the meter position;
- the consumer unit or distribution board to which the alternative or additional source is connected; and
- at all points of isolation of all sources of supply.

The labels are required to indicate the presence of the SSEG in the premises. The Health and Safety (Safety Signs and Signals) Regulations 1996 require that the labels should display the prescribed triangular shape and size using black-on-yellow colouring. A typical label is shown as Figure 12.8.

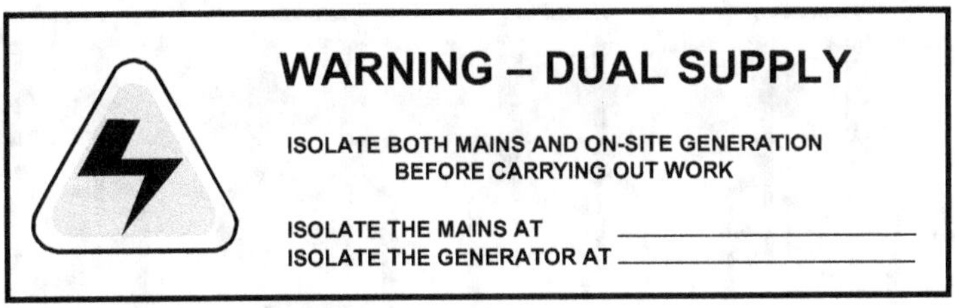

Figure 12.8 SSEG warning

Engineering Recommendation G83 also requires up-to-date information to be displayed at the point of connection with the distributor's network. The information required is:

(a) a circuit diagram showing the relationship between the embedded generator and the network operator's cut-out (see Figure 12.9); the diagram is also required to show by whom the generator is owned and by whom it is maintained; and
(b) a summary of the protection settings incorporated within the equipment.

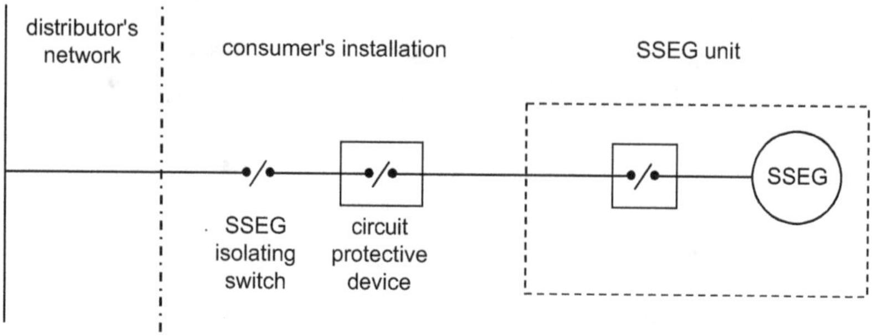

Figure 12.9 Circuit diagram for a SSEG installation

12.2.6 Domestic combined heat and power (DCHP)

Engineering Recommendation G83 specifies the particular requirements for combined-heat-and-power sets that have been agreed by the distributors. This will help manufacturers by specifying acceptable performance criteria, including that of the interface with the electricity supply.

DCHP generators can replace household gas central-heating boilers, so may become common. Most small-scale DCHP generators use a Stirling engine (see Figure 12.10). The Stirling engine does not burn the gas within the cylinder: the power to the engine is conducted heat from the combustion gases of the gas burner and the energy transfer achieved by the temperature difference between the burner exhaust gases and burner input air or circulating water.

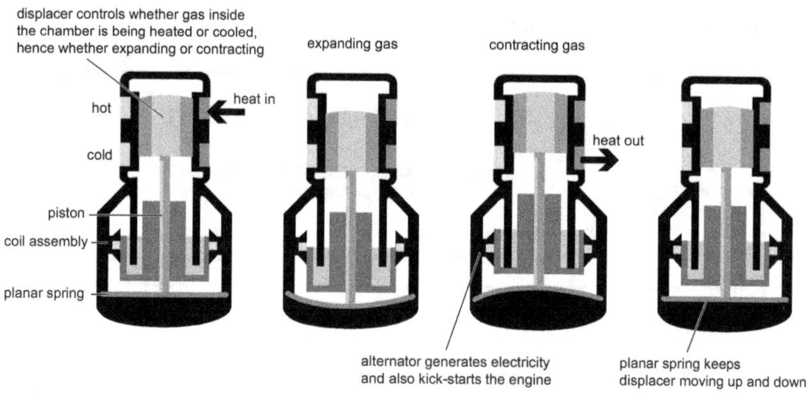

Figure 12.10 Stirling engine

12.2.7 Solar photovoltaic (PV) power supply systems

It is necessary for solar photovoltaic power systems to comply with the requirements of the ESQCR in the same way as any other small-scale embedded generator. Among other requirements, they must be configured to disconnect from power connection when the distributor's equipment disconnects the supply of electricity to the consumer's installation.

BS 7671 includes Section 712, 'Solar photovoltaic (PV) power supply systems', Guidance on the installation of photovoltaic systems is to be found in 'Guide to the installation of PV systems' report S/P2/00355/REP/1, published in 2002. Authors are BRE, EA Technology, Halcrows and Sundog. Another useful source of information is the British Photovoltaic Association on www.pv-uk.org.uk.

12.3 Static inverters
Regulation 551.4.3.3

Uninterruptible power supplies (UPS) are almost exclusively static inverters (BS 7671 calls these 'converters'). The high internal impedance of static inverters will result in high earth fault loop impedances and, as a consequence, the disconnection times of Regulation group 411.3.2 may not be met unless an RCD is installed. The use of supplementary bonding instead of an RCD is recognized in Regulation 551.4.3.3.1. See also Regulation 415.2.2.

When installing UPS equipment, thought needs to be given in particular to (i) isolation and (ii) the waveform (harmonics).

12.3.1 Isolation

The UPS will continue to supply equipment and circuits when the mains supply has been interrupted or switched off. Therefore, it is important to label distribution boards, including the main boards supplying equipment with UPSs, and the equipment supplied by a UPS, so that users will be aware of the arrangements. The hazard that could arise if a person working on the system was not aware of the standby supplies needs to be warned against. All equipment supplied by a UPS needs local isolation.

Some smaller UPS have a common neutral from the input to the load connections. This can be a hazard if not correctly isolated.

12.3.2 Harmonics

UPS equipment may generate a non-sinusoidal waveform and IT equipment itself tends to draw non-sinusoidal currents. Where the current contains third harmonics and multiples thereof, particular care has to be taken in sizing neutrals. In balanced three-phase systems, the line currents sum to zero in the neutral. This is not true of third (and multiples of) harmonics, which are additive in the neutral. The neutral current may exceed the line current and cables will need to be sized accordingly (see section 17.2).

12.4 Rotating machines
Section 552

Before making arrangements for installing large rotating machines, the distributor should be consulted to ensure that disturbances to the supply will be within acceptable limits. Guidance is given in the Energy Networks Association document P28, *Planning limits for voltage fluctuations caused by industrial, commercial and domestic equipment in the United Kingdom.*

Regulation 552.1.1 requires that account be taken of cumulative effects of starting or braking currents on the temperature rise of the equipment of the circuit. An approach to this is described in section 6.2.3.

12.5 Accessories
Section 553

The general requirement (Regulation 553.1.3) is that socket-outlets for household and similar use should comply with BS 1363; these are the 13 A socket-outlets suitable for the standard 13 A plugs of the Plugs and Sockets etc. (Safety) Regulations. These Regulations require all equipment to be supplied either with a standard plug to BS 1363, or a non-UK plug to IEC 884-1 complete with conversion plug to fit a BS 1363 socket-outlet.

It is likely that designers may be asked to design installations with socket-outlets in common use in Europe. These are generally 5 A or 16 A devices. The European plugs are unfused, so if such plugs are to be installed, unless there is some control over the sizes of the flexible cables, circuits supplying socket-outlets for these plugs should be fused at no more than 20 A (see section 6.7).

12.6 Water heaters and boilers
Regulation groups 554.1, 554.3

Regulation group 554.3 deals with a particular type of water heater having an immersed and uninsulated heating element. This is not a standard metal-sheathed heating element, but an element that is directly immersed in the water, and is encountered in certain industrial water heaters and makes of instantaneous shower (see Figure 12.11).

This type of heater is to be distinguished from an electrode boiler, which actually uses the water as a conducting medium. Requirements for electrode boilers are found in Regulation group 554.1 of BS 7671 (see Figure 12.12).

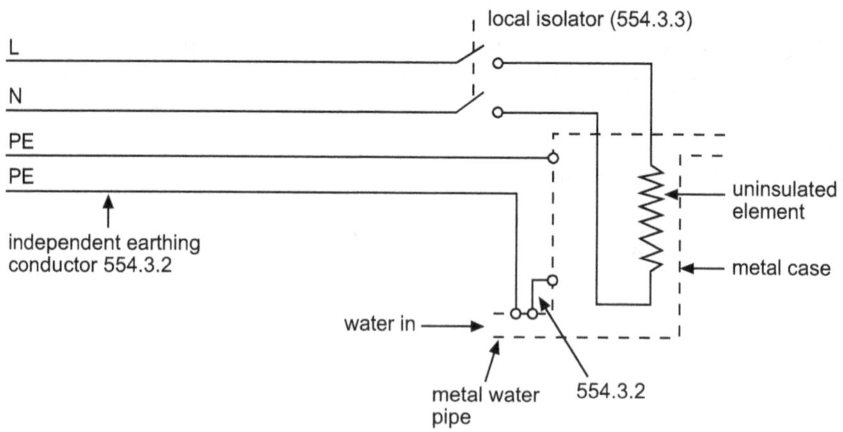

Figure 12.11 Uninsulated immersed-element water heater

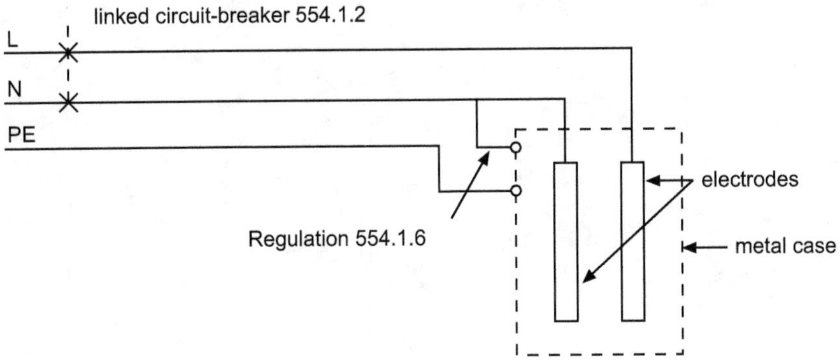

Figure 12.12 Single-phase electrode boiler

12.7 Autotransformers
Section 555

Where an autotransformer is connected to a circuit having a neutral conductor, the common terminal of the autotransformer must be connected to the neutral conductor. This arrangement is shown diagrammatically in Figure 12.13.

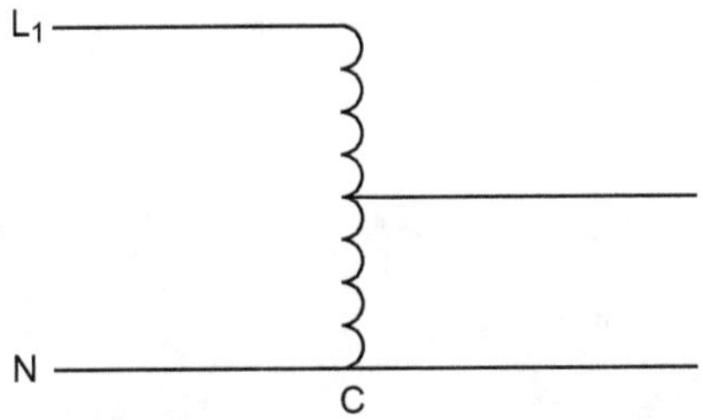

Figure 12.13 Connection of autotransformer to the neutral

Chapter 13
Safety services

(Adviser Mr Colin Reed)

Chapter 56

13.1 Scope
Regulation 560.1

The scope of Chapter 56 includes supplies for "emergency lighting" and "fire detection and alarm systems". BS 7671 does not include the detailed requirements of the HDs for "emergency lighting" and "fire detection and alarm systems" and instead refers to and requires compliance with the requirements of BS 5266 *Emergency lighting* and BS 5839 *Fire detection and fire alarm systems for buildings.* This results in compliance with the technical intent of the HD and avoids any conflict between the standards and BS 7671.

13.2 Classification
Regulation 560.4

Standby supplies are classified according to the changeover time (see Table 13.1).

Table 13.1 Classification of standby supplies

Classification	Standby supply available within (seconds)
No-break	provides a continuous supply
Very short break	up to 0.15
Short break	within 0.5
Normal (Lighting) break[1]	within 5
Medium break	within 15
Long break	more than 15

1. BS 5266 *Emergency lighting* requires a response time of 5 seconds from the failure of the normal lighting, but at the discretion of the enforcing authority allows the period to be extended to 15 seconds in premises that are likely to be occupied for the most part by persons who are familiar with them and the escape routes (clause 6.3.2).
2. BS 5839 *Fire detection and alarm systems for buildings* requires the standby supply to comprise a secondary rechargeable battery with an automatic charger.

13.3 Battery life
Regulation 560.6.10

The British Standard for emergency lighting (BS 5266) requires batteries to comply with BS EN 50171 *Central power supply systems,* and this, like BS 7671, requires a minimum battery life of ten years for central systems (clause 6.12.2).

BS 5839 *Fire detection and fire alarm systems for buildings* requires a minimum battery life of four years.

13.4 Segregation of circuits
Regulation group 560.7

Circuits of safety services are required to be independent of other circuits. The note to the regulation advises that this may necessitate separation by fire-resistant materials or different routes or enclosures. While this requirement is somewhat unclear, the segregation requirements for fire systems and emergency lighting are more specific.

13.4.1 Segregation requirements of BS 5839-1 Fire detection and fire alarm systems

The requirements of BS 5839 for segregation are given in clause 26.2:

26.2 Recommendations

(k) To avoid the risk of mechanical damage to fire alarm cables they should not be installed within the same conduit as the cables of other services. Where fire alarm cables share common trunking, a compartment of the trunking, separated from other compartments by a strong, rigid and continuous partition, should be reserved solely for fire alarm cables.

(l) To avoid electromagnetic interference with fire alarm signals, any recommendations by the manufacturer of the fire alarm equipment in respect of separation of fire alarm cables from other cables of other services should be followed.

(m) Where multicore cable is used for interconnection of fire alarm circuits, none of the conductors should be used for circuits other than those of the fire alarm system.
NOTE 12: This recommendation does not preclude the multiplexing of signals of other systems with those of the fire detection and fire alarm system. Guidance on such integrated systems is given in DD CLC/TS 50398.

(n) Fire alarm cables carrying power in excess of extra-low voltage should be segregated from extra-low voltage fire alarm circuits by use of cables complying with 26.2(b) and 26.2(l). In particular, the mains supply cable to any control, indicating or power supply equipment should not enter the equipment through the same cable entry as cables carrying extra-low voltage. Within the equipment, low voltage and extra-low voltage cables should be kept separate to the extent practicable.

(o) All fire alarm cables should be of a single, common colour that is not used for cables of general electrical services in the building, to enable these cables to be distinguished from those of other circuits.
NOTE 13: The colour red is preferred.

13.4.2 Segregation requirements of BS 5266-1 Emergency lighting

The requirements for segregation in BS 5266-1 are given in clause 8.2.6:

8.2.6 Segregation

It is essential that the wiring of emergency escape lighting installations is exclusive to the installation and separate from the wiring of any other circuits. Normal lighting dimming control circuits may have common elements, but should not interfere with the operation of the emergency lighting.

Where multicore cable is used for the emergency escape lighting, none of the conductors should be used for any other circuit.

Where cables for emergency lighting circuits are installed in a common containment system such as trunking, ducting, cable basket or cable tray, as cables of other services, they should be segregated from each other, in order to avoid the risk of mechanical damage to the emergency escape lighting cables, by a partition which:

1. is of the same base material as the common containment system, e.g. metal, and permanently and securely fixed by the same fastenings used to assemble the common containment system, e.g. metal screws and nuts, pop rivets, etc.;
2. is rigid, mechanically strong, and continuous for the whole length of the emergency lighting cable installation;
3. is without perforation and the same height as the sides of the common containment system.

Ducting, trunking or channel reserved for emergency escape lighting system cables should be marked to indicate this reservation.

Multicore cables should not be used to serve both emergency escape lighting and any other circuit.

13.5 Wiring systems
Regulation 560.8

Note 2 to Regulation 560.8.1 provides simple advice that cables to BS EN 60702-1, BS 7629-1 and BS 7846 as being suitable when appropriately selected for the application. Designers must make reference to the emergency lighting and fire detection and fire alarm system standards.

13.6 Emergency lighting systems
Regulation 560.9

The wiring system selected shall meet the requirements of the relevant code of practice appropriate to the application and shall be mounted and installed in such a way that the circuit integrity will not be impaired during a fire.

BS 5266-1 requires cables or cable systems with an inherently high resistance to attack by fire for connecting centrally supplied emergency escape lighting to the standby supply. Such cables and cable systems should have duration of survival of 60 minutes for standard systems and 120 minutes for enhanced systems, when tested in accordance with BS EN 50200:2006.

This corresponds for standard emergency lighting systems to a classification of PH 60 as detailed in BS EN 50200:2006, and a duration of survival of 30 minutes when tested in accordance with BS EN 50200:2006, Annex E. For enhanced emergency lighting systems, this corresponds to a classification of PH 120 as detailed in BS EN 50200:2006 and a duration of survival of 120 minutes when tested in accordance with BS 8434-2.

Cables

BS 5266-1, clause 8.2.2, advises that cables with an inherently high resistance to attack by fire that may be suitable include:

1. mineral insulated cables to BS EN 60702-1, with terminations complying with BS EN 60702-2;
2. fire-resistant electric cables having low emission of smoke and corrosive gases to BS 7629;
3. armoured fire-resistant electric cables having thermosetting insulation and low emission of smoke and gases to BS 7846.

13.7 Selection of cables for fire detection and fire alarm systems to BS 5839

13.7.1 Cable performance

BS 5839 recognises two levels of performance for cables as follows:

1. standard performance fire-resisting cables, and
2. enhanced performance fire-resisting cables.

Standard performance fire-resisting cables are to be used for all parts of the critical signal paths and for the final circuits to the fire system.

Note, critical signal path: all components and interconnections between every fire alarm initiation point (manual call point or automatic fire detector) and the input terminals on or within each fire alarm device.

Enhanced performance fire-resisting cables are recommended for systems in which cables might need to operate satisfactorily in:

1. unsprinklered buildings (or parts of buildings) in which the fire strategy involves evacuation of occupants in four or more phases;
2. unsprinklered premises of greater than 30 m in height;
3. unsprinklered premises and sites in which a fire in one area could affect cables of critical signal paths associated with areas remote from the fire, in which it is envisaged people will remain in occupation during the course of the fire. Examples may be large hospitals with central control equipment and progressive horizontal evacuation arrangements, and certain large industrial sites;
4. in any other buildings in which the designer, specifier or regulatory authority, on the basis of a fire risk assessment that takes fire engineering considerations into account, consider that the use of enhanced fire-resisting cables is necessary.

13.7.2 Standard performance fire-resistant cables (used for critical signal paths), BS 5839-6 clause 26.2

Standard fire-resistant cables should meet the PH 30 (or PH 120) classification when tested in accordance with BS EN 50200 and the 30 minutes survival time when tested to Annex E (water spray) of that standard, see note below, and be selected from one of the following:

1. Mineral insulated copper sheathed cables, with an overall polymeric covering, conforming to BS EN 60702-1, with terminations conforming to BS EN 60702-2.
2. Fire-resistant cables that conform to BS 7629-1.
3. Armoured fire-resistant cables that conform to BS 7846.
4. Cables rated at 300/500V (or greater) that provide the same degree of safety to that afforded by compliance with BS 7629-1.

NOTE 1: Standard fire-resisting cables are required to meet the PH 30 classification and maintain circuit integrity when tested in accordance with BS EN 50200 and additionally Annex E, which requires circuit integrity is maintained when subjected to a flame at a temperature of 830°C and mechanical shock for a period of 15 minutes and then a further 15 minutes with the addition of water spray as well as the flame and mechanical shock.

In practice, the specifier or installer needs confirmation from the cable supplier that cables selected meet or exceed the requirements of BS 5839-1 for standard fire-resistant cables.

NOTE 2: BS EN 50200 and Annex E only cover cables of 20 mm in diameter or smaller. The appropriate standard for cables of greater than 20 mm in diameter that tests for fire, mechanical shock and water spray is BS 8491:2008.

Table 13.2 One manufacturer's statements on compliance

FP200 Gold	BASEC and LPCB approved to BS 7629-1 and suitable for use according to BS 5839-1:2013 requirements for "standard" cables	
FP PLUS	BASEC and LPCB approved to BS 7629-1 and suitable for use according to BS 5839-1:2013 requirements for "enhanced" cables.	
FP600S	BASEC approved to BS 7846 category F 120 and LPCB approved to BS 8491 for 120 minute fire survival time. *Tested to BS 8491:2008 Method for assessment of fire integrity of large diameter power cables for use as components for smoke and heat control systems and certain other active fire safety systems for 120 minute fire survival time.*	

13.7.3 Enhanced fire-resisting cables

Enhanced fire-resisting cables are required to meet the PH 120 classification when tested in accordance with BS EN 50200 and the 120 minutes classification when tested to BS 8434-2, which requires circuit integrity is maintained when subjected to a flame at a temperature of 930°C and mechanical shock for a period of 60 minutes and then a further 60 minutes with the addition of water spray as well as the flame and mechanical shock. Cables are to be selected from one of the following:

1. Mineral insulated copper sheathed cables, with an overall polymeric covering, conforming to BS EN 60702-1, with terminations conforming to BS EN 60702-2.
2. Fire-resistant cables that conform to BS 7629-1.
3. Armoured fire-resistant cables that conform to BS 7846.
4. Cables rated at 300/500 V (or greater) that provided the same degree of safety to that afforded by compliance with BS 7629.

As the test requirements for cables of BS 5839-1 do not align fully with the requirements of the cable standards, specifiers and installers must confirm with the cable supplier that the requirements are met or exceeded.

NOTE: BS EN 50200 and BS 8434-2 only cove cables of 20 mm in diameter or smaller. The appropriate standard for cables of greater than 20 mm in diameter that tests for fire, mechanical shock and water spray is BS 8491:2008.

13.7.4 Cable colours

All fire system cables should be of a common colour different to that used for the cables of the general electrical installation, the preferred colour being red.

13.7.5 Cable installation

BS 5839 makes the following recommendations for the installation of cables:

1. Cables should be installed without external joints, wherever practicable.
2. Mineral insulated copper sheathed cables and steel wire armoured cables may be used throughout without additional mechanical protection (note: or protection by RCDs).
3. Other cables should be given mechanical protection in any area where physical damage or rodent attack is a risk. Particular attention is to be given to cables installed less than 2 m above floor level.

   NOTE: Mechanical protection includes use of: cable tray, conduit, ducting, trunking and burying in the structure of the building. In many cases the fire-resisting cable is contained within a trunking system. In such cases the cable must be securely fixed to the building structure through the trunking itself.
4. Cable fixings and supports must be such as to match the fire, water and shock withstands specified for the cables.
5. Penetrations of walls, ceilings, floors must be sealed to maintain fire and smoke integrity.
6. Fire system cables shall be segregated from other cables*
7. ELV cables shall be segregated from LV cables.

* The use of separate fire-resistant cables and separate conduits and trunking (or segmented trunking with metal partitions) is appropriate.

NOTE: Clause 37 of BS 5839-1 has good installation practice and workmanship recommendations that the installer will need to follow.

Chapter 14
Inspection and testing

Part 6

Note: (For detailed guidance on inspection and testing, reference should be made to IET Guidance Note 3: *Inspection & Testing.*)

14.1 Initial verification
Chapter 61

14.1.1 General

The purpose of initial verification of an installation is to verify in so far as it is possible by inspection and test:

(i) the general characteristics (see Part 3 of BS 7671);
(ii) that the installation has been carried out in accordance with the design; and
(iii) that the installation has been correctly constructed and is safe.

Many inspection and testing activities cannot be carried out properly if left until completion of the works. This is recognized in Regulation 610.1, requiring that every installation shall during erection and on completion before being put into service, be inspected and tested, so far as is reasonably practicable, to verify that the requirements of the Regulations have been met. 'Put into service' includes use by the builder, subcontractor etc.

A fundamental requirement of initial verification is that precautions shall be taken to avoid danger to persons, and to avoid damage to property and installed equipment during inspection and testing. Inspecting-and-testing personnel must never forget that other people may very well be at work on an electrical installation, or in contact with it, while they are testing, and nothing should be done to put these people at risk. Apart from the routine precautions well outlined in the Regulations with respect to the procedures for test etc., testers must be well aware of the particular hazards that might arise if there are installation errors or if earthing connections, or even neutral connections, are broken without isolating the supply.

14.1.2 Inspection
Section 611

14.1.2.1 Compliance with standards and manufacturers' instructions
Regulation 611.2

Regulation 611.2 requires that an inspection be made of an installation to confirm that all equipment complies with the relevant requirements of the applicable British

Standard or Harmonized Standard. Inspectors should be looking preferably for an approval body mark (see Figure 2.1) or a maker's CE mark and identification of the standard. Where the equipment is not covered by a British Standard or Harmonized Standard, the installer should be asked to confirm in writing what action has been taken to verify compliance with the general requirements.

More recently a requirement has been added for installers to take account of manufacturers' instructions.

14.1.2.2 Correct selection and erection

The inspector of an installation will not generally be expected to verify the design. It is normally expected that the installer/designer provide design data for the installation, so that it can be checked that the installation has been installed as designed. The designer and installer are required to sign the Electrical Installation Certificate where possible, confirming that the installation has been designed and constructed in accordance with the Regulations, and to list any deviations from Parts 3 to 7. Deviations from Part 1, the fundamental principles, are not allowed.

14.1.2.3 Damage

An important element of the verification of new installations is to check that the equipment installed is not visibly damaged or defective in some other way.

14.1.2.4 Inspection lists

Regulation 611.3 lists items for inspection. This list forms the basis of the 'Examples of items requiring inspection during initial verification' schedule given in Appendix 6 of BS 7671.

14.1.3 Testing
Section 612

14.1.3.1 General

The first regulation in Section 612 requires that the tests, where relevant, be carried out in the sequence of Regulation groups 612.2 to 612.6. The measurement of earth fault loop impedance and earth electrode resistance, and functional testing (Regulations 612.7 to 612.13), may consequently be effected out of sequence. Both polarity and earth fault loop impedance testing may very well be carried out with the installation live, as well as certain checks made before the installation is switched on. Before the installation is energized, compliance with the requirements for continuity of protective conductors, insulation resistance, etc. must be verified. Regulation 612.1 gives the tester the necessary reminder that, if an inspection or test identifies a failure to comply, after rectification the relevant parts of the installation shall be inspected and tested again. Even if certain checks of earth loop impedance and polarity have been made on an isolated installation, final checks, such as of loop impedance and polarity of sockets, phase rotation of three-phase machines, need to be made when the installation is live, before handing over to the customer.

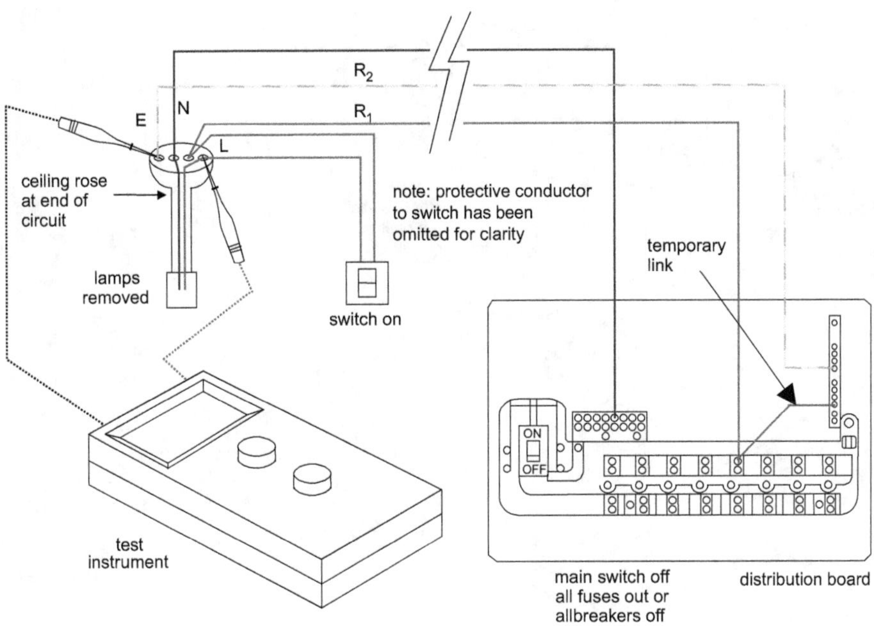

Figure 14.1 Connections for testing continuity of protective conductor, method 1

14.1.3.2 Continuity testing
Regulation 612.2

The regulation has specific requirements for the no-load voltage and short-circuit current (200 mA) of continuity testers. These requirements are included in clauses 4.1 and 4.2 of BS EN 61557-4, as below:

> The following requirements as well as those given in IEC 61557-1 shall apply.
>
> 4.1 The measuring voltage may be a d.c. or an a.c. voltage. The open-circuit voltage shall not exceed 24 V and shall not be less than 4 V.
>
> 4.2 The measuring current within the minimum measuring range according to 4.4 shall be not less than 200 mA (0.2 A).

It cannot be said that a current of 200 mA stresses the installation any more than, say, 20 mA, so neither really verifies that connections have been properly made. However, it is a most necessary test of continuity of all protective conductors, including earthing and main protective bonding conductors. If a connection is open-circuit, this will be shown up by the continuity test. Regulation 612.1 does allow other methods of testing provided that they give valid results.

Two methods of testing the continuity of protective conductors are shown in Figures 14.1 and 14.2. Method 1 is normally used for circuit protective conductors, though method 2 can be used, while method 2 must be used for testing bonding conductors.

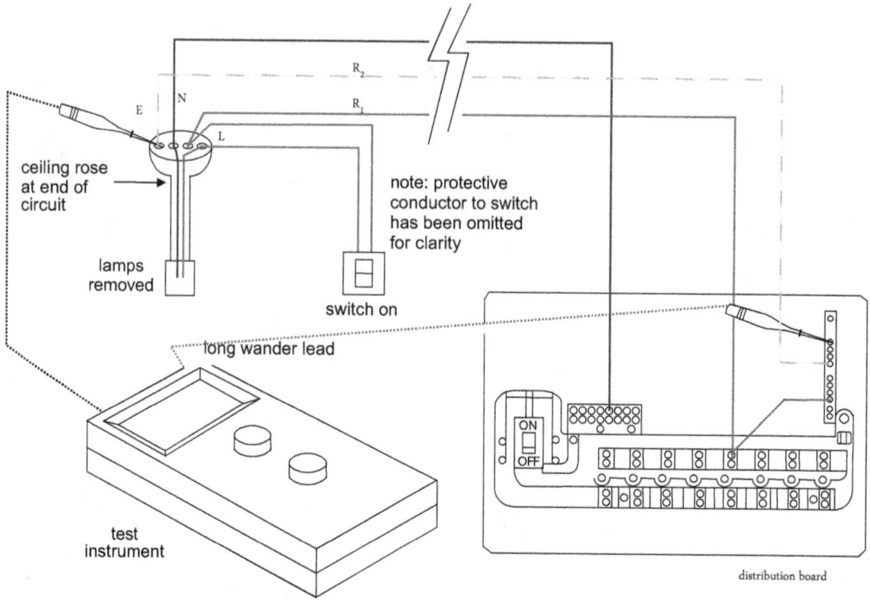

Figure 14.2 Connections for testing continuity of protective conductor, method 2

14.1.3.3 Insulation resistance
Regulation group 612.3

The 16th Edition requirement that the test instrument shall be capable of supplying the test voltage indicated when loaded with 1 mA has been deleted, since the minimum insulation resistance at 500 V is now 1 MΩ and this will draw only 0.5 mA. However, it is included in the requirements of the standard for insulation testers (BS EN 61557-1) as follows:

4.2 The open-circuit voltage shall not exceed 1.25 times the rated output voltage.
4.3 The rated current shall be at least 1 mA.
4.5 The maximum percentage operating uncertainty within the measurement range to be marked or stated shall not exceed ±30% with the measured value as fiducial value.

It is the major criterion for insulation testing that the test voltage be maintained.

A serious concern with reference to insulation resistance is the low insulation resistance of 1 MΩ allowed for installations up to 500 V. The guidance given in IET Guidance Note 3 is that, where an insulation resistance of less than 2 MΩ is recorded, the possibility of a latent defect exists; each circuit should then be tested separately and its measured insulation resistance should be greater than 2 MΩ.

BS 5839-1 *Fire detection and fire alarm systems for buildings,* clause 38.2, requires a minimum insulation resistance of 2 MΩ at 500 V d.c. as follows:

38.2 Recommendations

The following recommendations are applicable.

(a) All installed cables with a manufacturer's voltage rating suitable for mains use should be subject to insulation testing at 500 volts d.c. Prior to this test, cables should be disconnected from all equipment that could be damaged by the test.

(b) Insulation resistance measured in the above test, between conductors, between each conductor and earth, and between each conductor and any screen, should be at least 2 MΩ.

Note: Control and indicating equipment can have fault sensing for wiring insulation resistance to earth. If this is, for example, set at 1 MΩ, the combined effect of all wiring earth insulation resistance ought to be well above this to avoid nuisance fault indications. For large systems, the 2 MΩ minimum needs to be much higher than this to achieve something in excess of 1 MΩ overall. For a small non-addressable system up to about four zones, 2 MΩ might be acceptable.

Regulation 612.3.3 recognizes the possibility of there being electronic devices installed in a circuit, and requires only a measurement between live conductors connected together and the earthing arrangement. To avoid damage to equipment, it is most important that the line and neutral conductors are connected together before carrying out this test is undertaken.

Regulation 612.3.2 Disconnection of current-using equipment

If a line-to-neutral insulation test is to be carried out, prior disconnection of all current-using equipment is necessary, otherwise the test instrument will be short-circuited by the load resistance. There is no particular requirement in the 17th Edition to test appliances. Should an inspector wish to test an appliance, guidance is given in the IET *Code of Practice for In-Service Inspection and Testing of Electrical Equipment.*

14.1.4 Protection by SELV, PELV or by electrical separation Regulation 612.4, Table 61

14.1.4.1 General

The testing of SELV and PELV circuits is carried out at 250 V d.c., with readings in accordance with Table 61.

14.1.4.2 Protection by separation of circuits supplying more than one item of equipment

Regulations 612.4.3, 418.3

Regulation 612.4.3

Where electrical separation is used to supply circuits with more than one item of current-using equipment there are additional inspection and test requirements necessary to verify compliance with the requirements of Regulation 418.3.

14.1.5 Polarity
Regulation 612.6

Regulation 612.6 requires that a test for polarity is made to verify that:

(i) every fuse and single-pole control and protective devices is connected in the line conductor only, and
(ii) except for E14 and E27 (Edison screw) lampholders to BS EN 60238, in circuits having an earthed neutral conductor, centre contact bayonet and Edison screw lampholders have their outer or screwed contacts connected to the neutral conductor, and
(iii) the wiring has been correctly connected to socket-outlets and similar accessories.

The prime requirement is to test the polarity of the supply and, for a three-phase installation, the rotation of the supply. Testing the polarity of a PME supply might be complicated in the rather unusual circumstances of there being a cross-connection in the distribution system. If this is suspected, isolate the supply (line and neutral conductors), remove the main earth and test with a proprietary voltage-detection device; alternatively, contact the distributor.

14.1.6 Protection by automatic disconnection of supply
Regulation 612.8

This regulation simply confirms that for protection by automatic disconnection of supply in TN and TT systems:

(i) an earth fault loop impedance measurement should be made for circuits protected by both overcurrent devices and RCDs;
(ii) the rating of overcurrent devices should be checked and RCDs tested; and
(iii) RCD test instruments complying with BS EN 61557-6 are to be used.

14.1.7 Earth fault loop impedance
Regulation 612.9, Appendix 14

It is necessary to measure the earth fault loop impedance of the supply, Z_e, (Figure 14.3), as well as the earth fault loop impedance at the extremity of the circuits. The measurement of Z_e is important to verify that there is no fault on the electricity supply. The earth fault loop impedance of the supply should be below 0.35 Ω for PME supplies and below 0.8 Ω for a TN-S supply but, in practice, values higher than these can be experienced depending on how far the installation is from the substation. (The impedance (R_B) of a TT supply should be below 20 Ω, but this cannot be easily measured unless the impedance to earth of the installation electrode (R_A) is known – see Figure 3.3.) The test is carried

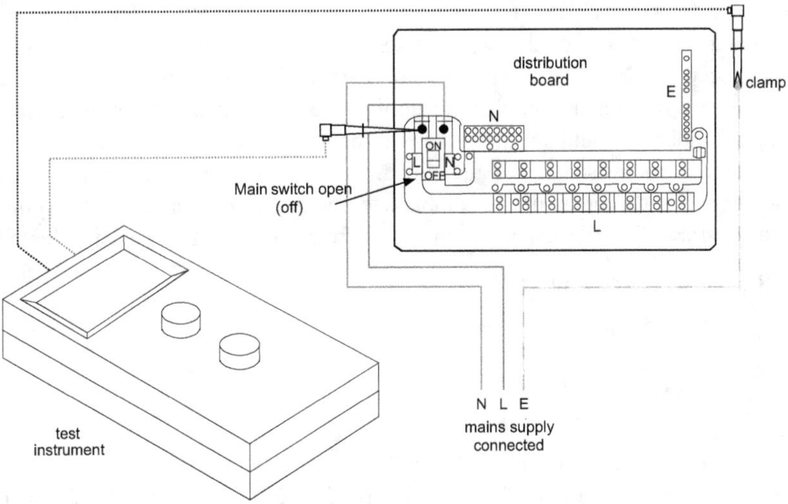

The diagram shows the test probes connected to phase and earth only. Three wire loop impedance testers require a connection to the neutral for the instrument to operate.

Figure 14.3 Measurement of Z_e at the origin

out with the earthing conductor temporarily disconnected, having verified that the method of test is such that no danger can occur. This will generally mean that the supply will need to be isolated before disconnecting the earthing conductor. Regulation 542.4.2 requires the facility to measure the resistance of the earthing arrangements, and means to be provided in an accessible position for disconnecting the earthing conductor. Regulation 543.2.4 does not allow gas, water utility or other services to be used as a protective earth electrode, so the earth electrode of TT supplies must be disconnected from the main bonding before measurement of the earth electrode resistance.

Regulation 612.9 allows the earth fault loop impedance of the installation to be measured or determined by an alternative method. Checks for compliance with the loop impedance requirements of Tables 41.2 to 41.4 for overcurrent devices and Table 41.5 for RCDs can be made by measurement. When overcurrent devices are used, another method would be having knowledge of the external supply impedance Z_e and the measured length of the circuits. The use of standard circuit length is described in Chapter 18. If the loop impedance Z_e is within the range assumed in the design, e.g. less than 0.35 Ω for a PME supply or 0.8 Ω for a TN-S supply, and the circuit lengths are less than the maximum allowed for those supplies, then it can be said that it has been determined that the earth fault loop impedances are satisfactory. Testing of the continuity of the protective conductors is still required (Regulation 612.2.1).

When earth fault loop impedance is checked by measurement, the readings obtained should be compared with the figures given by the designer. If such figures are not available, as almost certainly will be the case when carrying out periodic testing of an installation, the tester will have to determine the reference impedances to be used. Where protective conductors are not of reduced cross-sectional area,

Tables 41.2 to 41.4 of BS 7671 may be used after correcting for temperature. The notes at the bottom of the tables state that the loop impedances given should not be exceeded when the conductors are at their maximum permitted operating temperature. When testing, cables are unlikely to be at their maximum permitted operating temperature and are more likely to be at, say, 10 °C.

Corrections to resistance for the difference in temperature between maximum permitted operating temperature and ambient temperature can be made by correcting to 20 °C using Table C3 and then correcting again from 20 °C to ambient using Table C2.

$$R_{70} = 1.2.R_{20} \text{ from Table C3}$$

$$R_{10} = 0.96.R_{20} \text{ from Table C2}$$

$$\frac{R_{10}}{R_{70}} = \frac{0.96}{1.2} = 0.8$$

This is how the factor of 0.8 in Appendix 14 is calculated.

Then the values in Tables 41.2 to 41.4 of BS 7671 can simply be divided by 1.25 (or multiplied by 0.8). If loop impedances are near the limit, allowance for the supply external loop impedance not being zero can be made as follows:

$$Z_s = Z_e + 1.25(R_1 + R_2)$$

where:

Z_e = external impedance
R_1 = impedance of the line conductor at 10 °C
R_2 = impedance of the protective conductor at 10 °C
1.25 = correction factor from 10 °C to 70 °C.

Table 14.1 BS 88-2 fuses - maximum measured earth fault loop impedances Z_t (in ohms) at 10 °C where the overcurrent device is a fuse to BS 88-2 with 5-second disconnection

Protective conductor csa (mm²)	Device rating								
	20 A			32 A			63 A		
	Z_a	Z_s	Z_t	Z_a	Z_s	Z_t	Z_a	Z_s	Z_t
1.0	2.18	2.80	1.76	0.77	1.75	0.62	NP	0.78	NP
1.5	2.43	2.80	1.96	1.25	1.75	1.0	NP	0.78	NP
2.5	≥3.52	2.80	2.24	1.82	1.75	1.4	0.42	0.78	0.34
4.0	≥3.52	2.80	2.24	2.3	1.75	1.4	0.61	0.78	0.49
6.0	≥3.52	2.80	2.24	≥2.30	1.75	1.4	0.91	0.78	0.62

Notes:

Z_a *is from* Table 11.4.
Z_s *is from* Table 41.4 *of BS 7671.*
$Z_t = Z_a$ *or* Z_s *(whichever is less) ÷ 1.24 to correct to 10 °C.*
NP means that the combination of the protective conductor and the fuse is not permitted.

Now $Z_{test} = Z_e + (R_1 + R_2)$.
Substituting for $(R_1 + R_2)$ from the equation above, we have $Z_{test} \leq \frac{Z_s}{1.25} + \frac{Z_e}{5}$

$$Z_{test} \leq \frac{Z_s}{1.24} + \frac{Z_e}{5}$$

This indicates that the figures obtained by dividing Tables 41.2 to 41.4 by 1.24 are a little conservative and, if Z_e has been measured, an increase can be allowed of $\frac{Z_e}{5}$. (If an ambient temperature of 20 °C is considered the increase is $\frac{Z_e}{6}$.)

If the protective conductor is of a smaller cross-sectional area than the live conductors, then consideration has to be given to the adiabatic limitations given in Tables 11.2 to 11.6 of Chapter 11 of this Commentary. Where the protective conductor is of a reduced size, the neewd to limit the loop impedance, to prevent the conductor overheating under fault conditions, may be more limiting than the loop impedance limitation for shock protection purposes. Table 14.1 gives the limiting loop impedances Z_a (adiabatic) and Z_s (shock) for three BS 88-2 fuse ratings, and a range of protective conductor sizes.

The test value Z_t is corrected to 10 °C. It can be seen by looking at the tables that for a 32 A fuse, and a 1 mm² protective conductor, the loop impedance is severely limited by adiabatic constraints, as it is for 1.0 mm² to 4.0 mm² protective conductors with 63 A devices. As a general rule, for 0.4 s disconnection purposes with standard twin-and-earth cables, the adiabatic loop impedance will not be the limiting factor, but it may well be for 5 s disconnection times. As discussed earlier, a correction can be applied to these figures; Z_t can again be increased by $Z_e/5$.

Table 14.2 Semi-enclosed fuses – maximum measured earth fault loop impedance (in ohms) at ambient temperature where the overcurrent protective device is a semi-enclosed fuse to BS 3036

(i) 0.4 second disconnection (final circuits in TN systems)

Protective conductor (mm²)	Fuse rating			
	5 A	15 A	20 A	30 A
1.0	7.3	1.9	1.3	NP
≥1.5	7.3	1.9	1.3	0.83

(ii) 5 seconds disconnection (find circuits exceeding 32 A and distribution circuits in TN systems)

Protective conductor (mm²)	Fuse rating			
	20 A	30 A	45 A	60 A
1.0	2.3	NP	NP	NP
1.5	2.9	1.9	NP	NP
2.5	2.9	2.0	1.0	NP
4.0	2.9	2.0	1.2	0.85
≥6.0	2.9	2.0	1.2	0.85

Note: NP means that the combination of the protective conductor and the fuse is not permitted.

Note that Z_e must be the actual measured external loop impedance, and not the value assumed for design purposes. The tables for test purposes for the usual devices are reproduced in Tables 14.2 to 14.7. A value of k of 115 from Table 54.3 of BS 7671 is used. This is suitable for PVC insulated and sheathed cables to Table 4, 7 or 8 of BS 6004 and for thermosetting insulated and sheathed cables to Table 3, 5, 6 or 7 of BS 7211. The k value is based on both the thermoplastic (PVC) and thermosetting cables operating at a maximum temperature of 70 °C.

Regulation 434.5.2 and minimum protective conductor size (mm^2)

Regulation 434.5.2 of BS 7671:2008 requires the protective conductor csa meets the requirements of BS EN 60898-1,-2 or BS EN 61009-1 or the minimum quoted by the manufacturer. The values in Table 14.6 are for energy limiting Class 3 type B and C devices only. See section 11.4.2 for the derivation.

Table 14.3 BS88-2fuses–maximummeasuredearthfaultloopimpedance(inohms) at ambient temperature where the overcurrent protective device is a fuse to BS 88-2

(i) 0.4 second disconnection (final circuits not exceeding 32 A in TN systems)

Protective conductor (mm^2)	Fuse rating							
	2 A	4 A	6 A	10 A	16 A	20 A	85 A	32 A
1.0	26.5	12.5	6.2	3.7	1.9	1.3	1.0	0.71
1.5			6.2	3.7	1.9	1.3	1.0	0.80
≥2.5			6.2	3.7	1.9	1.3	1.0	0.80

(ii) 5 seconds disconnection (find circuits exceeding 32 A and distribution circuits in TN systems)

Protective conductor (mm^2)	Fuse rating							
	20 A	25 A	32 A	40 A	50 A	63 A	80 A	100 A
1.0	1.8	1.4	0.77	0.55	NP	NP	NP	NP
1.5	2.0	1.8	0.84	0.84	NP	NP	NP	NP
2.5	2.4	1.8	1.4	1.0	0.56	0.29	NP	NP
4.0	2.4	1.8	1.4	1.0	0.80	0.53	0.23	0.13
6.0	2.4	1.8	1.4	1.0	0.80	0.66	0.36	0.16
10.0	2.4	1.8	1.4	1.0	0.80	0.63	0.44	0.33
16.0	2.4	1.8	1.4	1.0	0.80	0.63	0.44	0.34

Note: NP means that the combination of the protective conductor and the fuse is not permitted.

Table 14.4 BS 88-3 fuses – maximum measured earth fault loop impedance (in ohms) at ambient temperature where the overcurrent protective device is a semi-enclosed fuse to BS 88-3

(i) 0.4 second disconnection (final circuits not exceeding 32 A in TN systems)

Protective conductor (mm^2)	Fuse rating			
	5 A	16 A	20 A	32 A
1.0	7.9	1.84	1.55	0.67
1.5	7.9	1.84	1.55	0.73
2.5 to 16	7.9	1.84	1.55	0.73

(ii) 5 seconds disconnection (find circuits exceeding 32 A and distribution circuits in TN systems)

Protective conductor (mm^2)	Fuse rating					
	20 A	32 A	45 A	63 A	80 A	100 A
1.0	2.13	0.59	NP	NP	NP	NP
1.5	2.57	0.76	NP	NP	NP	NP
2.5	2.57	1.25	0.55	0.24	NP	NP
4.0	2.57	1.25	0.79	0.32	0.19	NP
6.0	2.57	1.25	0.79	0.51	0.29	0.16
10	2.57	1.25	0.79	0.55	0.41	0.26
16	2.57	1.25	0.79	0.55	0.41	0.30

Note: NP means that the combination of the protective conductor and the fuse is not permitted.

Table 14.5 Circuit-breakers – maximum measured earth fault loop impedance (in ohms) at ambient temperature where the overcurrent device is a circuit-breaker to BS 3871 or BS EN 60898 or RCBO to BS EN 61009

For 0.1 to 5 seconds disconnection times (includes 0.4 second disconnection time)

Circuit-breaker type	Circuit-breaker rating (amperes)															
	3	5	6	10	15	16	20	25	30	32	40	45	50	63	100	125
1	14.6	8.74	7.28	4.37	2.91	2.73	2.19	1.75	1.46	1.37	1.09	0.98	0.87	0.69	0.44	0.35
2	8.39	4.99	4.16	2.50	1.68	1.56	1.25	1.00	0.83	0.78	0.62	0.56	0.50	0.40	0.25	0.20
B	11.7	7.05	5.83	3.50	2.33	2.19	1.75	1.40	1.17	1.10	0.87	0.78	0.70	0.55	0.44	0.28
3&C	5.87	3.52	2.94	1.76	1.17	1.10	0.88	0.70	0.59	0.55	0.44	0.39	0.35	0.28	0.18	0.14
D 0.1 to 3sec	2.91	1.76	1.46	0.87	0.59	0.55	0.44	0.35	0.29	0.27	0.22	0.20	0.18	0.14	0.09	0.07
D 5sec	5.87	3.52	2.94	1.76	1.17	1.10	0.88	0.70	0.59	0.55	0.44	0.39	0.35	0.28	0.18	0.14

Table 14.6 Circuit-breakers – minimum protective conductor sizes to meet let-through energy requirements

Energy-limiting Class 3 device rating	Fault level (kA)	Protective conductor csa (mm^2)	
		B type	C type
Up to and including 16 A	≥3	1.0	1.5
	≥6	2.5	2.5
Over 16 up to and including 32 A	≥3	1.5	1.5
	≥6	2.5	2.5
40 A	≥3	1.5	1.5
	≥6	2.5	2.5

Table 14.7 Ambient temperature correction factors

Ambient temperature (°C)	Correction factor (from 10 °C) Notes 1 and 2
0	0.96
5	0.98
10	1.00
20	1.04
25	1.06
30	1.08

Notes:

1. The correction factor is given by: {1 + 0.004 (Ambient temp – 10} where 0.004 is the simplified resistance coefficient per °C at 20 °C given by BS EN 60228 for both copper and aluminium conductors.
2. The ambient correction factor is applied as a multiplier to the earth fault loop impedances of Tables 14.2 to 14.5 if the ambient temperature is other than 10 °C.
   For example, if the ambient temperature is 25 °C the measured earth fault loop impedance of a circuit protected by a 32 A type B circuit-breaker to BS EN 60898 should not exceed 1.16 × 1.06 = 1.23 Ω.

14.1.8 *Earth electrode resistance* Regulation 612.7

For measurement of a low impedance earth, such as for the combined HV/neutral earth of an HV/LV substation, a proper earth electrode resistance tester must be used (see Figure 14.4). For testing the impedance of electrodes used with RCDs, where resistances as high as 100 or 200 Ω are acceptable, a standard earth loop impedance tester may be used with sufficient accuracy. Before the test is undertaken, all protective bonding should be disconnected from the earth electrode to ensure that all the test

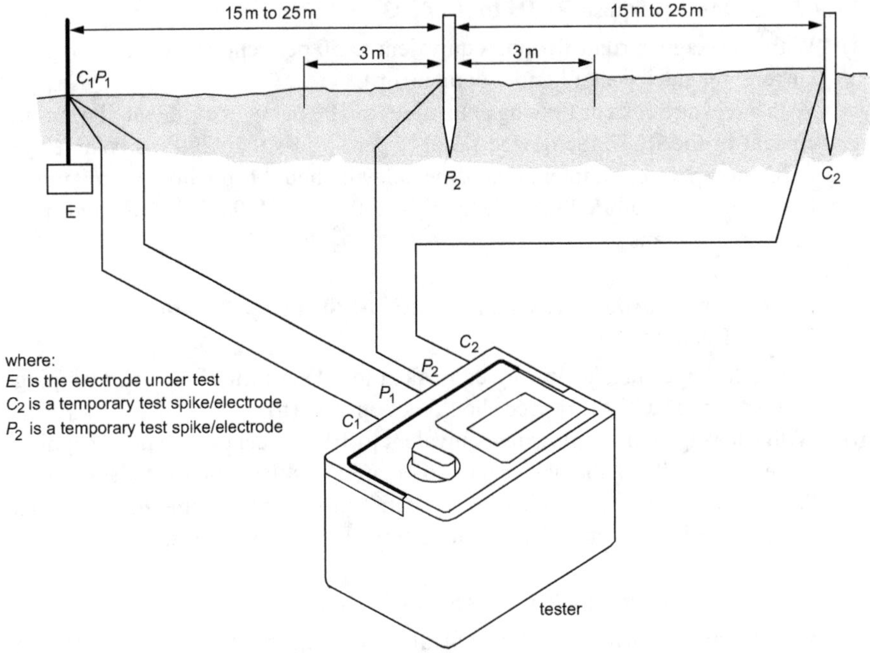

Figure 14.4 Measurement of earth electrode resistance to Earth

current passes through the earth electrode or electrodes only. The loop impedance tester is connected between the line conductor of the origin of the installation and the earth electrode with the test link open, and a test performed. This impedance reading can be treated as the electrode resistance, as the impedance of the supply (say, at most 20 Ω) will be small compared with the impedance of the electrode. During this test the installation supplied by the RCD will not be able to provide fault protection, so the supply to circuits must be switched off.

14.1.9 RCD testing
Regulation 612.10, 612.13.1

Residual current devices, whether for use for additional protection or used for fault protection, are required to be tested using equipment complying with BS EN 61557-6. The appropriate tests would be the type tests of the standard as listed below.

14.1.9.1 General test procedure

The tests are made on the load side of the RCD, as near as practicable to its point of installation, and between the line conductor of the protected circuit and the associated circuit protective conductor. The load supplied should be disconnected during the test.

14.1.9.2 General-purpose RCDs to BS 4293

(i) With a leakage current flowing equivalent to 50 per cent of the rated tripping current, the device should not operate (trip).

(ii) With a leakage current flowing equivalent to 100 per cent of the rated tripping current of the RCD, the device should open in less than 200 ms. Where the RCD incorporates an intentional time delay it should trip within a time range from '50% of the rated time delay plus 200 ms' to '100% of the rated time delay plus 200 ms'.

14.1.9.3 General-purpose RCCBs to BS EN 61008 or RCBOs to BS EN 61009

(i) With a leakage current flowing equivalent to 50 per cent of the rated tripping current of the RCD the device should not operate (trip).

(ii) With a leakage current flowing equivalent to 100 per cent of the rated tripping current of the RCD, the device should open in less than 300 ms unless it is of 'Type S' (or selective), which incorporates an intentional time delay. In this case, it should trip within a time range from 130 ms to 500 ms.

14.1.9.4 RCD protected socket-outlets to BS 7288

(i) With a leakage current flowing equivalent to 50 per cent of the rated tripping current of the RCD the device should not operate (trip).

(ii) With a leakage current flowing equivalent to 100 per cent of the rated tripping current of the RCD, the device should open in less than 200 ms.

14.1.9.5 Additional protection

Where an RCD or RCBO with a rated residual operating current $I_{\Delta n}$ not exceeding 30 mA is used to provide additional protection (against direct contact with live parts), with a test current of 5 $I_{\Delta n}$ the device should open in less than 40 ms. The maximum test time must not be longer than 40 ms, unless the protective conductor potential rises by less than 50 V. (The instrument supplier will advise on compliance.)

14.1.10 Prospective fault current testing
Regulation 612.11

Methods of determining prospective fault current are described in section 6.3.3 of this Commentary. It is preferable if the rating of switchgear is selected on the basis of the maximum declared value of prospective fault current provided by the distributor, attenuated only by the length of service cable installed on the customer's property. This is to be preferred to direct measurement, as fault levels are not constant over time but are dependent on the configuration of the electricity supply network but will not exceed the distributor's maximum declared value.

Where measurements are taken (see Figures 14.5 and 14.6), particular care is needed with those made close to transformers: there have been some questions asked as to the accuracy of earth loop impedance measuring instruments when fault levels are high and when the reactive component of the supply impedance predominates. In such circumstances it would be unreasonable to take other than the design value calculated on the basis of the transformer impedance or the figure declared by the distributor. For three-phase supplies, estimates of three-phase-to-earth fault levels can be made by doubling the single-phase-to-earth values provided by many test instruments.

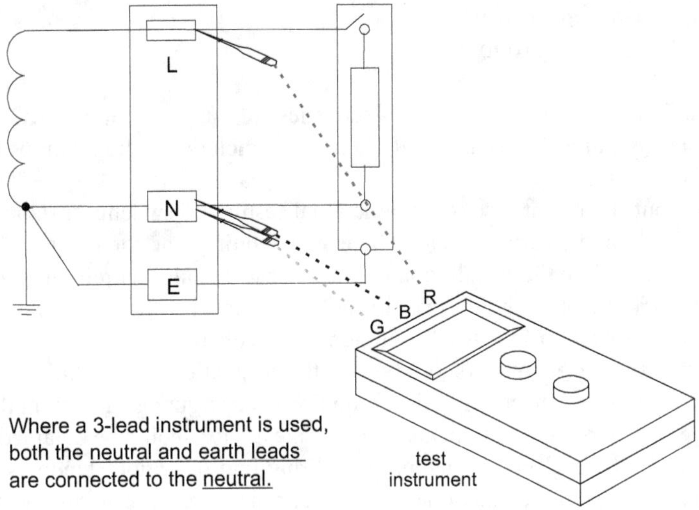

Figure 14.5 Measurement of prospective short-circuit current

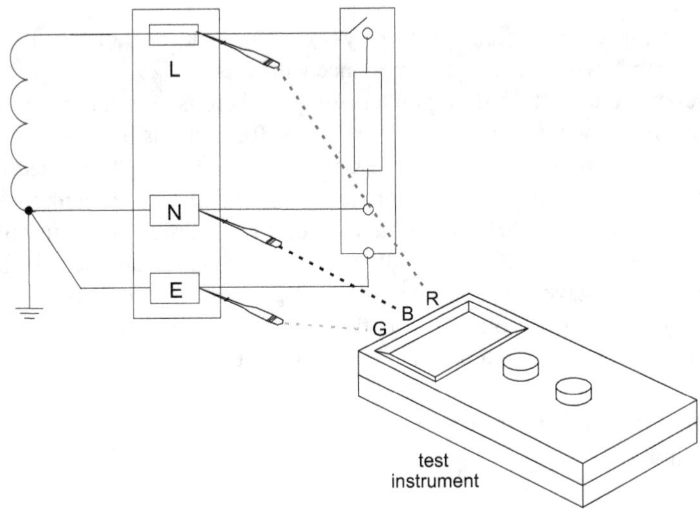

Figure 14.6 Measurement of prospective earth fault current

14.1.11 Check of phase sequence
Regulation 612.12

It is clearly necessary to check phase sequence at all three-phase distribution boards, particularly if rotating machines are to be installed. Confirmation of this process, where applicable, should now be possible on most commercially-available test schedules.

14.1.12 Functional testing
Regulation group 612.13

Regulation 612.13.1 requires the effectiveness of the test button of RCDs to be checked. The requirement to use an RCD tester is included in Regulations 612.8.1 and 612.10.

A final but important test is the functional testing of switchgear, controlgear, drives, controls and interlocks, etc. There is a requirement for practical tests to switch on and operate the installation to ensure that it is all securely mounted, and that it all works before handing over to the customer. The electrician who carries out the final inspection and test in conjunction with the installer and specialist plant-commissioning engineers should switch the installation on, checking for correct operation, phase rotation, etc. Care is required when energizing equipment that may be damaged by incorrect rotation, and appropriate advice should be obtained.

Switching off the electricity supply may bring into operation standby supplies, so it is important that these functional tests be carried out, e.g. fire alarms, emergency lighting, UPSs and standby generation, etc. with the specialist engineer in attendance.

14.1.13 Verification of voltage drop

There is no specific requirement to verify compliance with Section 525. Regulation 612.14 states that, where required, compliance with Section 525 may be verified.

The reference to voltage drop is interesting, as there is no requirement as such to limit the phenomenon. The requirement of the Regulations is to ensure that the voltage at the terminals of any fixed current-using equipment etc. is adequate and such as not to impair the safe functioning of that equipment (Regulations 525.1 and 525.201). See section 9.10 for calculation of voltage drop in large installations. The most practical way of checking voltage is actually to measure the voltage at the locations where voltage drop is expected to be greatest. By measuring voltage at the origin and the farthest point under both no -load and full load, voltage drops can be determined if this is something that needs pursuing.

14.2 Periodic inspection and testing
Chapter 62

14.2.1 The nature of inspection and testing

The purpose of periodic inspection and testing is to determine whether the installation is satisfactory for continued service. Regulation 621.2 requires inspection, comprising detailed examination of the installation, to be carried out without dismantling or with partial dismantling as required, supplemented with the appropriate tests of Chapter 61.

The inspection and testing is, as far as is reasonably possible, to provide for:

(i) the safety of persons and livestock against the effects of electric shock and burns;
(ii) protection against damage to property by fire and heat arising from an installation defect;
(iii) confirmation that the installation is not damaged or deteriorated so as to impair safety; and
(iv) the identification of installation defects and departures from the requirements of the Regulations, which may give rise to danger.

The scope or extent of the work is to be decided by a competent person taking into account the condition of the installation and records, if any, of previous inspections.

There is no requirement as such to carry out all the inspections and all the tests of the initial verification. What is required is inspection comprising detailed examination of the installation without dismantling or with partial dismantling as required, supplemented by testing to determine whether the installation is suitable and safe for continued service. The requirements are general: defects in the protection against shock or protection against fire, or identification of damage and installation defects or non-compliance with the Regulations, which may give rise to danger, will need to be identified and reported.

It is worth referring to the note by the Health and Safety Executive in BS 7671:2008, part of which states:

> Existing installations may have been designed and installed to conform to the standards set by earlier editions of BS 7671 or the IEE Wiring Regulations. This does not mean that they will fail to achieve conformity with the relevant parts of the Electricity at Work Regulations 1989.

What is required of a periodic inspection is not a simple reproduction of the tests in Chapter 61, but a more general approach, which fundamentally is a careful scrutiny looking for potential hazards and deterioration of the installation, supplemented by testing where appropriate. From the age of the installation and its use or abuse, an experienced inspector will know the particular deficiencies to look for. Previous periodic inspection reports will assist and may help identify deterioration of an installation.

14.2.2 Frequency of inspection and testing

Regulation 622.1, Chapter 34, Appendix 6 'Electrical Installation Certificate – Notes:'

The Regulations do not provide specific guidance on the frequency of periodic inspection and testing. Maintainability of an installation is specified in Chapter 34. This requires the designer to make an assessment of the frequency and quality of maintenance that the installation can be expected to receive during its intended life. The designer must plan the installation with an idea of the frequency and quality of the maintenance to be expected, and this must underlie the design. The frequency of inspection is not determined solely by the design, but also by the wear and tear the installation can be expected to receive. An electrical installation in a student's lodging or hotel room could be expected to receive somewhat harsher treatment, and defects less reliably reported, than, shall we say, in a home. An installation in an area open to the public can similarly expect to receive some harsh treatment from time to time, without any reliable reporting of defects.

The designer should initially advise the client of the frequency at which he recommends that the inspection of the installation should take place, and so provide some guidance on the nature of the inspection. It might be more appropriate for a routine inspection to be carried out frequently, perhaps once a year, or every six months without testing, unless defects are identified or reported. Testing is likely to identify faults that are generally associated with ageing and may be hidden, whereas inspections will identify the more likely problems, which are to be associated with breakages or other damage. Frequent inspections could perhaps be supplemented by thorough formal inspection with some dismantling and testing at a longer period.

The Electrical Installation Certificate requires a recommendation to be inserted regarding the initial period before the first periodic inspection and testing, which is endorsed by the designer/installer/tester. This should also be noted on the notice required by Regulation 514.12.1.

The guidance on the initial frequencies of inspection given in IET Guidance Note 3 is reproduced as Table 14.8.

14.2.3 Installations under effective supervision
Regulation 622.2

For installations under effective management system for preventative maintenance in normal use, a regime of continuous monitoring and maintenance program may replace periodic inspection and testing. Competent persons supervising installations in normal use will be able to determine any work (including inspection and testing) necessary to keep the installation in a condition suitable for continued use, i.e. to ensure that the fundamental safety requirements of Regulation 621.2 are met. Records of inspections and tests carried out routinely and following repairs, alterations or additions are to be retained.

14.3 Certification and reporting
Chapter 63, Appendix 6

BS 7671 provides model forms of the Electrical Installation Certificate, Minor Electrical Installation Works Certificate, Electrical Installation Condition Report, and inspection and test schedules. Samples of these are also to be found in Guidance Note 3 and the *On-Site Guide*.

14.3.1 Electrical Installation Certificate

The first page of the certificate requires particulars of the installation. These should have been determined by the designer before the installation reaches the inspection stage, as they are intrinsic to the design of the installation.

Table 14.8 Recommended initial frequencies of inspection of electrical installations

Type of installation	Routine check subclause 3.5	Maximum period between inspections and testing as necessary	Reference (see notes below)
General installation			
Domestic	-	Change of occupancy/10 years	
Commercial	1 year	Change of occupancy/5 years	1, 2
Educational establishments	4 months	5 years	1, 2
Hospitals	1 year	5 years	1, 2
Industrial	1 year	3 years	1, 2
Residential accommodation	at change of occupancy/1 year	5 years	1, 2
Offices	1 year	5 years	1, 2
Shops	1 year	5 years	1, 2
Laboratories	1 year	5 years	1, 2
Buildings open to the public			
Cinemas	1 year	1 to 3 years	2, 6,7
Church installations	1 year	5 years	2
Leisure complexes (excl. swimming pools)	1 year	3 years	1, 2, 6
Places of public entertainment	1 year	3 years	1, 2, 6
Restaurants and hotels	1 year	5 years	1, 2, 6
Theatres	1 year	3 years	2, 6,7
Public houses	1 year	5 years	1, 2, 6
Village halls/Community centres	1 year	5 years	1, 2
Special installations			
Agricultural and horticultural	1 year	3 years	1,2
Caravans	1 year	3 years	8
Caravan parks	6 months	1 year	1, 2, 6
Highway power supplies	as convenient	6 years	
Marinas	4 months	1 year	1,2
Fish farms	4 months	1 year	1,2
Swimming pools	4 months	1 year	1, 2, 6
Emergency lighting	Daily/monthly	3 years	2, 3, 4
Fire alarms	Daily/weekly/monthly	1 year	2, 4, 5
Launderettes	1 year	1 year	1, 2, 6
Petrol filling stations	1 year	1 year	1, 2, 6
Construction site installations	3 months	3 months	1,2

Notes:

1. Particular attention must be taken to comply with SI 2002 No. 2665, the Electricity Safety, Quality and Continuity Regulations 2002 (as amended).
2. SI 1989 No. 635, the Electricity at Work Regulations 1989 (Regulation 4 and Memorandum).
3. See BS 5266-1:2011 *Emergency lighting – Part 1: Code of practice for the emergency escape lighting of premises*
4. Other intervals are recommended for testing operation of batteries and generators.
5. See BS 5839-1:2008 *Fire detection and fire alarm systems for buildings – Part 1: Code of practice for system design, installation, commissioning and maintenance.*
6. Local Authority Conditions of Licence.
7. It is recommended that a caravan be inspected and tested annually if it is used frequently (see Regulation 721.514.1 and Figure 721 – instructions for electricity supply).
8. The person carrying out subsequent inspections may recommend that the interval between future inspections be increased or decreased as a result of the findings of their inspection.
9. Medical locations shall have their isolating transformer equipment inspected and tested for functionality as well as alarms etc.; every third year the output leakage current of the IT isolating equipment shall be measured.
10. The Landlord & Tenant Act 1985 requires that properties under the Act have their services maintained. Periodic inspection and testing is the IET recognised method of demonstrating this.
11. The Management of Houses in Multiple Occupation Regulation (England and Wales).

The model electrical installation certificate in BS 7671 has facilities for certification by the designer, the constructor and those carrying out the inspection and test. This is necessary, as often three different parties carry out the work. This may not be the case in, say, a domestic installation, but is almost certainly the case in a large installation designed by consultants and tendered for competitively. Those carrying out inspection and testing should not sign in respect of construction or design unless they are responsible for them; similarly, the constructor should not sign for the design unless responsible for that design.

14.3.2 Electrical Installation Condition Report form

Those carrying out periodic inspection and testing are required to make recommendations to the client with respect to the results of the inspections and tests. The inspector must consider each of the departures from the current edition and give each departure a code as follows:

- Code 1: Danger present. Risk of injury. Immediate remedial action required;
- Code 2: Potentially dangerous - urgent remedial action required;
- Code 3: Improvement recommended,

If no further advice is provided in BS 7671, the inspector is finally required to decide whether the installation is satisfactory for continued use, and he or she should advise what work, if any, is required to make it so.

Electrical Safety First has prepared a best practice guide, *Electrical installation condition reporting: Classification Codes for domestic and similar electrical installations* (BPG4), in association with, and with the assistance of, other industry bodies. BPG4 may be downloaded from the Electrical Safety First website http://www.electricalsafetyfirst.org.uk/electrical-professionals/best-practice-guides/

The advice given on coding is reproduced below:

Only one of the standard recommendation codes should be attributed to each observation.

If more than one recommendation code could be applied to an observation, only the most serious one should be made (Code 1 being the most serious). In general terms, the recommendation codes should be used as follows:

Code C1 Danger present. Risk of injury. Immediate remedial action required

This code is to be recorded for installations that are a risk, require immediate attention. The situation must be made safe before leaving the site, either by carrying out the repair or isolating the defective part of the installation. Examples include accessible bare live parts, badly damaged equipment with risk of access to live parts, incorrect polarity, arcing found in switchgear.

Code C2 Potentially dangerous – urgent remedial action required

This code is to be recorded for installations that are potentially dangerous but do not require immediate remedial action. Examples include a non-earthed installation (this requires a further fault to manifest injury), fundamentally undersized cables, earth fault loop impedance values greater than BS 7671 requirements, a 'borrowed' neutral, equipment with inappropriately selected IP (this may be C1 if severe), insulation readings under 1MΩ, connections not terminated within appropriate enclosures.

Code C3 Improvement required

This code is to be recorded where C1 or C2 do not apply. Examples include the absence of most warning notices, absence of the required diagrams and charts, no or incorrect marking of conductors at terminations, absence of an RCD specified for additional protection (where the circuit otherwise tests as normal).

The inspector who carries out the periodic inspection and test must indicate on the report when he or she considers the installation should next be inspected. This will obviously depend on the age, condition and use of the installation. The proposed time period should be indicated also on the notice required by Regulation 514.12.1. The existing notice should be suitably amended or replaced during the periodic inspection.

Chapter 15
Special installations or locations

Part 7

15.1 General
Section 700

The introduction to Part 7 is important in that it makes it clear that the requirements of the sections of Part 7 either supplement or modify the general requirements in the remainder of the Regulations, and that, in the absence of a particular reference to the exclusion of any chapter, section or regulation, the corresponding general regulations in the main body of the Regulations are applicable.

The numbering of the regulations in the sections of Part 7 are not sequential. The number of a regulation in Part 7 links that regulation to the particular regulation in Parts 3 to 6 that it modifies. For example, the numbering of Regulation 701.415 identifies that:

- 701: concerns locations containing a bath or shower; and
- 415: modifies 415 of Chapter 41 concerning additional protection.

Similarly the numbering of Regulation 701.411.3.3 identifies that:

- 701: concerns locations containing a bath or shower; and
- 411.3.3: modifies Regulation 411.3.3 of Chapter 41 concerning additional protection by RCDs.

The requirements of Part 7 generally indicate that there is an increased risk, e.g. a reduction in body-contact resistance due to water and lack of clothing, which requires the regulations to be modified.

In each section of this chapter, an attempt is made to summarize the key requirements of the section concerned, and the regulations that give rise to debate are discussed.

The sections of Part 7 have a scope statement, which, as well as providing guidance on the application of the section, in a number of instances gives an indication of the particular risk. For example, the scope of Section 702 'Swimming pools and other basins' notes that in these areas in normal use the risk of electric shock is increased by a reduction in body resistance and contact of the body with Earth potential.

15.2 Locations containing a bath or shower
Section 701

15.2.1 The particular risk

The particular risk associated with locations containing a bath tub or shower is that the risk of electric shock is increased by a reduction in body resistance and contact of the body with earth potential. The reduction in body-contact resistance arises from lack of clothing and the presence of water. In determining the 0.4 s disconnection times for general dry situations (see Chapter 4), an impedance of 1000 Ω for the presence of footwear and floor resistance is added to the body impedance of say 1600 Ω. In bathrooms, swimming pools and similar locations, body impedance is reduced by immersion in water and there is often no footwear and an earthy floor. As a consequence, touch currents are greatly increased and special precautions need to be taken.

15.2.2 Key requirements

(a) The protective measures of obstacles or placing out of reach are not allowed – Regulation 701.410.3.5.
(b) The protective measures of non-conducting location and earth-free local equipotential bonding are not allowed – Regulation 701.410.3.6.
(c) Switches and other means of electrical control or adjustment are not allowed in zones 0, 1 and 2 (shall not be within reach of a person using the bath or shower) apart from specified exemptions – Regulation 701.512.3 (see Table 15.1).
(d) Minimum degrees of protection are specified for equipment in the zones (see Table 15.1).
(e) There are limitations on the installation of current-using equipment (see Table 15.1).

There are four significant changes in this section introduced by the 17th Edition:

(i) All low voltage circuits serving the location and those passing through zones 1 and/or 2 not serving the location are required to be protected by an RCD having a rated residual operating current not exceeding 30 mA.
(ii) Supplementary bonding is not required provided that all the other requirements of the Regulations are met. Particularly itemised are the requirements that:
30 mA RCD protection shall be applied to all circuits of the location and all final circuits complying with the requirements for automatic disconnection, and all extraneous-conductive-parts shall be main bonded.
(iii) 230 V socket-outlets are allowed provided they are at least 3 m horizontally from the boundary of zone 1, i.e. from the edge of the bath or shower basin (and, of course, protected by a 30 mA RCD).
(iv) Accessible metal conduit, trunking or metal sheathed wiring systems are allowed.

Table 15.1 Requirements for equipment (current-using and accessories) in locations containing a bath or shower (Regulations 701.3, 701.5)

Zone	Minimum degree of protection	Current-using equipment	Switchgear and accessories[1]
0	IPX7	Only 12 V a.c. rms or 30 V ripple-free d.c. SELV, the safety source installed outside the zones.	None allowed.
1	IPX4 (IPX5 if water jets)	Items having a nominal voltage not exceeding 25 V a.c. rms or 60 V ripple-free d.c, SELV or PELV, the safety source installed outside the zones. The following mains voltage fixed permanently connected equipment is allowed: whirlpool units, electric showers, shower pumps, ventilation equipment, towel rails, water heating appliances, luminaires.	Only 12 V a.c. rms or 30 V ripple-free d.c SELV switches, the safety source installed outside the zones.
2	IPX4[2] (IPX5 if water jets)	Fixed and permanently connected equipment allowed - general rules apply.	Only switches and sockets of SELV circuits allowed, the source being outside the zones, and shaver supply units complying with BS EN 61558-2-5 if fixed where direct spray is unlikely.
Outside zones	No requirement	General rules apply.	Accessories allowed and SELV socket-outlets and shaver supply units to BS EN 61558-2-5 allowed. Socket-outlets allowed 3 m horizontally from the boundary of zone 1.

Notes:

1. Luminaires installed above zone 1 should be so installed or constructed (e.g. IP44) so as to protect the lamp of the luminaire from splashed or sprayed water.
2. The space under a bath or shower basin is considered to be zone 1 unless the space is only accessible with a tool, when it is considered to be outside the zones.

15.2.3 Discussion
Switches

Regulation 701.512.3 requires that switchgear or accessories shall not be installed in the zones (0, 1 and 2) with some exceptions for SELV circuits. Ordinary light switches can be installed in a bathroom, although they must be outside the zones. While it is always prudent to use pull-cord light switches in a bathroom, whether or not they are accessible to a person using a bath or shower, there are situations where pull-cord switches are not the correct solution. Those with impaired vision or otherwise handicapped may have some difficulty in using pullcord switches, and it may then be appropriate to use a rocker switch outside the zones.

15.2.3.1 Current-using equipment
Regulation 701.55

While only SELV equipment at a voltage not exceeding 12 V a.c. rms or 30 V ripple-free d.c. is allowed in zone 0, i.e. where it may be immersed, in zone 1, specified 230 V fixed current-using equipment meeting the IP requirement is allowed (see Table 15.1). In zone 2 any fixed current using-equipment meeting the IP requirement is allowed.

15.2.3.2 Washing machines

Home laundry equipment such as washing machines may be installed outside zone 2 provided it is supplied by a switched, fused flex outlet (connection unit) installed outside zone 2 or a socket-outlet installed at least 3 m horizontally from the boundary of zone 1.

15.2.3.3 Socket-outlets
Regulation 701.512.3

230 V, 13 A socket-outlets are allowed 3 m horizontally from the boundary of zone 1. While this is unlikely to be able to be applied in most bathrooms, because they are too small, it is applicable in bedrooms with a shower cubicle and in the very large bedrooms in some houses in which a bath is installed. Such a bedroom is a location containing a bath or shower and consequently all circuits serving that location must be protected by an appropriately rated RCD. In most domestic premises this is likely to be as cables installed in most walls or partitions at a depth of less than 50 mm are unlikely to be protected by earthed metal conduit.

15.3 Swimming pools and other basins
Section 702

The risks associated with swimming pools, fountains, hot tubs and other basins are similar to those of bathrooms: the risk of electric shock is increased by reduction in body resistance from the presence of water, including immersion and contact of the body with true Earth.

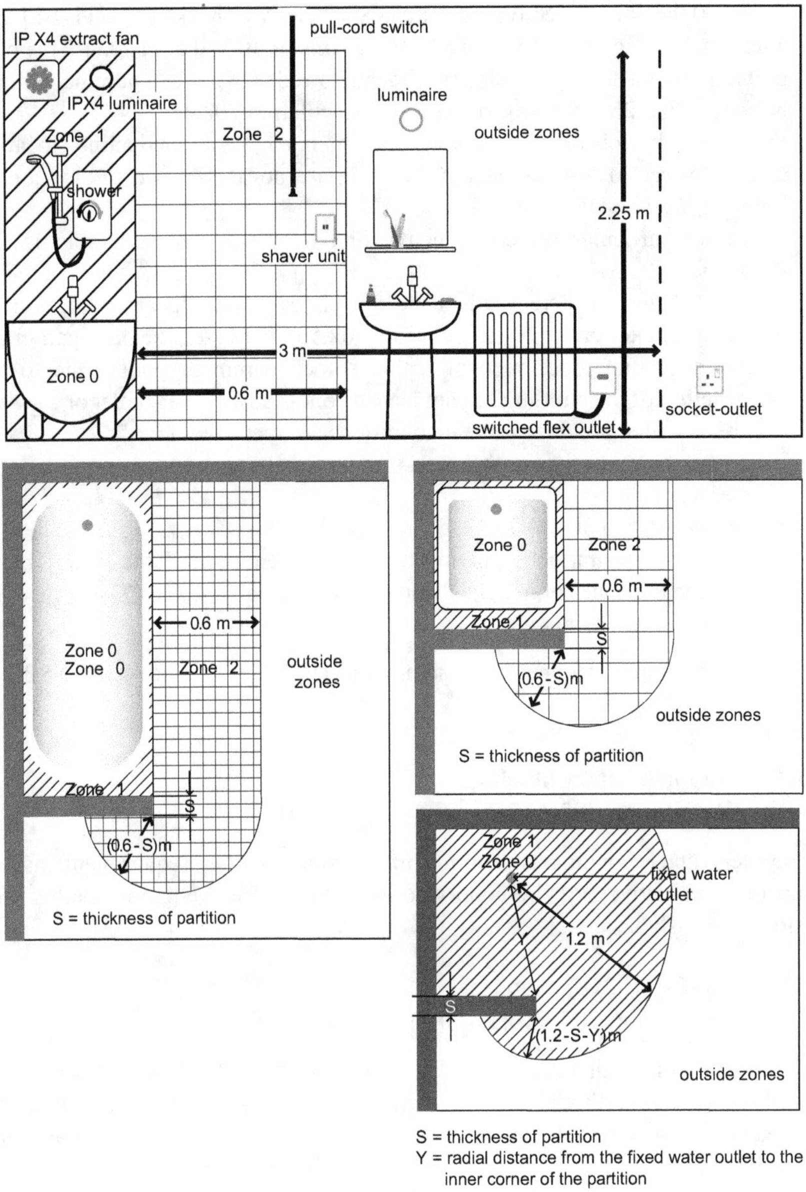

Figure 15.1 Bathroom zones

15.3.1 Key requirements for swimming pools and similar basins

(a) The swimming pool and similar basin surrounds are divided into three zones, 0, 1, and 2 (Regulation 702.32).
(b) Supplementary bonding is required in all zones (Regulation 702.415.2).

(c) In zone 0 the only protective measure against electric shock allowed is SELV at a nominal voltage not exceeding 12 V a.c. rms or 30 V d.c. In zone 1 the only protective measure against electric shock allowed is SELV at a nominal voltage not exceeding 25 V a.c. rms or 60 V d.c. Regulation 702.410.3.4.1.
(d) Protection by obstacles, placing out of reach, non-conducting location and earth-free local equipotential bonding are not to be used in any zone – Regulations 702.410.3.5 and 6.
(e) There are particular IP requirements for all three zones (see Table 15.2 – Regulation 702.512.2).
(f) In zones 0 and 1, there shall be no switchgear, controlgear or accessories including socket-outlets – Regulation 702.53. However the last paragraph of this regulation does recognize that some swimming pools, due to the construction of the building, do not have a zone 2 so a socket-outlet or a switch may be installed provided certain conditions are met – see 15.3.2.7.
(g) In zone 2, socket-outlets are allowed provided that they are:
  - SELV; or
  - protected by a 30 mA RCD; or
  - electrical separation supplying one item of equipment, the source being outside the zones or in zone 2 and protected by a 30 mA RCD and meeting the IP requirements.

The requirements for fountains are summarised in Table 15.3 and section 15.3.2.5.

15.3.2 Discussion

15.3.2.1 Supplementary bonding
Regulation 702.415.2

Supplementary bonding is required in the zones with a requirement that all extraneous-conductive-parts be bonded to exposed-conductive-parts, whether they are simultaneously accessible or not.

15.3.2.2 TN-C-S (PME) supplies
Regulation 702.410.3.4.3(ii)

There is a note with a recommendation that, where a PME facility is used for the installation, an earth electrode or earth mat with a suitably low resistance be installed (and connected to the protective equipotential bonding). This has particular benefits in the event of an open-circuit PEN or if there is a long service cable (see section 11.5.4 for the calculation).

15.3.2.3 Metal grids
Regulation 702.55.1

While there is no particular requirement for a metallic grid in the floor, there is a requirement for electric heating elements in the floor, other than SELV, to have an earthed metallic sheath or be covered by an earthed metallic grid (both with 30 mA RCD protection).

Table 15.2 Requirements for equipment in swimming pools and other basins

Zone	Minimum degree of protection	Current-using equipment	Switchgear and accessories[(1)]
0	IPX8	Only 12 V a.c. rms or 30 V ripple-free d.c. SELV, the safety source installed outside the zones. Only fixed equipment designed for swimming pools. Luminaires to comply with BS EN 60598.	None allowed.
1	IPX4 (IPX5 if water jets)	25 V a.c. rms or 60 V ripple-free d.c. SELV, the safety source installed outside the zones. Fixed equipment designed for swimming pools – Note 1.	Only 12 V a.c. or 30 V ripple-free d.c. SELV switches, the safety source installed outside the zones. A socket-outlet shall not be installed. Note 2.
2	IPX2 for indoor IPX4 for outdoor (IPX5 if water jets)	SELV, the safety source installed outside the zones. Automatic disconnection of supply using a 30 mA RCD. Electrical separation (Section 413), the source for electrical separation supplying only one item of current-using equipment and being installed outside the zones. It is permitted to install the source in zone 2 if its supply circuit is protected by a 30 mA RCD. Fixed, permanently connected equipment allowed. General rules apply.	Sockets and switches permitted.
Outside zones		General rules apply	

Notes:

1. Fixed equipment designed for use in swimming pools and other basins (e.g. filtration systems, jet-stream pumps) is allowed in zone 1 provided it is:
   (i) inside an insulating enclosure providing at least Class II or equivalent insulation and providing protection against mechanical impact of medium severity (AG2); and
   (ii) accessible only via a hatch (or a door) by means of a key or a tool. The opening of the hatch (or door) shall disconnect all live conductors. The supply cable and the main disconnecting means shall be installed in a way which provides protection of Class II or equivalent insulation; and
   (iii) the supply circuit of the equipment is protected by:
   (a) SELV at a nominal voltage not exceeding 25V a.c. rms or 60V ripple-free d.c., the source of SELV being installed outside the zones; or
   (b) a 30mA RCD; or
   (c) electrical separation (Section 413), the source for electrical separation supplying a single fixed item of current-using equipment and being installed outside the zones.
2. For a swimming pool where it is not possible to locate a socket-outlet or switch outside zone 1, a socket-outlet or switch, preferably having a non-conductive cover or coverplate, is permitted in zone 1 if it is installed outside (1.25 m) from the border of zone 0, is placed at least 0.3 m above the floor and is protected by:
   (a) SELV (Section 414), at a nominal voltage not exceeding 25V a.c. rms or 60V ripple-free d.c., the source for SELV being installed outside zones 0 and 1; or
   (b) automatic disconnection of supply, using a 30mA RCD; or
   (c) electrical separation (Section 413) for a supply to only one item of current-using equipment, the source for electrical separation being installed outside zones 0 and 1.

Table 15.3 Requirements for equipment in fountains

Zone	Minimum degree of protection	Current-using equipment	Switchgear and accessories
0	IPX8	SELV, the safety source installed outside zones 0 and 1. Automatic disconnection using a 30 mA RCD. Electrical separation, source supplying one item of equipment and installed outside zones 0 and 1. Mechanical protection to AG2 only removable with a tool. A luminaire shall be fixed and shall comply with BS EN 60598-2-18. An electric pump shall comply with the requirements of BS EN 60335-2-41.	None allowed.
1	IPX4 (IPX5 if water jets)	As zone 0	None allowed
2	IPX2 for indoor locations, IPX4 (for outdoor locations, IPX5 where water jets are likely to occur for cleaning purposes)	There is no zone 2 for fountains	There is no zone 2 for fountains

The metallic sheath or grid must be connected to the supplementary equipotential bonding mentioned in 15.3.2.1 above.

15.3.2.4 Lighting

For swimming pools where there is no zone 2, 230V Class II (all-insulated) AG2 luminaires may be installed on a wall or on a ceiling provided that the circuit is protected by automatic disconnection with the additional protection of a 30mA RCD and the height from the floor is at least 2 m.

This is a relaxation from a minimum height otherwise of 2.5 m.

15.3.2.5 Fountains

Automatic disconnection of supply with a 30 mA RCD is allowed for the protection of equipment in zones 0 and 1. There is no zone 2 for fountains.

15.3.2.6 Ponds and water features

There is no definition of fountain, so it might be reasonable to treat any pool, pond or water feature in which people are not going to immerse themselves as a fountain.

15.3.2.7 Socket-outlets in zone 1

Regulation 702.53 does not allow socket-outlets in zone 0 or 1 unless the size of the surrounds makes it impossible to install a socket-outlet outside zone 1. Then a socket-outlet (and/or switch), preferably having a non-conductive enclosure, is permitted if it is installed 1.25 m outside the border of zone 0, is placed at least 0.3 m above the floor and is protected by:

(i) SELV (Section 414), at a nominal voltage not exceeding 25 V a.c. rms or 60 V ripple-free d.c., the source for SELV being installed outside zones 0 and 1; or
(ii) automatic disconnection of supply (Section 411), using a 30 mA RCD; or
(iii) electrical separation (Section 413) for a supply to only one item of current-using equipment, the source for electrical separation being installed outside zones 0 and 1.

Regulation 702.410.3.4 concerning zones 0 and 1 recognizes that a socket-outlet is necessary to clean the pool area; it is not there for general use, and consequently it must be labelled for use only when the pool is not occupied by persons, but it must be IPX4 or IPX5 protected. However, the advantage of an industrial socket is that is does prevent casual use and discourages use when the pool is in use.

15.4 Saunas

Section 703

15.4.1 Key requirements

(a) There shall be no switchgear, accessories or equipment other than equipment that forms part of the sauna heater installed in zone 2 (Regulation 703.537.5).
(b) There are particular temperature requirements for equipment and cabling (Regulation 703.512.2).
(c) The sauna heating equipment shall be in accordance with BS EN 60335-2-53 (Regulation 703.55).
(d) All low voltage circuits of the sauna must be protected by an ≤ 30 mA RCD.
(e) Fault protection by non-conducting location or earth-free local equipotential bonding is not allowed (Regulation 703.410.3.6).
(f) Basic protection against electric shock by obstacles or placing out of reach is not allowed (Regulation 703.410.3.5).

15.4.2 Discussion

The basic requirement for saunas is that no equipment shall be installed within the sauna area other than that necessary for its operation. Cables and flexes have to be appropriate for the temperatures within the zones.

15.5 Construction and demolition site installations
Section 704

15.5.1 General

The particular risk associated with a construction site is the onerous environment. This can lead to damage to equipment and cables and as a consequence there are particular requirements for shock protection and equipment, including cables.

The other particular problem is the difficulty of bonding all extraneous-conductive-parts to the main earthing terminal and as a consequence there is a particular requirement for TN-C-S systems. Regulation 704.411.3.1 specifically notes that a PME earthing facility shall not be used unless the means of earthing all extraneous-conductive-parts are reliably bonded to the main earthing terminal.

15.5.2 Key requirements

(a) Connecting an installation to a PME earthing facility is not allowed unless all extraneous-conductive-parts are reliably connected to the main earthing terminal (Regulation 704.411.3.1).

(b) Circuits supplying socket-outlets or hand-held equipment up to a rating of 32 A must be protected by either:

   (i) reduced low voltage (with 5 s disconnection); or
   (ii) automatic disconnection of supply with additional protection by a ≤30 mA RCD; or
   (iii) electrical separation, each socket-outlet and item or hand-held electrical equipment being supplied by an individual transformer or separate winding of a transformer; or
   (iv) SELV or PELV (Regulation 704.410.3.10).

(c) Circuits supplying socket-outlets with a rating equal or greater than 32 A are required to be protected by a 500 mA RCD (Regulation 704.411.3.2.1).

(d) A plug and socket-outlet with a rating exceeding 16 A are required to comply with BS EN 60309-2 (Regulation 704.511.1).

(e) All switchgear assemblies are required to comply with BS EN 60439-4 (Regulation 704.511.1).

(f) There are particular requirements for the cables used on the site (Regulation 704.52).

15.5.3 Notable changes made by the 17th Edition

(a) There is no longer a requirement to use SELV hand lamps and reduced low voltage (RLV) for supplies to socket- outlets and hand-held equipment. However, two notes to Regulation 704.410.3.10 advise that RLV is strongly preferred for the supply to portable handlamps for general use and portable hand tools and local lighting up to 2 kW, and a SELV system is strongly preferred for portable handlamps in confined damp locations.

However automatic disconnection of supply with additional protection by means of a ≤30 mA RCD is allowed.

This brings the requirements in the UK into line with practice on the Continent. See Health and Safety Executive Guidance HSG141 *Electrical safety on construction sites*.

(b) There are no particular requirements for socket-outlets up to 16A.
(c) Reduced disconnection times are no longer required.

15.5.4 Discussion
PME supplies

The Energy Networks Association Engineering Recommendation G 12/4, *Requirements for the application of protective multiple earthing to low voltage networks*, advises that it is usually impractical to comply with the bonding requirements of the statutory ESQCR and, consequently, that a PME earthing terminal (TN-C supply) should not be provided.

This means that generally a TT system has to be adopted. However, for large sites with their own transformer, as advised in G12/3, an earthing terminal connected directly to the neutral will usually be possible (TN-S system).

13 A socket-outlets

13 A plugs and sockets to BS 1363 may be used; however, they will need to be suitable, i.e. robustly water-protected (say IP44).

15.6 Agricultural and horticultural premises
Section 705

15.6.1 General

The requirements for agricultural and horticultural premises do not apply to those locations and areas for household use such as farmhouses, restrooms, perhaps workshops. The particular risks perceived are those due to damage of equipment by livestock, damage to equipment as a result of the harsh environment and nature of the activities undertaken, a risk of fire and difficulty in bonding all extraneous-conductive-parts.

15.6.2 Key requirements

(a) In locations intended for livestock, all accessible (to livestock, remembering herding effect) exposed-conductive-parts and extraneous-conductive-parts shall be supplementary bonded. (Regulation 705.415.2.1).

(b) Socket-outlet circuits with a rating up to 32 A are required to be protected by a ≤30 mA RCD (Regulation 705.411.1).
(c) Socket-outlet circuits with a rating of more than 32 A are required to be protected by a ≤100 mA RCD (Regulation 705.411.1).
(d) All other circuits to be protected by a ≤300 mA RCD (Regulation 705.411.1).
(e) Electrical equipment generally shall be inaccessible to livestock (Regulation 705.513.2).
(f) A minimum degree of protection of IP 44 is required (Regulation 705.512.2).
(g) The electrical installation of each building or part of a building shall be isolated by a single isolation device (Regulation 705.537.2).
(h) Isolation generally shall include all live conductors (Regulation 705.537.2).
(i) There are requirements for the supply to animal life-support systems, including standby generation and alarms (Regulation 705.560.6).

15.6.3 Discussion

15.6.3.1 Fire

The requirement to protect all circuits with an RCD meets the requirement for devices for protection against the risk of fire in Section 532.

The electrical heating appliances commonly used for the breeding and rearing of livestock are required to comply with the European standards and to be fixed so as to maintain a sufficient distance from livestock and combustible materials. For radiant heaters the clearance required is not less than 0.5 m, or that distance recommended by the manufacturer. Regulation 705.422.6.

15.6.3.2 PME supplies

The Energy Networks Association Engineering Recommendation G12/3 makes particular recommendations for farms and horticultural premises (see below), a typical installation arrangement being shown in Figure 15.2.

15.6.3.3 Farms and horticultural premises

Where in remote buildings all extraneous conductive parts cannot be bonded to the earthing terminal, the pipes and metalwork of isolated buildings, whether or not they have an electricity supply, shall be segregated from metalwork connected to the PME earthing terminal. Any supplies to such buildings should be controlled by a residual current operated device and the associated earth electrode and protective conductor shall be segregated from any metalwork connected to the PME earthing terminal.

Where segregation is not possible then the alternative of using suitable earth electrodes and rods for the whole of the installation should be considered. Alternatively if a dedicated transformer is used to supply the premises then protective neutral bonding (PNB) may be used.

Particular care must be taken in areas where livestock are housed as they are sensitive to very small voltages. A suitable metallic mesh shall be installed in the concrete bed of a dairy and bonded in accordance with the PME requirements.

If PME is to be applied to an existing dairy the steel reinforcement in the floor should be bonded. Alternatively if small voltage differences are unacceptable the area concerned should be protected by an RCD and the associated earthing system segregated electrically from the remainder of the installation.

NOTE: If PME is to be used and the steel reinforcing mesh of the concrete cannot be bonded or does not exist the customer must be advised that in the case of dairies the small voltage differences referred to above may adversely affect livestock feeding at milking and also milk output. For details consult the current edition of BS 7671.

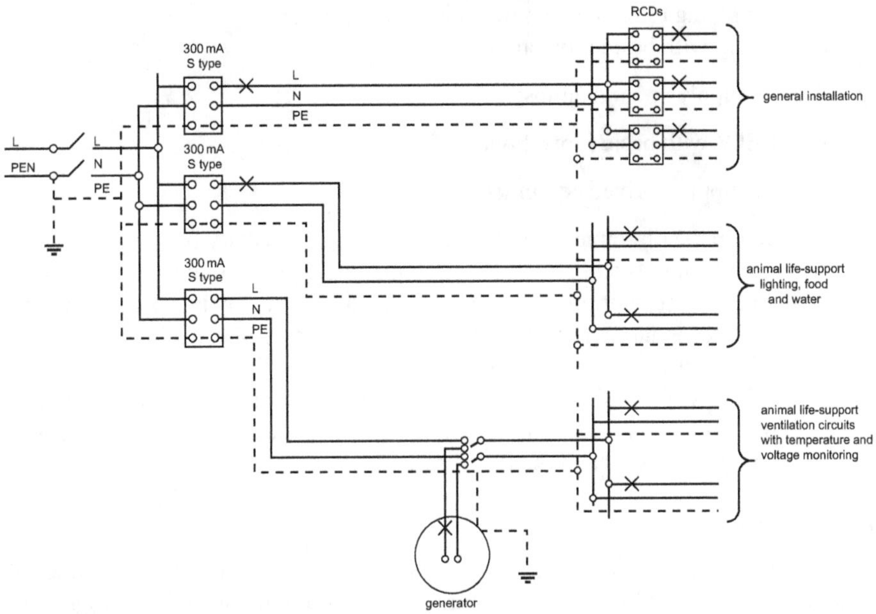

Figure 15.2 Agricultural and horticultural premises schematic

15.6.3.4 The farm environment

The farm environment is very harsh. Not only is there a considerable risk of mechanical damage, but the use of fertilisers and presence of urine and often wet locations can result in severe corrosion. Metal conduit or trunking is often not suitable for such locations and the use of non-metallic wiring enclosures may be preferred.

15.7 Conducting locations with restricted movement
Section 706

15.7.1 General

Conducting locations with restricted movement are particularly dangerous. The conductive location means that any person within the location will have a low resistance to Earth (and high touch currents). The restricted or confined location will mean that persons may be hot and sweaty, decreasing body impedance, and the restrictive nature of the location might prevent escape from any hazard.

15.7.2 Key requirements

Protective measures are specified for the supply to current-using equipment.

(a) For the supply to hand-held tools and mobile equipment:

- (i) electrical separation, the source being located outside the location and supplying only one item of equipment; or
- (ii) SELV with basic protection.

(b) For the supply to hand lamps:

- (i) SELV with basic protection.

(c) For the supply to fixed equipment:

- (i) automatic disconnection of supply with supplementary bonding; or
- (ii) use of Class II equipment plus ≤30 mA RCD protection; or
- (iii) electrical separation with one piece of equipment only being connected to each secondary winding, with the sources located outside the location; or
- (iv) SELV with basic protection; or
- (v) PELV with basic protection and with equipotential bonding within the conducting location of all exposed-conductive-parts, extraneous-conductive-parts and the PELV earth connection.

15.7.3 Discussion

Conducting locations with restricted movement would include the inside of metal tanks or boilers; by their very nature such locations are inherently dangerous. Where functional earths are required for instruments, equipotential bonding is required to the functional earth (Regulation 706.411.1.2).

15.8 Caravan/camping parks
Section 708

15.8.1 General

The particular risk associated with caravan parks is the accessibility of the metal body of the caravans to a person outside the caravan and in contact with Earth. As a result they merit special attention in the ESQCR (see section 15.8.3). Connection by a flexible cable via a plug and socket is less reliable than the connection of a fixed installation.

15.8.2 Key requirements

(a) The protective measures of obstacles and placing out of reach are not allowed (Regulation 708.410.3.5).

(b) The protective measures of non-conducting location and earth-free local equipotential bonding are not allowed (Regulation 708.410.3.6).
(c) For PME supplies, the protective conductor of all pitch socket-outlets used to supply caravans and tents must not be bonded to the PME terminal, and a TT system installed (Regulation 708.553.1.14).
(d) Each pitch socket-outlet shall be protected individually by a ≤30 mA RCD (Regulation 708.553.1.13).
(e) Each pitch socket-outlet shall be protected individually by an overcurrent device (Regulation 708.553.1.12).
(f) Caravan pitch equipment shall preferably be supplied by underground cables (Regulation 708.521.1).
(g) Underground cables shall be installed at a depth of at least 0.6 m and unless provided with additional protection shall be installed outside any caravan pitch area (Regulation 708.521.1.1).
(h) If overhead conductors are used, particular requirements with respect to clearance etc. are to be met (Regulation 708.521.1.2).

15.8.3 Discussion

PME supplies

Regulation 9(4) of the Electricity Safety, Quality and Continuity Regulations 2002 (ESQCR) requires a distributor not to provide a PME earth to any metalwork in a caravan or boat. The caravan installation must form part of a TT system. The particular hazard anticipated by the ESQCR is the loss of the supply neutral, and the voltage difference between a loaded PEN conductor and true Earth. A person standing outside a metal caravan would be particularly at risk. Figures 15.3 and 15.4 illustrate typical site distribution arrangements.

Regulation 9(4) applies to the caravans, so the earth connection of any buildings on the caravan site can be connected to the neutral/earth connection of the supply. For large sites the distributor could provide a TN-S supply.

The advice in Engineering Recommendation G12/4 is as follows:

> **6.2.6.1 Caravans**
>
> The Electricity Safety, Quality and Continuity Regulations 2002, as amended, preclude the provision of a PME earthing terminal to a caravan. Supplies to caravans should be from a TT system utilising a separate earth electrode segregated from the PME earth and protected by an RCD which must be provided by the consumer or site owner.
>
> **Note:** This does not preclude the provision of PME to mobile homes which are not of caravan construction (See 6.2.7)
>
> **6.2.6.2 Caravan sites, campsites and amenity shower blocks**
>
> Due to the higher probability of persons being barefooted at toilet and amenity shower blocks (including sports pavilions with shower facilities) the extension of PME earthing is not recommended unless a buried bonded metal

grid has been installed. This requirement also applies to other locations where toilet or shower blocks have been provided for general public use where people are likely to be barefoot e.g. beachside locations, parks etc.

Where outside showers have been provided, provision of PME earthing is not recommended as providing a reliable equipotential cage may prove impractical.

Where no shower area exists nor is likely to exist in a sports pavilion, PME may be offered provided the appropriate metalwork is bonded and due consideration is given to the construction of the building, i.e. wooden or brick construction.

Note: This does not preclude a PME earthing terminal being provided for use in permanent buildings on a caravan site such as the site owner's living premises and any bars or shops.

Supplies to caravans and boats should be two-wire line and neutral supplied through an appropriately rated RCD which must be provided by the customer or site owner. This method of supply is also recommended for toilet or amenity blocks. An independent earth electrode is required (see Figure 4 for the recommended method of giving a supply to a caravan site).

15.9 Marinas

Section 709

15.9.1 The particular hazards

Installations in marinas and pleasure craft are characterized by the risk of corrosion, movement of jetties, mechanical damage due to moving of the craft and the jetties, presence of fuel and the generally arduous environment. The risk of electric shock is

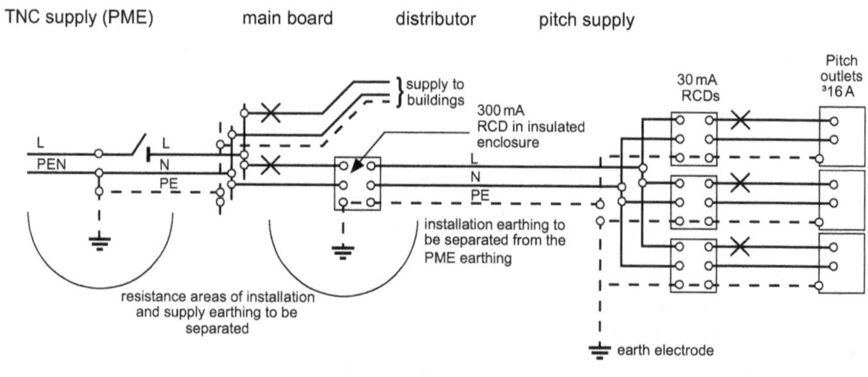

Figure 15.3 Typical caravan site distribution for a PME supply, with separation from PME earth at the main distribution board

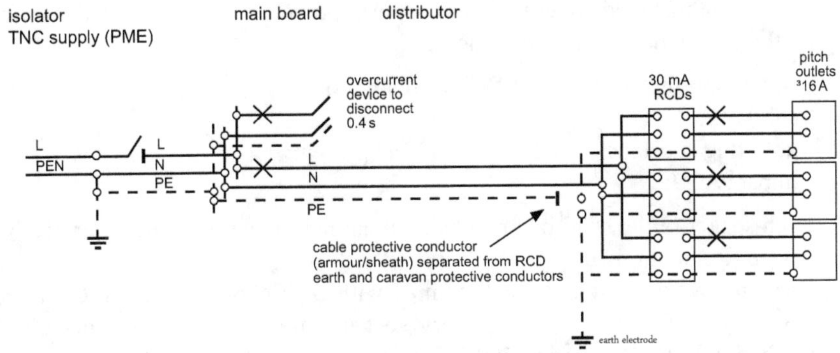

Figure 15.4 Typical caravan site distribution for a PME supply, with separation from PME earth at pitch supply point

increased by a reduction in body resistance and good contact with Earth due to the presence of water and the nature of the leisure activity.

15.9.2 Key requirements

(a) Metalwork in boats are not permitted to be supplied from a PME supply (see Regulation 9(4) of the ESQCR). In the case of TN systems, only TN-S systems may be used to supply ‘pleasure craft and houseboats’. TN-C-S can be used for buildings in the marina (offices, dwellings, shops etc.). In view of the general nature and size of many marinas, it may be practicable to install a TN-S system. This obviously is reasonable if the marina is of such a size that it has its own transformer (Regulation 709.411.4).

(b) The protective measures of obstacles, placing out of reach or non-conducting location must not be used (Regulation 709.410.3.5,.6).

(c) ≤30 mA RCDs shall be installed to protect each boat supply socket-outlet individually, and be able to disconnect all poles, including the neutral (Regulation 709.531.2).

(d) The following wiring systems, among others, are suitable for distribution circuits of marinas:

  (i) copper cables with thermoplastic or elastomeric insulation and sheath installed; within an appropriate cable management system taking into account external influences such as movement, impact, corrosion and ambient temperature or

  (ii) mineral-insulated cables with a PVC covering; or

  (iii) cables with armouring and serving of thermoplastic or elastomeric material; or

  (iv) other materials no less suitable (Regulation 709.521.1.4).

(e) Cable management systems (conduit and trunking) shall have suitable apertures for drainage (Regulation 709.521.1.6).

(f) Equipment installed on jetties, wharfs, piers, pontoons etc. shall be suitable for all the following external influences:

water: splashes IPX4, jets IPX5, waves IPX6;
solid objects IP3X;
polluting substances AF2, or AF3 if fuel present;
impact AG2.

(g) Each socket-outlet shall be provided with an individual overcurrent protective device (Regulation 709.533).
(h) Boat supply socket-outlets shall comply with BS EN 60309-2 up to 63 A and BS EN 60309-1 above 63 A. Every socket-outlet shall meet the degree of protection IP44 (Regulation 709.553.1.8).

15.9.3 Discussion

Marinas require particularly specialist knowledge. Many of the practical precautions that need to be taken are not included within the specific electrical requirements of Section 709. The chapter 'Marinas and similar locations' in IET Guidance Note 7: *Special Locations* contains further advice. Particular care must be taken in the selection of cables. Some of the notes from Guidance Note 7 are included below:

Wiring systems
Notes

(1) Cables should be installed in locations where they are protected from physical damage and wherever practicable out of water.
(2) Many cable types including PVC insulated and sheathed cables are not suitable for continuous immersion in water. The suitability of the cable types should be checked with the manufacturers. Floating pontoons are usually manufactured with a service void in them, enclosed and accessible from above, to accommodate cables and water piping.
(3) Fixed cables installed permanently under water at a depth of more than 4 m will normally need to be metal sheathed, e.g. lead. Fixed cables not permanently immersed or at a depth of less than 4 m should be armoured and incorporate extruded MDPE (medium density polyethylene) outer sheath.
(4) Due to the possibility of corrosion, the galvanized steel armouring of cables must not be used wholly or in part as a circuit protective conductor (cpc) on the floating section of marinas. A separate protective conductor should be used which, when sized in accordance with Regulation 543.1.2, can be common to several circuits if necessary. The armour must still, however, be connected to protective earth.
(5) Protective bonding connections must be single-core PVC insulated to BS 6004 (HAR reference – H07V-R and H07Z-R), or BS 6007 (flexible type), or with an oversheath or further mechanical protection as applicable to the particular location.

(6) Conductor colour coding should be in accordance with the requirements of BS 7671 Table 51. Terminations should be protected against corrosion either by the selection of suitable materials or covering with grease or water-resistant mastic or paint.
(7) Care should be exercised when installing cables to prevent damage from abrasion due to movement between pontoon sections, etc. Cables must be adequately fixed, protected and supported, and if necessary cable types suitable for the flexing movement must be used.
(8) Where cables are installed at onshore locations due consideration should be given to the routing, depth of lay and protection, especially where heavy traffic and point loads are experienced. Cables should normally be laid above the water table, or cable types suitable for continual immersion used. It is not usually practicable for buried cable duct systems to be made totally watertight. The watertight termination of ducts into drawpits and cable trenches below switchboards is also difficult to achieve.

Distribution boards, feeder pillars and socket-outlets

Note: While 2 P & E and 4 P & E plugs and sockets are generally used, other configurations may be necessary as in the case of special security circuits indicating unauthorized use of particular socket-outlets on remote monitoring systems.

[63 A socket-outlets, or larger, are required for some craft. Where necessary, 63 A sockets and greater should have a pilot isolating circuit or mechanical interlock in order to ensure that the load is disconnected before the plug is inserted or withdrawn. BS 7671 allows a plug and socket-outlet to be used as an on-load isolator only for loads up to 16 A.]

General notes

1. Pontoon amenity lighting

It is important that the routes of pontoons and their termination points are clearly delineated. The lighting may be controlled by either automatic photoelectric cells or time switches, the former being preferred as they sense poor conditions caused by fog, etc. when natural light is waning. Luminaires should be of rugged and watertight construction and should preferably be mounted at low level with the light source facing the walkway, not omnidirectional.

2. Navigation lighting

The local waterway authority should be consulted in order that all necessary and suitably coloured navigation lighting is provided. The light sources should have an extended life expectancy. Photoelectric cell control is preferred to time switches.

3. Fuelling stations

The relevant local authority should be consulted in order to ensure that the completed installation complies with its requirements. Where applicable special emergency control facilities should be established onshore. Fuel hoses are required to be non-conducting. Ship/shore bonding cables are not to be used – see *The International Safety Guide for Oil Tankers and Terminals*. Electrical equipment in the proximity of fuelling stations should comply with APEA/IP *Guidance for the Design, Construction, Modification, Maintenance and Decommissioning of Filling Stations*.

4. Metering systems

Metering systems are outside the scope of this guidance note and must be agreed between the designer and the marina owner to provide all necessary electricity consumption information for accurate billing. The meters may be required to be installed locally in the feeder pillars for local direct reading, or may be part of a site wide data network system. The metering system must be fit for the installation and type of use. Functional and safety earthing must be adequate.

Check metering for the various main sections of the distribution system may be required in order that the marina operator can use this data in establishing tariffs for the resale of electricity. Such equipment must be installed within the main switchgear and feeder pillars, and must be of adequate rating and quality for the duty required. The increased use of items of electrical equipment exhibiting low power factor characteristics, e.g. dehumidifiers, refrigerators, battery chargers, etc., requires that electricity metering should record suitable data to ensure the marina operator does not suffer a loss of revenue. (This particularly applies when kVAh metering is installed by the electricity distributor.)

5. Location of equipment

Due consideration should be given to the location of items of equipment so that they are, as far as practicable, not vulnerable to damage either on or offshore at the marina. In the case of onshore areas there will be the need for clear vehicular movement including large mobile boat hoists, transit lorries and cars, etc. The location therefore of feeder pillars and lighting columns requires special attention. For marina areas, the lighting columns and power supply feeder pillars should be so positioned that the risk of contact with luggage trolleys etc. and such items as the bowsprit of craft are, as far as practicable, reduced to a minimum. This is particularly important where lighting and power supply equipment has moulded enclosures which are unable to withstand such mechanical forces and impact and may be damaged.

Site investigations should be carried out at an early stage to determine maximum wave heights which can be experienced. This is of particular importance in exposed coastal sites. Where marinas have breakwater type pontoons, it is likely that under certain conditions waves will pass over the structure.

6. Routine maintenance and testing

Initial inspection and testing of all electrical systems should be carried out on completion of the installation, in accordance with the requirements of Part 6 of BS 7671, the recommendations of the IET's Guidance Note 3 and this guidance note. A periodic inspection and test of all electrical systems should be carried out annually and the necessary maintenance work implemented. If the site is considered to be exposed, or operational experience shows problems (i.e. misuse), the inspection frequency should be increased to cater for the particular conditions experienced.

All RCDs should be tested regularly by operating the test button and periodically by a proprietary instrument to ensure they conform with the parameters of their relevant product standards e.g. BS 4293, BS EN 61008.

All tests should be tabulated for record purposes and the necessary forms required by Part 6 of BS 7671 must be provided by the contractor or persons carrying out the inspection and tests to the person ordering the work.

15.10 Exhibitions, shows and stands
Section 711

15.10.1 Introduction

The requirements in Section 711 apply to the electrical installations to exhibitions and shows both erected on a greenfield site and also those exhibitions, shows and stands erected within an exhibition building. The requirements do not apply to the exhibition building itself.

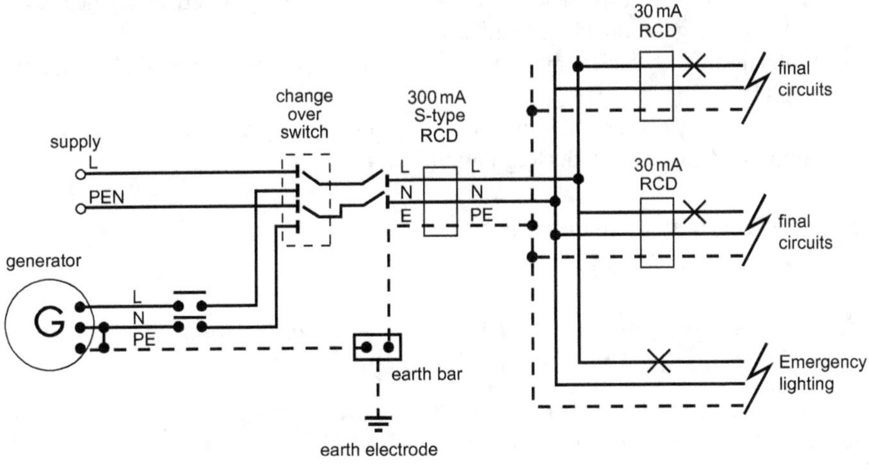

Figure 15.5 Typical show installation

15.10.2 Particular risks

The particular risks associated with exhibitions, shows and stands are those of electric shock and fire. The risk of electric shock is increased because users may be in contact with true Earth and equipment, particularly cables, are at risk of mechanical damage. The lack of permanent structures and the temporary nature of the works puts equipment at risk.

Risks are also increased by access of the public.

15.10.3 Key requirements

(a) Distribution cables are to be protected at its origin by a ≤300 mA RCD, time delay or S-type (Regulation 711.410.3.4).
(b) Each socket-outlet not exceeding 32 A and all final circuits other than emergency lighting are required to be protected by ≤30 mA RCDs (Regulation 711.411.3.3).
(c) Accessible metalwork of the stands (vehicles, caravans, containers, etc.) is to be connected to the main earthing terminal of the unit or stand (Regulation 711.411.3.1.2).
(d) Except for an installation within a building, a TN-C-S system cannot be installed; if a PME supply is provided, the installation must be TT (Regulation 711.411.4). See Figure 15.5.
(e) Armoured cables or cables protected against mechanical damage shall be used wherever there is a risk of mechanical damage (Regulation 711.52).
(f) Where no fire alarm system is installed in the building used for exhibitions etc. the cable systems shall be flame-retardant and low-smoke, or enclosed in metallic or non-metallic conduit or trunking providing fire protection and ingress protection of at least IP4X (Regulation 711.521).
(g) Every separate temporary structure, such as a vehicle, stand or unit shall be provided with its own readily accessible and properly identifiable means of isolation (Regulation 711.537.2.3).
(h) Protection by obstacles and placing out of reach is not permitted (Regulation 711.410.3.5).
(i) Protection by non-conducting location and earth-free local equipotential bonding is not permitted. (Regulation 711.410.3.6).

15.11 Electric vehicle charging installations
Section 722

15.11.1 The risk – Open circuit PEN conductor in the supply
Regulation 722.411.4 TN installations

The risk that is of concern for electric vehicle charging is an open-circuit of the supply PEN conductor (see sections 11.5.3 and 11.5.4). Precautions are required for outdoor installations, to keep the voltage between the main earthing terminal of the installation and earth below 70V in the event of an open circuit supply PEN conductor.

The ESQC Regulations do not permit a PME supply to be made available to a caravan or boat because of this concern.

15.11.2 Domestic installations

The requirements of Section 722 can be met by making all charging point installations part of a TT system. However, a TT installation would need to be separated from the house TN installation (or the house converted to TT as well) and an earth electrode installed. The concern is that the very reliable multiply earthed TN-C-S system with low touch voltages is exchanged for a TT system with high touch voltages and dependent upon a sensitive device (the RCD). The IET commissioned the Health and Safety Laboratory to calculate and compare the risks of the two approaches and to make recommendations.

The main findings of the report "Charging of Electric Vehicles at Domestic Dwellings Using Protective Multiple Earthing (PME) – Risk Analysis PE01417/2012/1" are copied below as they are not definitive. The UK National Committee had some difficulty in applying the results.

Main Findings

The increase in individual risk of a person being seriously injured or killed associated with permitting the use of PME supplies for charging electric vehicles outside domestic dwellings was assessed to be within the range 1.1×10^{-6} *to* 7.8×10^{-9} *per year, depending on the assumptions made. This increase in risk will not apply to those individuals who will not come into contact with an electric vehicle, and the number of these individuals will itself depend on the number of electric vehicles in use in society.*

The upper increase of risk of 1.1×10^{-6} *(just over one in a million) chance per year of injury or fatality is the order of magnitude of the 'Tolerable' region in risk tolerability criteria published by the Health and Safety Executive. However, given the large amount of uncertainty in the values used to calculate these risks then a large amount of conservatism has been built into these figures. It is therefore considered that the increase in risk seen from the use of PME supplies for charging electric vehicles at domestic dwellings is unlikely to increase the risk from the 'Broadly Acceptable' region, assuming that the base level of risk from electrical hazards to a normal household is already within this region.* ***With this level of risk additional control measures to reduce risk should be implemented if it is reasonably practicable to do so****, but a less onerous justification would be required than if the risks were to fall in the higher risk 'Tolerable' region.*

This finding is also based upon a number of fixed assumptions about the nature of faults on electricity distribution networks, in particular the reliability of their detection and the speed of rectification.

The report concerned domestic dwellings and its main finding was "With this level of risk, additional control measures to reduce risk should be implemented if it is reasonably practicable to do so".

This leaves a difficult decision as to what is reasonably practicable in each particular installation. The expression used in the report "reasonably practicable" has been copied into Regulation 722.411.4.1:

The requirement for single-phase domestic installations is to either:

1. adopt a TT system; or
2. take reasonable precautions to keep the voltage between true earth and the main earthing terminal below 70V in the event of an open-circuit PEN in the supply; or
3. install a device that trips the supply if an open-circuit PEN in the supply occurs.

A device to trip the supply (this will be similar to the voltage-operated trips commonly installed before RCDs took over,) is not yet available. When they are they will probably be installed in every domestic charging point installation.

So what is reasonably practicable for an outside charging point for a dwelling? The author suggests that

1. the TN system be retained because of the inherently low touch voltages;
2. at least two earth electrodes are installed, preferably with a combined resistance to earth of not more than 20 ohms; and
3. all obviously accessible metal parts in contact with earth are bonded to the earthing terminal of the car-charging installation.

15.11.3 Industrial and commercial installations

For outdoor non-domestic installations touch voltages have to be kept below 70 V (in the event of an open-circuit PEN), either because the load is balanced or by installing sufficient earthing. Indoors, bonding will ensure touch voltages are low.

The voltage ($V_{oo'}$) to be kept low is that between the star point of the load and true earth, see Figure 15.6. This can be readily calculated using Millman's theorem.

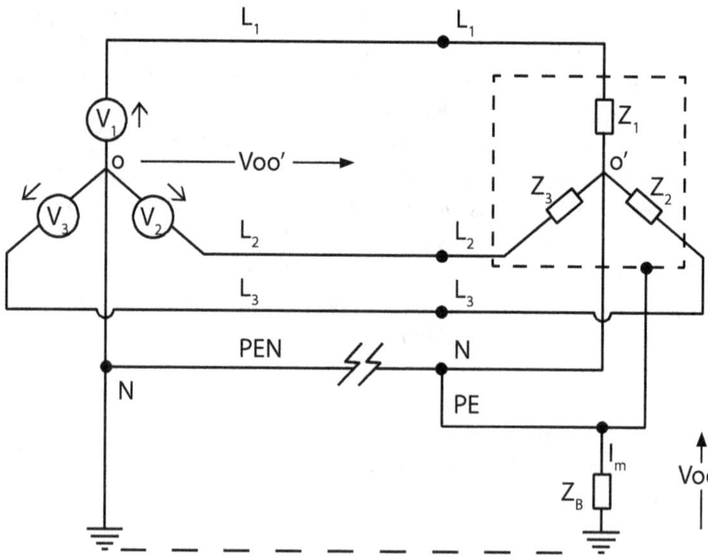

Figure 15.6 TN-C-S system with open-circuit supply PEN conductor

Millman's theorem states:

$$V_{oo'} = \frac{Y_{L1}V_{L1} + Y_{L2}V_{L2} + Y_{L3}V_{L3}}{Y_{L1} + Y_{L2} + Y_{L3} + Y_{ZB}}$$

where:

$V_{oo'}$ is the voltage between the star point (o) of the source and the star point of the load (o') when this point is not earthed or connected to neutral.

Y_{L1} is the load admittance in line 1 (and the corresponding is true for Y_{L2} and Y_{L3})

Y_{ZB} is the admittance of the installation earthing

V_{L1} is the voltage relative to point o at the free end of the admittance Y_{L1} (and the corresponding is true for V_{L2} and V_{L3})

$V_{oo'}$ is the voltage we wish to calculate ∠

$$Y_{L1} = \frac{I_{L1}\angle\,\theta_{L1}}{U_0}$$

$$Y_{L2} = \frac{I_{L2}\angle\,\theta_{L2}}{U_0}$$

$$Y_{L3} = \frac{I_{L3}\angle\,\theta_{L3}}{U_0}$$

$$Y_{ZB} = \frac{1}{R_B}$$

Equations (5), (6) and (7)
where: I_{L1}

I_{L1} is the load on line 1
U_o is the nominal line voltage to earth

Hence,

$$V_{oo'} = \frac{\frac{I_{L1}\angle\theta_{L1}}{U_0}U_0 < 0 + \frac{I_{L2}\angle\theta_{L2}}{U_0}U_0 < 240 + \frac{I_{L3}\angle\theta_{L3}}{U_0}U_0 < 120}{\frac{I_{L1}\angle\theta_{L1}}{U_0} + \frac{I_{L2}\angle\theta_{L2}}{U_0} + \frac{I_{L3}\angle\theta_{L3}}{U_0} + \frac{1}{Z_B}}$$

hence,

$$V_{oo'} = \frac{I_{L1}\angle\,\theta_{L1} \times I_{L2}\angle\,\theta_{L2} \times I_{L3}\angle\,\theta_{L3} \times U_0\angle\,120}{I_{L1}\angle\,\theta_{L1} + I_{L2}\angle\,\theta_{L2} + I_{L3}\angle\,\theta_{L3} + \frac{U_0}{Z_B}}$$

Example calculation

Consider a three-phase 400 V installation with loads as follows

Line 1 100 A, line 2 100 A at 0.8 pf, line 3 120A at 0.8 pf and a resistance to earth of the earthing terminal (ZB) of 20 ohms.

Calculate the maximum voltage between the main earthing terminal of the installation and Earth in the event of an open-circuit fault in the PEN conductor of the low voltage network.

$$Y_{L1} = \frac{100\angle 0}{230},$$

$$Y_{L2} = \frac{100\angle 36.86}{230},$$

$$Y_{L3} = \frac{120\angle 36.86}{230},$$

$$Y_{L1ZB} = \frac{1}{20}$$

$$V_{L1} = 230 \angle 0,$$

$$V_{L2} = 230 \angle 240,$$

$$V_{L3} = 230 \angle 120$$

$$V_{oo'} = \frac{100\angle 0 \times 230\angle 0 + 100 \angle 36.86 \times 230\angle 240 + 120 \angle 36.8 \times 230\angle 120}{100\angle 0 + 100\angle 36.86 + 120\angle 36.86 + \frac{230}{20}}$$

$$V_{oo'} = \frac{230.100\angle 0 + 230.100 \angle 276.86 + 230.120 \angle 156.8}{100\angle 0 + 100\angle 36.86 + 120\angle 36.86 + 11.5}$$

$$V_{oo'} = \frac{230(1.9 + j51.7)}{287 + j132}$$

$$V_{oo'} = 37.7\,V$$

This is less than 70V and is satisfactory.

15.11.4 Annex A722 of section 722

The equations given in Annex A722 of Section 722 may be used if the power factor of the loads on each phase is similar. However, they are basically no quicker although slightly simpler than carrying out the full calculation described above.

The proof:

Using

$$V_{oo'} = \frac{\frac{I_{L1}\angle\theta_{L1}}{U_0} U_0 < 0 + \frac{I_{L2}\angle\theta_{L2}}{U_0} U_0 < 240 + \frac{I_{L3}\angle\theta_{L3}}{U_0} U_0 < 120}{\frac{I_{L1}\angle\theta_{L1}}{U_0} + \frac{I_{L2}\angle\theta_{L2}}{U_0} + \frac{I_{L3}\angle\theta_{L3}}{U_0} + \frac{1}{Z_B}}$$

If $Ø_{L1}=Ø_{L2}=Ø_{L3}=Ø$ and Z_B is assumed to be large then

$$V_{oo'} = \frac{(I_{L1}\angle\,\theta_{L1} \times U_0\angle\,0) + (I_{L2}\angle\,\theta_{L2} \times U_0\angle\,240) + (I_{L3}\angle\,\theta_{L3} \times U_0\angle\,120)}{(I_{L1} + I_{L2} + I_{L3})\angle\,\theta}$$

$$V_{oo'} = \frac{U_0(I_{L1}) + (I_{L2}\angle\,240) + (I_{L3}\angle\,+120)}{(I_{L1} + I_{L2} + I_{L3})}$$

the neutral current $I_m = (I_{L1} + (I_{L2}\angle 240) + (I_{L3}\angle + 120)$

Hence:

$$V_{oo'} = \frac{U_0 I_m}{(I_{L1} + I_{L2} + I_{L3})}$$

15.12 Operating and maintenance gangways
Section 729

15.12.1 Introduction

CENELEC agreed requirements for operating or maintenance gangways specifying spacing between switchgear panels and minimum distances from walls to facilitate operational access in HD 60364-7-729: *Low-voltage electrical installations – Part 7-729: Requirements for special installations or locations – Operating or maintenance gangways*. However there is a UK special national condition not allowing unprotected live parts and making reference to the Electricity at work Regulations and Health and Safety Executive Guidance Note HSR 25, the memorandum of guidance on the Electricity at Work Regulations.

United Kingdom

Clause 729.410.3.7: In the UK where the gangway has unprotected live parts refer to UK government guidance on the Electricity at Work Regulations 1989, Statutory Instrument 1989, number 635, guidance document HSR25.

In the Health and Safety Executive's view, work activities in a gangway of such construction (unprotected live parts) would be counter to the requirements of EWR (Regulations 4, 7 and 14). It is important to note that the HSE objection (and consequently the UK note) relates to gangways with exposed live parts, not those that have the live parts enclosed and inaccessible in normal operation to persons in the gangway.

For this reason requirements for gangways with live parts was not included in BS 7671 as they are not allowed. Requirements for gangways where the protective measures of barriers or enclosures or obstacles only were included.

15.12.2.1 Requirements for operating and maintenance gangways (Regulation 729.513.2)

The width of gangways and access areas shall be adequate for work, operational access, emergency access, emergency evacuation and for the movement of equipment. Gangways shall permit equipment doors or hinged panels to be opened to at least 90°.

15.12.2.2 Restricted access areas where the protective measure of barriers or enclosures applies (Regulation 729.513.2.1)

For restricted access areas the following apply:

(i) restricted access areas shall be clearly and visibly marked by appropriate signs;
(ii) unauthorized persons shall not have access to restricted access areas; and
(iii) door(s) provided for closed restricted access areas shall allow easy evacuation to the outside by opening without the use of a key, tool or any other device that is not part of the opening mechanism.

Where the protective measure is provided by barriers or enclosures in accordance with Chapter 41, the following minimum distances apply (see Figure 15.7):

(a) width of gangways with barriers or enclosures between switch handles and circuit-breakers in position 'isolation' or switch handles and the wall: 700 mm;
(b) width of gangway between barriers or enclosures and other barriers or enclosures, or barriers or enclosures and the wall 700 mm;
(c) height of panelling above the floor 2000 mm;
(d) height of live parts out of reach (Regulation 417.3) 2500 mm.

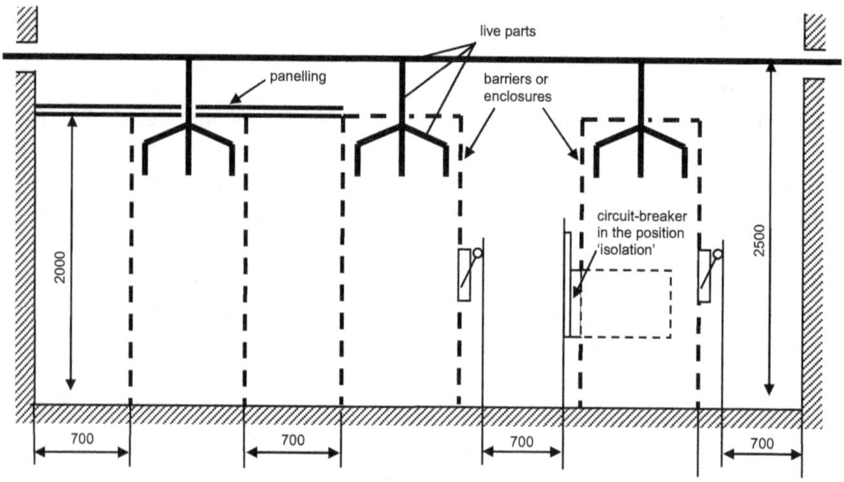

Note: Where additional workspace is needed, e.g. for special switchgear and controlgear assemblies, larger dimensions may be required.
Note: The above dimensions apply after all parts of the panelling have been mounted and closed and to circuit-breakers in the position 'isolation'.

Figure 15.7 Gangways in installations where the protective measure of barriers or enclosures applies [Figure 729.1 of BS 7671]

15.12.2.3 Restricted access areas where the protective measure of obstacles applies (Regulation 729.513.2.2)

Regulation 729.513.2.2 provides requirements for restricted access areas where the protective measure of obstacles is the protective measure for basic protection (against direct contact with live parts). This protective measure (and placing out of reach) can be used only in locations where access is restricted to (a) skilled persons or (b) instructed persons under the supervision of skilled persons.

Where the protective measure of obstacles is used, the minimum distances are:

(a) width of gangway between obstacles and switch handles, obstacles and the wall, or switch handles and the wall 700 mm;
(b) height of panelling above the floor 2000 mm; and
(c) height of live parts above the floor 2500 mm.

However, for a new installation it would be difficult to demonstrate that the protective measure of obstacles, even in a location restricted to skilled or instructed persons, complied with Regulation 4 of the Electricity at Work Regulations.

A drawing of gangways where the protective measure of obstacles is used is not included to avoid giving the impression to a casual reader that the method is considered appropriate.

15.12.2.4 Emergency exits

Gangways longer than 10 m are required to be accessible from both ends to aid entry and, importantly, exit (see Figure 15.8). It is also recommended for gangways longer than 6 m. This may be accomplished by placement of the equipment a minimum of 700 mm from the end walls or by providing an access door, if needed, on the opposite end wall. Closed, restricted access areas with a length exceeding 20 m should have doors at both ends to aid exit and access.

To assist evacuation, the doors of any equipment inside the location are required to close in the direction of the evacuation route. Gangways are to be wide enough for equipment doors or hinged panels to be opened to a minimum of 90° (see Figure 15.10).

Doors giving access to gangways in closed restricted access areas shall open outwards (see Figure 15.8) and have the following minimum distances: width 700 mm; height: 2000 mm. For doors that can be fixed in the open position or circuit-breakers or equipment that is withdrawn fully for maintenance (position: completely extracted), a minimum distance of 500 mm shall be provided between the door edge or circuit-breaker/equipment edge and the opposite side of the gangway (see Figures 15.9 and 15.10).

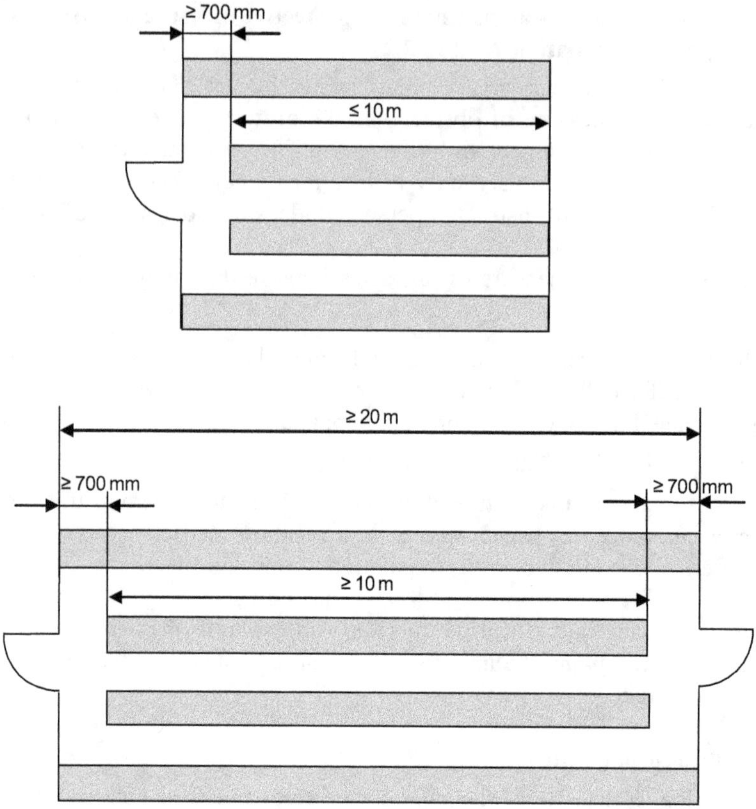

Figure 15.8 Examples of positioning of doors in long, closed, restricted access areas [Source: IEC 60364-7-729]

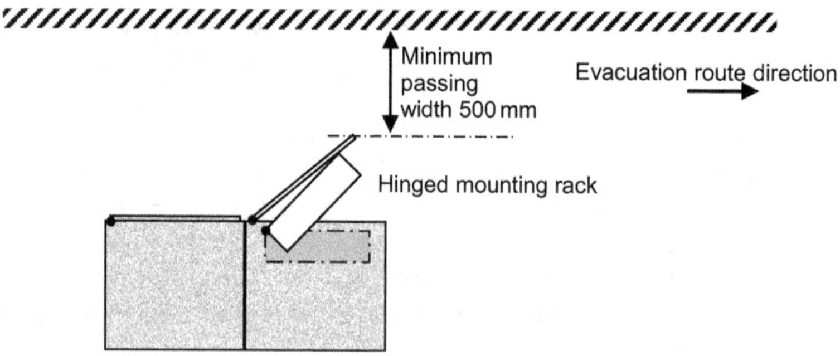

Figure 15.9 Minimum passing width in case of evacuation [Source: IEC 60364-7-729]

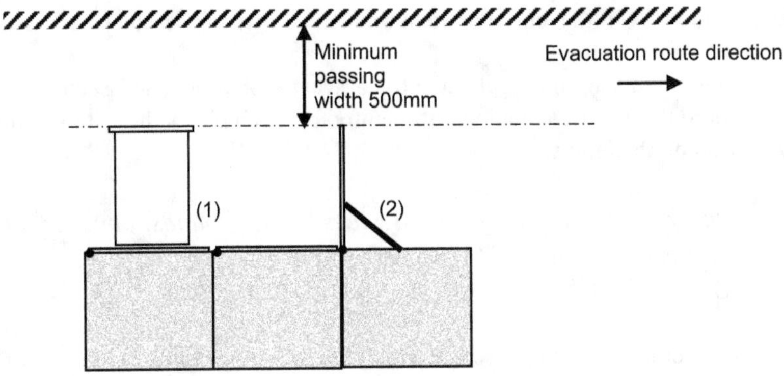

* A minimum-width gangway of 500 mm shall be provided between the wall and the circuit-breaker in the 'fully withdrawn' position or the door which has been fixed in the open position.

(1) Completely withdrawn circuit-breaker.
(2) Door fixed in open position.

Figure 15.10 Minimum passing width in case of evacuation [Source IEC 60364-7-729]

15.13 Other Section 7s

In addition to the special locations and installations discussed above, BS 7671:2008 includes particular requirements for the types of electrical installation summarised below.

Designers and installers should, as necessary, make a careful study of the requirements of the Regulations and seek guidance from equipment manufacturers and suppliers.

15.13.1 Solar photovoltaic (PV) power supply systems Section 712

The requirements of Section 712 apply to the electrical installations of PV power supply systems, other than systems intended for stand-alone operation, including systems with a.c. modules.

Key requirements include protective measures for protection against electric shock, and the provision of suitable devices for isolation to allow safe maintenance of the PV converter.

15.13.2 Mobile or transportable units Section 717

The types of unit (unit being defined as 'a vehicle and/or mobile or transportable structure') to which the requirements of this section apply include: technical and facilities vehicles for the entertainment industry, medical services, advertising, catering units, etc.

The requirements are not applicable to caravans (for which, see Section 721 of BS 7671), or to the electrical equipment required by a vehicle to allow it to be driven safely or used on the highway.

15.13.3 Structures and amusement devices at fairgrounds, amusement parks and circuses
Section 740

The requirements of this section aim both to protect the operators of the fairground equipment and the general public against electric shock, fire, burns and other injury, taking into account the repeated assembly and dismantling of machines and structures.

Key requirements: RCD protection is required at the origin of every temporary electrical installation; every structure or amusement device shall have its own readily accessible means of isolation and switching; and detailed precautions for wiring systems and lighting installations.

15.13.4 Floor and ceiling heating systems
Section 753

Key requirements for these systems: ≤30 mA RCD protection of circuits; a bonded conductive covering above floor (or below ceiling) heating elements; protection against overheating; provision of heating-free areas; a plan with comprehensive information for the user to be fixed adjacent to the distribution board.

Chapter 16

Current-carrying capacity and voltage drop for cables and cords

Appendix 4 of BS 7671
(Adviser: Mark Coates B.Eng)

16.1 Ambient conditions
Tables 4B1, Tables 4B2, 4B3

The cable rating tables in Appendix 4 of BS 7671 assume the following ambient conditions:

- Air temperature 30 °C
- Ground temperature 20 °C
- Ground thermal resistivity 2.5 K.m/W.

These are conservative conditions. BS 7769-3.1 (IEC 60287-3-2) provides standard conditions for the UK as follows:

4.14 United Kingdom

(1) Standard conditions

Soil thermal resistivity 1.2 K.m/W
Ground ambient temperature 15 °C

(2) Depth of burial

(a) For 1 kV cables – depth from ground surface to centre of cable, to centre of duct or to centre of trefoil group of cables or ducts 500 mm
(b) For 3.3 kV to 33 kV cables (unless pressure assisted) – depth from ground surface to centre of cable, to centre of duct or to centre of trefoil group of cables or ducts 800 mm
(c) For 33 kV pressure assisted cables – depth from ground surface to uppermost surface of cable or duct or of group of cables or ducts 900 mm

(3) Air ambient temperature

Out of doors 25 °C
In buildings 30 °C

(4) This temperature is also assumed as a standard condition in tabulating ratings of cable for use in certain indoor situations such as cable basements in power stations and other industrial plants.

More specifically, average annual soil temperatures at 0.5 m depth are usually assumed to be 8 °C in Scotland and 12 °C in the South. Peak summer temperatures at this depth can be expected to be 6.5 °C higher than the annual averages. It should also be noted that the above values are for natural soil conditions. If the cables are to be installed in made-up ground, thermal resistivity values may be higher than in the natural ground. Also, if cables are installed under tarmac or other impervious surfaces, both the ambient soil temperature and the thermal resistivity of the ground may increase.

Tables 16.1 and 16.2 provide data from BS 7769-3.1 for various countries and climatic conditions.

Table 16.1 Ambient conditions in a range of countries

Country	Soil thermal resistivity K.m/W	Ground ambient temperature °C	Air ambient temperature °C
United Kingdom	1.2	15	Outdoors: 25 Indoors: 30
Australia	1.2	Summer: 25 Winter: 18	Summer: 40 Winter: 30
Austria	0.7	0–20	Ave: 20
France	Summer: 1.2 Winter: 0.85	Summer: 20 Winter: 10	Summer: 30 Winter: 20
Germany	Ave: 1.0 Dry zone: 2.5	10–20 Ave: 10	–20 to 30 Ave: –10
Japan	1.0*	Summer max: 25 Winter max: 15 Min: 10	Summer max: 40 Winter max: 30

Source: BS 7769-3.1
*One manufacturer uses 1.2, 0.8, 0.4K.m/W for dry, normal, wet soil conditions.

Table 16.2 Thermal resistivity of soil (from clause 3.2.2 of BS 7769-3.1)

Thermal resistivity of soil K.m/W	Soil conditions	Weather conditions
0.7	Very moist	Continuously moist
1.0	Moist	Regular rainfall
2.0	Dry	Seldom rains
3.0	Very dry	Little or no rain

Table 16.3 Ratings comparison (for three-/four-core SWA with thermosetting insulation)

Conductor csa	BS 7671 (Table 4E4)	ERA 69-30 in ducts	ERA 69-30 buried
mm^2	A	A	A
50	135	177	217
70	167	218	266
95	197	263	319
120	223	300	363
150	251	338	406
185	281	382	458

The use of conservative base ambient conditions has resulted in the tabulated current ratings for cables laid in the ground being less than those given in ERA rating tables (ERA Reports 69-30, Parts III and V) and the ratings previously published by cable suppliers. Corrections for likely ambient conditions can be made using the appropriate rating factors in Tables 4B1 and 4B2 for air and ground ambient temperatures and Table 4B3 for soil thermal resistivity.

A comparison of current ratings from BS 7671 Table 4E4 A, column 7, and those given in ERA Report 69-30, Part V, for cables in ducts and buried direct is given in Table 16.3. This comparison shows the difference in ratings due to different ambient conditions and the use of 'generic' ratings against ratings calculated for specific types of cable.

16.1.1 BS 7769 Electric cables, calculation of current ratings

Tables 4B3, 4C2, and 4C3 advise that rating factors can be calculated by the methods in the various parts of BS 7769. However, the calculations are long and require full data on the cable constructions and as a consequence are not usually of practical use to a design engineer working on LV systems.

16.2 Ambient air temperature rating factor, C_a

16.2.1 Cables in air
Paragraph 2.1 and Table 4B1 of Appendix 4 of BS 7671

Table 4B1 provides rating factors (C_a) for ambient air temperature. For ambient temperatures above 30 °C the limiting factor is the maximum conductor operating temperature t_p and the following formula is used:

Ambient above 30 °C:

$$C_a = \sqrt{\frac{t_p - t_a}{t_p - t_o}}$$

For ambient (air) temperatures below 30 °C the limiting factor is the maximum conductor temperature t_e when the current to cause operation of the protective device in the conventional time is flowing and the following formula is used: Ambient below 30 °C:

$$C_a = \sqrt{1 + \frac{(t_o - t_a)}{1.45^2(t_p - t_o)}}$$

where:
C_a = rating factor for ambient temperature
t_p = conductor operating temperature given in the rating tables in Appendix 4
t_o = ambient temperature assumed for the tabulated rating, that is 30 °C
t_a = ambient temperature for which a rating factor is required.

16.2.1.1 Derivation of formulae

The above formulae are derived from the basic concept that temperature rise is proportional to I^2R.

$$\Delta T \propto I^2R$$

where:
I = current in the conductor
R = the resistance of the conductor.

or $\Delta T = KI^2R$

hence $(t_p - t_a) = KI_a^2R$ (a)

and $(t_p - t_o) = KI_o^2R$ (b)

Divide a by b, $\frac{t_p - t_a}{t_p - t_o} = \frac{I_a^2}{I_o^2}$

where:
t_p = conductor operating temperature
I_a = rating at ambient of t_a
I_o = rating at ambient of t_o

hence $C_a = \frac{I_a}{I_o} = \sqrt{\left(\frac{t_p - t_a}{t_p - t_o}\right)}$ for ambient above 30 ° C. (c)

This is not totally correct, but is reasonably accurate over the temperature rises considered. The proof assumes that the limiting factor is the maximum permissible sustained operating temperature (t_p) from Table 52.1 of BS 7671.

For ambient temperatures below 30 °C, the basis is that, under overload conditions, the conductor temperature is limited to the limiting final temperature t_e that a cable at an ambient of 30 °C will reach if a current of 1.45 I_o flows (that required to trip the device in conventional time).

The rating factor C_a for less than 30 °C ambient temperatures is given by

$$C_a = \sqrt{\frac{t_e - t_a}{t_e - t_o}} \qquad \text{(d)}$$

Now $I_e = 1.45\ I_t$ by definition

where I_e is the current to produce the limiting final temperature t_e

$$\text{and } \frac{t_e - t_o}{t_p - t_o} = \frac{I_e^2}{I_t^2} = 1.45^2$$

$$\text{and } t_e = 1.45^2\ (t_p = t_o) + t_o \qquad \text{(e)}$$

hence, for 70 °C cables $t_e = 1.45^2(70 - 30) + 30 = 114$ °C

and for 90 °C cables $t_e = 1.45^2(90 - 30) + 30 = 156$ °C

where:

I_t = current to produce the maximum conductor operating temperature t_p

t_o = ambient temperature for the conductor current ratings I_t e.g. 30 °C

t_a = new ambient temperature

t_e = final limiting temperature.

The substitution of (e) in (d) gives the equation for C_a:

$$C_a = \sqrt{1 + \frac{(t_o - t_a)}{1.45^2(t_p - t_o)}}$$

or, for 70 °C cables

$$C_a = \sqrt{\frac{(114 - t_a)}{(114 - 30)}}$$

16.2.2 Cables in the ground

16.2.2.1 Ambient ground temperature rating factor C_a

Paragraph 2.2 and Table 4B2 of Appendix 4 of BS 7671

The rating factors for ambient temperatures above and below 20 °C are calculated using the formula:

$$C_a = \sqrt{\frac{t_p - t_a}{t_p - t_o}}$$

where:
C_a = rating factor for ambient temperature
t_p = conductor operating temperature given in the rating tables in Appendix 4
t_o = ambient temperature assumed for the tabulated rating, that is 20 °C
t_a = ambient temperature for which a rating factor is required.

This is the same as for Table 4B1, but with $t_o = 20$ and not 30.

Hence the equation for thermoplastic cables t_p = 70 °C and a base ambient t_o = 20 °C is:

$$C_a = \sqrt{\frac{70 - t_a}{70 - 20}}$$

This is to limit the temperature to 70 °C, the assumed maximum overload temperature for 70 °C cables.

The equation for 90 °C cables is:

$$C_a = \sqrt{\frac{90 - t_a}{90 - 20}}$$

This method of calculating the effect of reduced ambient temperature while limiting the conductor temperature to the maximum conductor operating temperature t_p under steady load conditions could, under overload, result in the cable temperature exceeding 114 °C for thermosetting cables (t_p = 70 °C) and 156 °C for thermosetting cables (t_p = 90 °C).

Note: The ultimate conductor temperature (t_e) for a current of 1.45 I_{t70-30} with an ambient t_o of 30 °C and a conductor operating temperature (t_p) of 70 °C is:

$$t_e = 1.45^2(70 - 30) + 30$$
$$t_e = 114.1\ °C$$

However, this is corrected by the use of the rating factor C_c (see 16.4.2) for cables laid in the ground. This approach allows enhanced cable ratings when overload protection is not required.

16.3 Rating factor for grouping, C_g

Paragraph 2.3 and Tables 4C1 to 4C6 of Appendix 4

The group rating factors (C_g) in BS 7671 have been derived from experimental results on single-core cables in circular bundles of up to 37 cables. The general variation of these factors with numbers of cables has also been derived from theoretical considerations. Note that the tabulated grouping factors assume that all of the cables in the group are of a similar conductor size. Where there is a wide range of cable sizes in a group it is more appropriate to use a conservative grouping factor for cables that are enclosed or bunched and clipped direct based on $C_g = (n)^{-1/2}$ where n is the number of circuits or multicore cables, i.e.

$$C_g = \frac{1}{\sqrt{n}}$$

It is interesting to note that the cube root of n gives almost exactly the factor for bunched cables in Table 4C1:

$$C_g = \frac{1}{\sqrt[3]{n}}$$

Note 1 to Table 4C1 states that the published factors apply to uniform groups of cables (all one size), each carrying the same current. These conditions occur only rarely in practice; however, factors that are not limited in this way call for more complex tables. The factors are universally used even when the cables are not of the same size, and there are reportedly no problems with the general application. Presumably, the reason they work quite effectively (noting that all the cables in the group are unlikely to be fully loaded at the same time, or ever) is that, if the rating factors are applied, all the conductors in the group will be at approximately the same temperature (this is not so for, say, eight 50 mm^2 circuits in a group with one 2.5 mm^2 circuit and the 2.5 mm^2 circuit is in the middle: it will overheat if it is loaded to the rating given by applying the rating factor for nine circuits, assuming the other circuits are also fully loaded).

The factors are obviously not applicable when mixing cables with different conductor maximum operating temperatures (see Table 52.1 of BS 7671). 90 °C thermosetting insulated cables cannot be grouped with 70 °C thermoplastic insulated cables unless the thermosetting cables are derated to 70 °C thermoplastic current ratings.

16.4 Overload protection

Paragraph 4 of Appendix 4 of BS 7671, Regulation 433.1.202

16.4.1 Rating factor for semi-enclosed fuse, C_f

Paragraph 5.1 of Appendix 4

The overload protection of a cable does not, as such, affect the current-carrying capacity of the cable, since the maximum operating temperature is determined simply by the load current and the resistance of the cable. However, if the overcurrent device is providing overload protection, and the conductor temperature is to be limited to the limiting final temperatures t_e of 114 °C for thermoplastic cables and 156 °C for thermosetting cables, then the choice of overcurrent device will affect cable selection. BS 7671 presumes that, for fuses to BS 88 or BS 1361 or circuit-breakers to BS 3871 or BS EN 60898, the operating current (tripping or fusing) of the protective device I_2 will not exceed 1.45 I_n, and that, as a consequence, if I_z>I_n, then, for the devices mentioned, the overcurrent device will protect the cable against overload. For devices where I_2 exceeds 1.45 I_n, such as for semi-enclosed fuses to BS 3036, the cable size selected needs to be increased to provide a higher current-carrying capacity (you cannot increase the current-carrying capacity of a cable unless you change the installation conditions). When non-standard devices are being used, the operating current of the device

I_2 must be obtained from the manufacturer. The rating factor (C_f) to be applied to the current-carrying capacities of the cables = 1.45 I_n/I_2. For example, for a semi-enclosed fuse to BS 3036 the fusing factor I_2/I_n is 2 and hence the rating factor C_f is:

$$C_f = 1.45 \times \frac{I_n}{I_2} = 1.45 \times \frac{1}{2} = 0.725$$

16.4.2 *Rating factor for cables laid in the ground, C_c* Paragraph 5.1.1 of Appendix 4, Regulation 433.1.203

16.4.2.1 70 °C cables

The ambient temperature used to determine the rating of cables buried in the ground either direct or in ducts is 20 °C. This of course results in higher-rated currents I_{t20}, as the current to raise the conductor operating temperature to, say, 70 °C is greater than if an ambient ground temperature of 30 °C were assumed, as for air. However, under overload conditions it results in a final cable conductor temperature (t_e) greater than that reached for an ambient temperature of 30 °C, if the overload is the same multiple of the current rating.

Using: temperature rise (δt) ∞ I^2 and hence

$$\frac{\textit{temp rise } 1}{\textit{temp rise } 2} = \frac{I_1^2}{I_2^2}.$$

The cable rating tables and cable protection system of BS 7671 indicate that overcurrent devices will disconnect an overload of 1.45 I_n in the conventional time for the device – say 1 hour (see Table 6.2). Hence, the temperature rise of a cable when the tabulated (I_t) current rating exactly matches the overcurrent device rating (I_n) for an overload current of 1.45 I_n for a cable with an operating temperature of 70 °C and an ambient of 30 °C approximates to

$$\frac{\textit{temp rise for } 1.45 I_t}{70 - 30} = \left(\frac{1.45 I_t}{I_t}\right)^2$$

Table 16.4 Installation condition factor (C_c) for various ground temperatures

Ground temperature (°C)	Installation condition factor 70 °C thermoplastic, C_c	Installation condition factor 9 °C thermosetting, C_c
≥ 30	1	1
20	0.95	0.96
15	0.93	0.94
10	0.91	0.93

Hence the ultimate conductor temperature (t_e) for a current of 1.45 $I_{t\,70-30}$ with an ambient t_o of 30 °C and a conductor operating temperature (t_p) of 70 °C is

$$t_e = 1.45^2(70 - 30) + 30$$

$$t_e = 114.1\ °C$$

Also, the ultimate conductor temperature (t_e) for a current of 1.45 $I_{t\,70-20}$ with an ambient t_o of 20 °C and a conductor operating temperature (t_p) of 70 °C is

$$t_e = 1.45^2(70 - 20) + 30$$

$$t_e = 125.1\ °C$$

If the ultimate temperature under overload is to be limited to 114 °C and the temperature rise to (114 – 20) °C, a factor C_c needs to be applied to the tabulated current rating I_t, given by

$$\frac{(114–20)}{(125–20)} = \left(\frac{C_c 1.45 I_t}{1.45 I_t}\right)^2$$

$$C_c = 0.95$$

The factor given in paragraph 5.1.1 of Appendix 4 of BS 7671 has been rounded down to 0.9. The factor of 0.9 is more appropriate for a ground temperature of 10 to 15 °C, temperatures most likely to be assumed for the UK. Table 16.4 summarizes the installation condition rating factors.

16.4.2.2 90 °C cables

Similarly, the ultimate conductor temperature (t_e) for a current of1.45 $I_{t\,90-30}$ with an ambient t_o of 30 °C and a conductor operating temperature (t_p) of 90 °C is

$$t_e = 1.45^2(90 - 30) + 30$$
$$t_e = 156\ °C$$

Table 16.5 Effect of ambient temperature and installation condition factors

I_t for I_b = 100 A with and without overload protection (I_n = 100 A) for thermoplastic cables laid in the ground

Ambient temperature (°C)	40	30	20	15	10
C_a from Table 4B2	0.77	0.89	1	1.05	1.10
C_c from Table 16.4	1	1	0.95	0.93	0.91
No overload protection $I_t = I_b/C_a$ (amperes)	129	112	100	95	91
Overload protection $I_t = I_n/C_aC_c$ (amperes)	129	112	105	102	100

and ultimate conductor temperature (t_e) for a current of 1.45 $I_{t\,90-20}$ with an ambient t_o of 20 °C and a conductor operating temperature (t_p) of 90 °C is

$$t_e = 1.45^2(90 - 20) + 20$$
$$t_e = 167\ °\text{C}$$

If the ultimate temperature under overload is to be limited to 156 °C and the temperature rise to (156 − 20) °C, a factor C_c needs to be applied to the tabulated current rating I_t, given by

$$\frac{(156–20)}{(167–20)} = \left(\frac{C_c 1.45 I_t}{1.45 I_t}\right)^2$$

$$C_c = 0.96$$

Again, if overload protection is not required because, say, the load is fixed, the factor need not be applied, nor the rating factor for I_2 being greater than 1.45. These factors allow the appropriate ratings of cables to be determined for both when overload protection is required and when not (see Table 16.5).

16.5 Determination of the size of cable
Paragraph 5 of Appendix 4

16.5.1 *Application of rating factors with overcurrent device providing protection against overload and fault currents*
Paragraph 5.1.2 of Appendix 4

When an overcurrent device provides protection against both overload and fault currents, the cable must be selected on the basis of the rating of the device. The tabulated current-carrying capacity I_t may be determined as follows (Equation 2 of Section 5 of Appendix 4):

Current-carrying capacity I_t must be equal to or greater than:

$$I_t \geq \frac{I_n}{C_a C_g C_i C_f C_s C_c C_d}$$

where:

I_t = is the tabulated current-carrying capacity of a cable found in Appendix 4 of BS 7671

C_a = is the rating factor for ambient temperature, see Tables 4B1 and 4B2, of BS 7671

C_g = is the rating factor for grouping, see Tables 4C1 to 4C6 of BS 7671

C_i = is the rating factor for conductors surrounded by thermal insulation, see Table 52.2 of BS 7671

C_f = is a rating factor applied when overload protection is being provided by an overcurrent device with a fusing factor greater than 1.45 (for rewirable fuses)

$$C_f = 1.45 \times \frac{\text{device rating}}{\text{device fusing current}} = 1.45\frac{I_n}{I_2}$$

I_2 is the operating current (that is, the fusing current) of the device providing protection against overload), see Table 6.2

C_s is the rating factor for thermal resistivity of soil, see Table 4B3 of BS 7671

C_c is the rating factor for buried circuits, C_c is 0.9 for cables buried in the ground, otherwise is 1

C_d is a correction factor for depth of buried cable, see Table 4B4.

The rating factor C_f is unity for overcurrent devices to BS 88, BS 1361, BS 3871 and BS EN 60898, and C_f is 0.725 for semi-enclosed fuses to BS 3036. For other devices, the rating factor C_f (or I_2) must be obtained from the manufacturer.

Only the simultaneously applicable values are applied together. If there is grouping at one point of the cable route, but not thermal insulation, and thermal insulation at another part but not grouping, then C_g and C_i are not applied simultaneously. The worst situation with either C_g or C_i applied (with C_a and C_f, if appropriate) is determined and used to calculate I_t.

If the device is providing protection against overload, the rating of the overload device I_n must be greater than or equal to the load current, I_b, i.e.

$$I_n \geq I_b$$

16.5.2 Application of rating factors where overload protection is not required
Paragraph 5.2 of Appendix 4

When the overcurrent device is not providing protection against overload, the required I_t is determined by the load I_b, and not the nominal rating of the device I_n. (Equation 5 of section 5 of Appendix 4):

$$I_t \geq \frac{I_b}{C_a C_g C_i C_f C_s C_d} \qquad (5)$$

The rating factor C_c for cables laid in the ground (in ducts or direct) is 0.9. However, when overload protection is not required, and for cables installed above ground, it is 1. This approach may be taken if the load is fixed or has other overload protection.

16.5.3 Groups of cables
Paragraph 5.1.2 of Appendix 4

If it is to be presumed that all the cables in a group may overload simultaneously, and the protective device is providing protection against overload then

$$I_t \geq \frac{I_n}{C_g C_a C_s C_d C_i C_f C_c} \qquad \text{Equation 2}$$

In most circumstances the circuits of a group are not liable to simultaneous overload. If only one cable is likely to overload at any particular time, it will not increase the temperature of the group of cables to the same degree as if all the cables simultaneously overloaded. Section 5 of Appendix 4 provides equations numbered (3) and (4) to be used for devices other than semi-enclosed fuses to BS 3036, and equations (7) and (8) for semi-enclosed fuses.

Equations (3) and (4) are as follows:

$$I_t \geq \frac{I_b}{C_g C_a C_s C_d C_i C_f C_c} \qquad \text{Equation 3}$$

$$I_t \geq \frac{1}{C_a C_s C_d C_i}\sqrt{\left[\frac{I_n}{C_f C_c}\right]^2 + 0.48 I_b^2\left(\frac{1-C_g^2}{C_g^2}\right)} \qquad \text{Equation 4}$$

where:

I_b is the design current of the circuit, i.e. the current intended to be carried by the circuit in normal service.

I_n is the nominal current or current setting of the device protecting the circuit against overcurrent.

Equation 3 above ensures that the tabulated values of current (I_t) are adequate for when all the grouped cables are carrying the designated load current I_b, and (4) ensures that the tabulated values are satisfactory when one cable only is overloaded The two equations need to be applied to the hottest cable, i e to those cables nearest to their current-carrying capacity I_z, and this is generally not found by simple inspection; consequently, the equations have to be applied to every cable in the group, if they are of differing sizes.

Equation 4 is derived by assuming that only one cable is liable to overload in the group at any time. In the derivation, C_a and C_i are taken as 1. The temperature rise (Δt_g) of all the cables in the group as a result of the load current I_b in each cable is

$$\text{temperature rise } \Delta t_g = \left(\frac{I_b}{C_g I_t}\right)^2 (\Delta t_P - \Delta t_0) \qquad \text{(f)}$$

where:

Δt_g = temperature rise of all the cables in the group
t_p = conductor operating temperature
t_o = ambient temperature
I_t = tabulated current rating
I_b = load current

If all the cables are unloaded except one, the temperature rise Δt_1 of that one cable is:

$$\Delta t_1 = K\left(\frac{I_b}{I_t}\right)^2 (\Delta t_P - \Delta t_0) \qquad \text{(g)}$$

where K is a constant greater than unity and depends on the ratio

$$K = \frac{\text{heat emitting capability of a cable as part of a group}}{\text{heat emitting capability in isolation}}$$

The contribution of the other cables in the group to the temperature rise Δt_c of the one cable it is assumed will fault is:

$$\Delta t_c = \Delta t_g - \Delta t_1$$

or Δt_c = equation (f) – equation (g)

$$\Delta t_c = \left\{\left(\frac{I_b}{C_g I_t}\right) - K\left(\frac{I_b}{I_t}\right)^2\right\}(t_P - t_0) \qquad \text{(h)}$$

If the overloaded cable has a current PI_n then its temperature rise Δt_{1u} when grouped, with the other cables when they are unloaded in a similar manner to equation (g), will be:

$$\Delta t_1 = K\left(\frac{PI_n}{I_t}\right)^2 (t_P - t_0) \tag{i}$$

To determine the actual temperature rise ΔT of the one overloaded cable, add its temperature rise when overloaded with the other cables unloaded [Δt_{1u} or (i) above] to its temperature rise due to the other cables in the group carrying only their load current [Δt_c or (h)]:

$$\Delta T = \Delta t_{1u} + \Delta t_c \text{ or}$$

$$\Delta T = (i) + (h)$$

$$\Delta T = \left\{\left(\frac{PI_n}{I_t}\right)^2 K + \left(\frac{I_b}{I_t}\right)^2 \left(\frac{1-KC_g^2}{C_g^2}\right)\right\}(t_p - t_o) \tag{j}$$

This temperature rise ΔT must not exceed the limiting temperature rise of the cable type given by Table 52.1

$$\Delta T \leq (t_p - t_o)E^2 \tag{k}$$

where:

$$E = \frac{\text{current to raise cable to limiting temperature of Table 52.1}}{\text{tabulated current rating } I_t}$$

hence substituting (j) in (k):

$$E^2(t_s - t_o) = \left\{\left(\frac{PI_n}{I_t}\right)^2 K + \left(\frac{I_b}{I_t}\right)^2 \left(\frac{1-KC_g^2}{C_g^2}\right)\right\}(t_s - t_o)$$

and rearranging for I_t

$$I_t \geq \left\{\left(\frac{PI_n}{E}\right)^2 K + \frac{I_b^2(1-KC_g^2)}{E^2C_g^2}\right\}^{\frac{1}{2}}$$

If:
$K = 1$ (a conservative value),
$E = 1.45C_aC_sC_dC_i$ (see Regulation 433.1.1), and
$P = 1.45/C_c$ (as for devices to BS 88, BS 1361, BS 3036 etc.), *then*

$$I_t \geq \frac{1}{C_aC_sC_dC_i}\sqrt{\left(\left[\frac{I_n}{C_fC_c}\right]^2 + 0.48I_b^2\frac{(1-C_g^2)}{C_g^2}\right)} \qquad \text{Equation 5}$$

16.6 Tables of voltage drop
Paragraph 6 of Appendix 4

The voltage drop (mV/A/m) given in Appendix 4 is tabulated in the complex form $r + jx$ for conductor sizes over 16 mm^2. For smaller conductor sizes the inductance is not significant in the estimation of voltage drop. The inductance of cables reduces only slightly with increase in conductor size, as inductance depends on the interaction of the line and neutral currents, or line currents in a three-phase system, which is mostly determined by the physical configuration of the cable. As the conductor size increases and the resistance per metre (r) reduces, the inductance (x), which is fairly constant, becomes significant. The tabulated values of r relate to the a.c. conductor resistance at its maximum operating temperature.

The figures are given in the tables in the following form:
r is the resistance element $(\text{mV/A/m})_r$
x is the inductive element $(\text{mV/A/m})_x$
z is the impedance $(\text{mV/A/m})_z$.

$$\text{The impedance } z = \sqrt{(r^2 + x^2)}$$

For simple calculations the value of z is used to calculate the voltage drop:

$$\text{voltage drop} = \frac{I_b(\text{mV/A/m})_z L}{1000}\,\text{V}$$

where:
L = length of the conductor
I_b = load current
$(\text{mV/A/m})_z$ = tabulated voltage drop.

16.6.1 Correction for operating temperature
Paragraph 6.1 of Appendix 4

An equation for estimating the reduction in conductor operating temperature as a result of the load current being less than the maximum current-carrying capacity is given in section 9.5.3 of this Commentary.

The voltage drop tables of Appendix 4 assume that the conductors are at the conductor operating temperature given for each table. If the loading of the cable is such that the temperature is less than this, i.e. after allowances have been made for grouping, ambient temperature and thermal insulation, then the voltage drop will be less than that given in the tables. Equation (6) of Appendix 4 provides a correction factor that can be usefully applied if cables are underloaded. Note that the correction factor C_t is applied only to the resistive element of the voltage drop, as the reactive element of the voltage drop is unaffected by temperature, its value being determined generally by the configuration of the conductors.

$$C_t = \frac{230 + t_p - \left[C_a^2 C_g^2 C_s^2 C_d^2 - \frac{I_b^2}{I_t^2}\right](t_p - 30)}{230 + t_p} \qquad \text{Equation 6}$$

This correction factor may be determined as described below. The temperature of a cable is determined by:

(i) the ambient temperature;
(ii) the contribution from other cables, if it is grouped;

(iii) its own current; and
(iv) if buried, the depth and soil resistivity.

Let I_x be the rating of the cable after allowance for ambient temperature and grouping, that is:

$$I_x = C_a C_g C_s C_d I_t \tag{l}$$

where:
C_a = correction factor for ambient
C_g = correction factor for grouping
C_s = is the rating factor for thermal resistivity of soil
C_d = is correction factor for depth of buried cable
I_t = tabulated current-carrying capacity of a single circuit at an ambient temperature of 30 °C

The current I_x will cause the cable in the group to rise to its maximum operating temperature t_p.
The temperature rise ΔT_1 due to the current in the cable alone is given by:

$$\Delta T_1 = \frac{I_x^2}{I_t^2}(t_p - 30) \tag{m}$$

The temperature of the group when all cables are loaded is t_p, so that, if we wish to determine the contribution of the other cables T_2 to the temperature of the one conductor we are considering, we can say that

$$T_2 = t_p - \Delta T_1$$
$$= t_p - \frac{I_x^2}{I_t^2}(t_p - 30) \tag{n}$$

The rise in the conductor we are considering due to its current I_b is given by:

$$\Delta T_3 = \frac{I_b^2}{I_t^2}(t_p - 30) \tag{o}$$

The actual temperature (T_b) of a cable carrying a current I_b with ambient and grouping factors C_a and C_g is then given by the sum of T_2 and ΔT_3:

$$T_b = T_2 + \Delta T_3$$
$$= t_p - \frac{I_x^2}{I_t^2}(t_p - 30) + \frac{I_b^2}{I_t^2}(t_p - 30) \tag{p}$$

$$\text{Now } I_x = C_a C_g C_s C_d I_t \tag{q}$$

Substituting (q) in (p)

$$T_b = t_p - \left(C_a^2 C_g^2 C_s^2 C_d^2 - \frac{I_b^2}{I_t^2}\right)(t_p - 30) \tag{r}$$

Now using the relationship

$$\frac{R_1}{R_2} = \frac{\beta + t_1}{\beta + t_2}$$

where:
β = reciprocal of temperature coefficient of resistance of the conductor
t = conductor temperature
R_1 = conductor resistance at temperature t_1
R_2 = conductor resistance at temperature t_2

$$\text{then } C_t = \frac{R_{tb}}{R_{tp}}$$

where:
R_{tb} = resistance at temperature T_b
R_{tp} = resistance at temperature t_p

$$\text{hence } C_t = \frac{\beta + T_b}{\beta + t_p} \qquad \text{(s)}$$

Substituting (r) in (s) provides for the equation (6) above (also of Appendix 4) (assuming that β, the reciprocal of the temperature coefficient of resistance of the conductor, is 230). Reference to Table 6.14 will give values for β for the usual materials. Note that the value of 230 in equation (6) is an average for copper and aluminium. The formula may be more accurately calculated using the specific value of β for the material being considered:

$$C_t = \frac{\beta + t_p - \left(C_a^2 C_g^2 C_s^2 C_d^2 - \frac{I_b^2}{I_t^2}\right)(t_p - 30)}{\beta + t_p}$$

16.6.2 *Correction for load power factor* Paragraph 6.2 of Appendix 4

If account is to be taken of load power factor for cables of cross-sectional area of 16 mm^2 or less, the design value of voltage drop is given by

$$\text{voltage drop} = \frac{L(\text{mV/A/m})_z I_b \cos\varnothing}{1000}$$

where:
L = length of the conductor
I_b = modulus of the load current
$\cos\varnothing$ = power factor of the load current
(mV/A/m) = tabulated value of voltage drop.

For cables larger than 16 mm^2 the design value of (mV/A/m) is given as follows:

$$\cos\varnothing(\text{mV/A/m})_r + \sin\varnothing(\text{mV/A/m})_x$$

Figure 16.1 demonstrates how the equation cos ø [tabulated $(\text{mV/A/m})_r$] + sin ø [tabulated $(\text{mV/A/m})_x$] is derived. While the equation is not an accurate method of estimating the actual voltage drop, it is an effective way of estimating the reduction in supply voltage. What is important is the voltage applied to the terminals of the

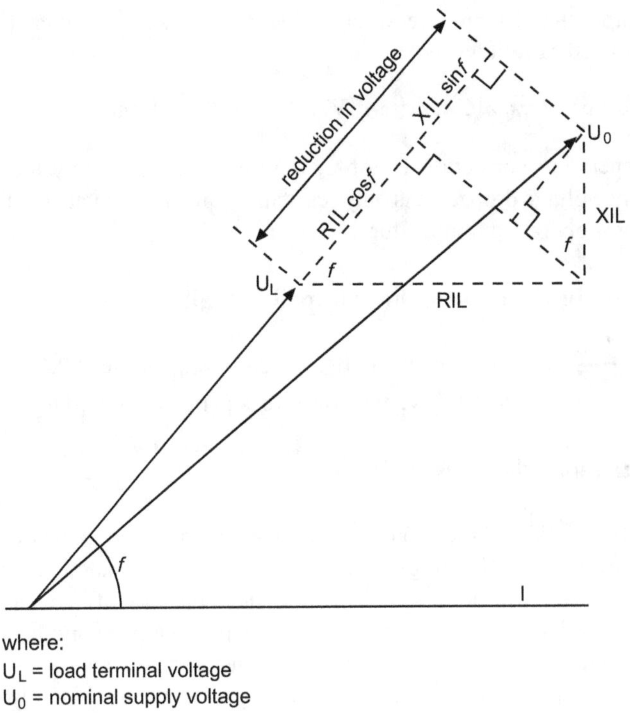

Figure 16.1 Derivation of correction of voltage drop for power factor

equipment, and this equation effectively allows this to be estimated. The accuracy of the method is also increased by the fact that the voltage drop is small compared with the voltage.

Voltage drop is calculated as follows:

$$\text{voltage drop} = \frac{LI_b}{1000}\{\cos\phi(\text{mV/A/m})_r + \sin\phi(\text{mV/A/m})_x\}$$

16.6.3 Correction for temperature and load power factor Paragraph 6.3 of Appendix 4

The calculation of voltage drop for conductors of cross-sectional area 16 mm^2 or less, where account is to be taken of reduction in conductor operating temperature due to the load being less than the maximum current-carrying capacity of the conductor, but including allowance for power factor, is given by

$$\text{voltage drop} = \frac{LI_bC_t\cos\phi(\text{mA/V/m})}{1000}$$

For cables larger than 16 mm^2, i.e. where inductive effects are significant, the voltage drop is calculated as follows:

$$\text{voltage drop} = \frac{LI_b}{1000}\left\{C_t \cos\phi (\text{mA/V/m})_r + \sin\phi (\text{mA/V/m})_x\right\}$$

The temperature correction factor C_t is in effect applied only to the resistive element of the voltage drop, as it is only resistance (and not inductance) that changes with conductor operating temperature.

16.7 Cables in closed trenches clipped to walls

Use the $\frac{W_{tot}}{3P}$ equation given in IEC 60287-2-1, Clause 2.2.6.2, Note ERA Report 69-30, Parts III and V, give the power dissipations required to calculate W_{tot}.

16.8 Unarmoured cables in ducts

For multicore cables, the tabulated ratings for armoured cables can be used. They come from the IEC ratings that were originally calculated for unarmoured cables. The ratings for armoured cables are reduced because of eddy current losses in the armour but increased because they have a larger diameter and hence a greater surface area to dissipate heat. These two factors just about balance.

For unarmoured single-core cables the ratings will be significantly better than those for armoured cables because there are no circulating currents. Thus armoured ratings used for unarmoured cables will be on the safe side. Ratings for unarmoured cables in ducts are not usually given by cable manufacturers as they would not recommend unarmoured cables in ducts because of the risk of damage to the cables when they are pulled into the ducts.

Chapter 17
Harmonics

Appendix 4 sections 5.5 and *5.6*
(Adviser: Mark Coates B.Eng)

17.1 Introduction

Appendix 4 sections 5.5 and 5.6, Section 523, Regulations 431.2.3, 533.2.2, 551.5.2

The electronic control and electronic power supplies to much equipment can result in non-linear or non-sinusoidal load current. The basic waveform can be considered to have further waveforms superimposed on it with frequencies that are multiples of the basic or fundamental 50 Hz waveform. These additional waveforms are called *harmonics*. Often these harmonics can be disregarded in the design of electrical installations; however, third harmonics or multiples of the third harmonic cannot. The third harmonic content of discharge lamps can be of the order of 25 per cent with a total harmonic distortion of 30 per cent, and the switch-mode power supplies of computers can produce third harmonics of the order of 70 per cent with a total harmonic distortion of 77 per cent, and 100 per cent is not unknown. Normal, that is fundamental, 50-hertz, three-phase load currents if balanced, cancel out in the neutral. This is a natural consequence of the 120-degree electrical displacement of each phase (see Figure 17.1).

Third and other triple harmonics do not cancel in the neutral but sum so that the neutral current equals the sum of the third and multiple of third harmonics of each

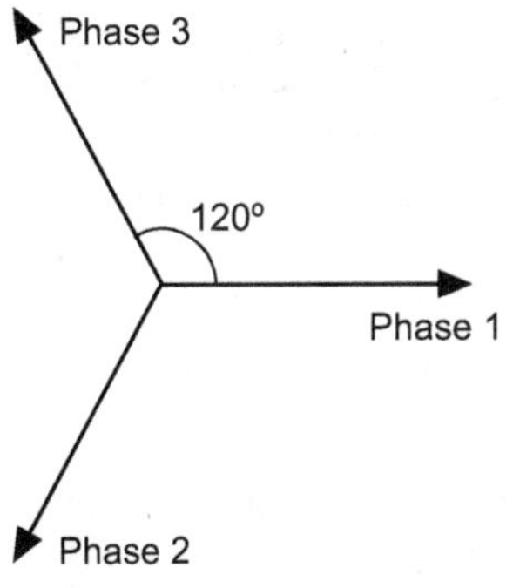

Figure 17.1 Phase displacement

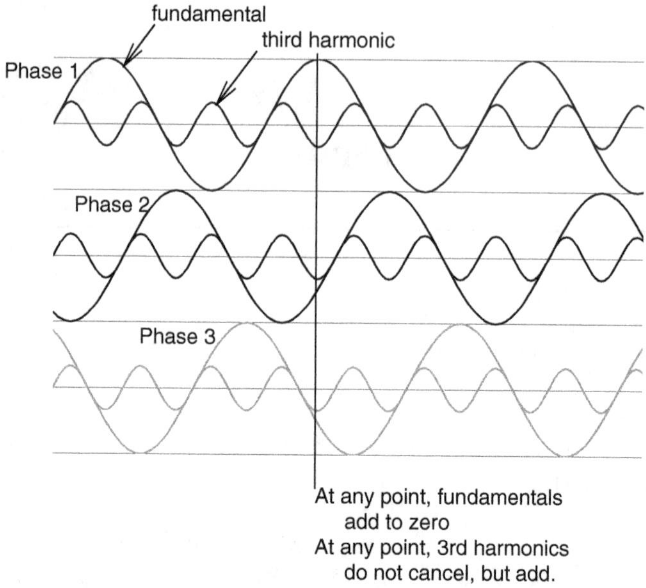

Figure 17.2 Three-phase, fundamental, 50 Hz waveforms and 3rd harmonic 150 Hz waveforms

phase (see Figure 17.2). These harmonic currents affect cable ratings, voltage drop, overcurrent protection and other equipment, e.g. UPS and transformers.

17.2 Cable ratings
Regulations 523.1, 523.6

The tabulated current-carrying capacities in Appendix 4 of BS 7671 are the same for both three- and four-core cables. This is because it is assumed that each line is carrying the rated current and the neutral current is zero; or, if the neutral current is not zero, as a result of imbalance between the three lines the increased current in the neutral will be balanced by a reduction in one of the lines. This understanding is not valid if there are harmonics in the line currents, particularly if there are triple harmonics, i.e. 3rd, 9th, 27th, etc., as these do not cancel out in the neutral but sum. Please see the figure showing that the third harmonics in each of the three phases in the drawing are in phase and will sum in the neutral.

The neutral current I_{bn} is given by

$$I_{bn} = \frac{3h}{100} I_{nL}$$

where:
I_{bn} = neutral current due to third harmonic currents
I_{nL} = fundamental line current
h = third harmonic as a percentage of the fundamental line current

Table 17.1 Harmonic current correction factors

3rd harmonic content of line current (%)	Neutral current as a percentage of line current[1] (%)	Cable selection[2]	Correction factor[3]
0–15	0–45	Line current -	1
15–33	45–99	Line current -	0.86
33–45	99–135	Neutral current	0.86
>45	>135	Neutral current	1

Notes:

1. Neutral current for 3rd harmonics is three times the phase (line) 3rd harmonic.
2. When the neutral current exceeds the line current, selection is based on the neutral current.
3. Harmonic currents reduce the rating of a cable. Where the neutral current exceeds 135% of the line current, selection is based on the neutral current and no correction factor is applied.

Table 17.2 Calculation of current-carrying capability

Load current, I_{bph}(A)	Triple harmonic content, h (%)	Neutral current, I_{bn}(A)	Rating selection basis	Current-carrying capacity required, I_z(A)
100	0–15	0–45	I_{nL}	100
100	15–33	45–99	$I_{nL}/0.86$	115
100	33–45	99–135	$I_{bn}/0.86$	115–157
100	>45	>135	I_{bn}	>135

Where:
Neutral current, $I_{bn} = 3h/100 I_{nL}$ for triple harmonics.
I_{nL} *is the fundamental line current*

That is, the neutral current is three times the triple harmonic currents in the lines.

The effect of this harmonic neutral current is to derate the cable. If the triple harmonic content exceeds 10 per cent the neutral conductor should not be of reduced cross-sectional area. If the harmonic content exceeds 15 per cent the cable must be derated (see Tables 17.1 and 17.2).

Example:

Consider a 3-phase load of 200 A with a harmonic content of 20 per cent to be supplied by a four-core armoured thermoplastic insulated cable on a cable tray. Table 17.2 advises a cable of rating 200/0.86 be selected – that is 233 A. Table 4D4A of BS 7671 says a 95 mm^2 cable must be selected. Without harmonics a 70 mm^2 cable would be suitable.

17.3 Voltage drop

Triple harmonics have a compound effect on voltage drop. As well as producing increased voltage drop because of the current in the neutral, the voltage drop is also increased because the triple harmonics increase the effective inductance of the cable. Inductive reactance ($2\pi fL$) is proportional to frequency: the higher the frequency the higher the inductive reactance.

The general equation for voltage drop from Appendix 4 of BS 7671 is as below:

$$\text{Voltage drop} = \frac{LI_b}{1000}\left\{\cos\o\left(\frac{mv}{\frac{A}{m}}\right)_r + \sin\o\left(\frac{mv}{\frac{A}{m}}\right)_x\right\}$$

where:
$L$is the length of the cable
I_b is the load
cos ø is the power factor
$(\text{mV/A/m})_r$ and $(\text{mV/A/m})_x$ are voltage drop values (in mV per amp per metre) given in Appendix 4 for the appropriate cable.

For cable sizes up to 16 mm^2, the simplified formula can be used as follows:

$$\text{Voltage drop} = \frac{LI_b}{1000}\cos\o\{(\text{mV/A/m})_r\}$$

These formulae, when used for three-phase circuits, assume a balanced load, that is a negligible neutral current. For a load with a high third harmonic content the neutral current is not zero and the inductance of the cable increases. The revised voltage drop formulae are given by the following.

17.3.1 For cable sizes larger than 16 mm^2

$$\text{Voltage drop} = \frac{LI_b}{1000}\left\{\cos\o\left(\frac{mv}{\frac{A}{m}}\right)_r\left[1+\frac{3h}{100}\right] + \sin\o\left(\frac{mv}{\frac{A}{m}}\right)_x\left[1+\frac{9h}{100}\right]\right\}$$

where:
L is the length of the cable
I_b is the load
cos ø is the power factor
$(\text{mV/A/m})_r$ and $(\text{mV/A/m})_x$ are voltage drop values (in mV per amp per metre) given in Appendix 4 for the appropriate cable.

(*Note:* Readers proving this equation for themselves will note that the neutral current is three times the line third harmonic, the neutral inductive voltage drop is trebled and, not to be forgotten, the line voltage drop is also slightly increased as its inductance to the third harmonic element is trebled.)

17.3.2 For cable sizes 16 mm² and smaller

$$\text{Voltage drop} = \frac{LI_b}{1000}\left\{\cos\phi\left(\frac{\frac{mv}{A}}{m}\right)_r\left[1+\frac{3h}{100}\right]\right\}$$

17.3.3 Example with cable size over 16mm²

Consider again the load of 200 A with a harmonic content of 20 per cent to be supplied by a four-core cable on a cable tray. Table 17.2 advises a cable of rating 200/0.86 be selected – that is 233 A. Table 4D4A of BS 7671 says a 95 mm² cable must be selected (and not 70 mm² if the third harmonic was neglected). Assume 50 m length and a power factor of 0.8.

$$\text{The voltage drop} = \frac{50\times 200}{1000}\left\{0.8\times 0.41\left(1+\frac{3\times 20}{100}\right)\right.$$

$$\left. +0.6\times 0.135\left(1+\frac{11\times 20}{100}\right)\right\}$$

The voltage drop = 7.8 V compared with 4.1 volts if third harmonics are neglected

The effect is most pronounced for large single-core cables, as they have a relatively high inductance.

Consider four 630 mm² single-core copper thermosetting insulated cables (BS 7671 Table 4E1) of length 50 m, laid flat and touching, supplying a load of 1000A with a third harmonic content of 20 per cent at a power factor of 0 9.

$$\text{The voltage drop} = \frac{50\times 1000}{1000}\left\{0.9\times 0.071\left(1+\frac{3\times 20}{100}\right)\right.$$

$$\left. +0.43\times 0.160\left(1+\frac{11\times 20}{100}\right)\right\}$$

The voltage drop = 16.1 V

If harmonics are neglected voltage drop is 6.6 V.

17.4 Overcurrent protection

High harmonic currents in the load do not affect fault current calculations, as fault currents are generally determined by circuit characteristics and not load characteristics, on the presumption that the fault currents are significantly higher than load currents. If this is not the case, allowances will need to be made. However, for overload protection this is not so. The usual formula

$$I_z \geq I_n \geq I_b$$

is applicable for a triple harmonic content up to 15 per cent. For greater harmonic contents, selection can be made as follows, where the device rating I_n installed in the line conductors has been selected on the basis of the line current I_b:

for triple harmonic content	0-15%	$I_z \geq I_n$
for triple harmonic content	>15-30%	$I_z \geq I_n/0.86$
for triple harmonic content	>33-45%	$I_z = 3hI_n/86$
for triple harmonic content	above 45%	$I_z = 3hI_n/100$

and $I_n > I_b$

where:
I_n is the rating of the overcurrent device in the line conductors
h is the percentage triple harmonic
I_z is the current-carrying capacity of the cable under particular installation conditions.

Overcurrent protection may be provided by devices in the line conductors; however, it may be appropriate to fit overcurrent detection in the neutral, which must disconnect the line conductors, but not necessarily the neutral (see Regulation 431.2.1).

For PEN conductors in TN-C or TN-C-S systems, the PEN conductor must not be switched (Regulation 537.1.2). It may be appropriate to fit an overcurrent device in the neutral. However, this must disconnect the line conductors and will not necessarily provide overload protection unless carefully selected with knowledge of the harmonic current.

With a triple harmonic content exceeding 33 per cent of the fundamental, neutral currents can exceed the line currents. There are then certain attractions in fitting the overcurrent detection in the neutral conductor; however, this overcurrent detection must disconnect the line conductors and care must be taken in adopting this approach, as there is a presumption that the harmonic content will remain over the life of the installation. It is perhaps preferable to degrade the overcurrent protection in the line conductors accordingly; this is more of a fail-safe approach.

Example:

Consider a load of 100 A with a harmonic content of 50 per cent.

This load can be protected by 100 A devices in the line conductors. The neutral current I_{bn} is given by:

$$I_{bn} = \frac{3h}{100} . I_{bph}$$

$$\text{hence } I_{bn} = \frac{3\times 50}{100} .100 = 150\text{ A.}$$

The cable rating I_z should be at least 150 A.

17.5 Harmonics in line conductors

Paragraph 5.6 of Appendix 4

The rms value I of a complex wave is given by

$$I = \sqrt{\frac{I_1^2 m}{2} + \frac{I_2^2 m}{2} + \cdots \ldots \ldots \frac{I_n^2 m}{2}}$$

equation 17.6

where I_{nm} is the peak value of wave m

or by

$$I = \sqrt{I_1^2 + I_2^2 + \cdots \ldots I_n^2}$$

equation 17.7

where I_n is the rms value of wave m.

If the design current I_b includes all the harmonics, that is it is given by equation 17.7 after making allowances for triple harmonic currents in the neutral, no further allowances for harmonics with respect to current rating need be made. If however the rms value of design current I_{fb} does not include the harmonics (that is, it is just the fundamental) an allowance must be made for the effect on the line current.

True rms design current $I_b = C_h I_{fb}$

$$\text{where } I_b = \sqrt{\frac{I_f^2 + I_2^2 + \cdots \ldots I_n^2}{I_f^2}}$$

Chapter 18
Calculations

18.1 Introduction

This chapter provides examples of typical calculations, including those for:

(i) cables buried in the ground;
(ii) short-circuit and earth fault loop impedances in a large installation; and
(iii) derivation of the standard circuits in the Electrician's Guide to the Building Regulations and the *On-Site Guide*.

The equipment data used is tabulated in Appendix C.

18.2 Calculation of tabulated current-carrying capacity, I_t

Consider a copper thermosetting cable to be laid underground at a depth of 0.7m, in one of a group of three touching ducts. Also consider for ease of comparison a load estimated at less than 80 A with a 100 A overcurrent device. Calculations of the minimum tabulated current-carrying capacity I_t necessary are made below assuming three different scenarios:

(i) simultaneous overload of all cables;
(ii) overload but not simultaneous; and
(iii) no overload.

(i) Simultaneous overload

Assuming:

(a) overload protection is required
(b) simultaneous overload is possible

then from paragraph 5.1.2 of Appendix 4 of BS 7671, current-carrying capacity (I_t) must be:

$$I_t \geq \frac{I_n}{C_g C_a C_s C_d C_i C_f C_c} \qquad \text{equation 2}$$

where:

I_t is the tabulated current-carrying capacity of a cable found in Appendix 4 of BS 7671

C_a is the rating factor for ambient temperature, see Tables 4B1 and 4B2 of BS 7671

C_g is the rating factor for grouping, see Tables 4C1 to 4C6 of BS 7671

C_i is the rating factor for conductors surrounded by thermal insulation, see Table 52.2 of BS 7671

C_f is a rating factor applied when overload protection is being provided by overcurrent devices with fusing factors greater than 1.45 (for rewirable fuses)

C_s is the rating factor for thermal resistivity of soil, see Table 4B3 of BS 7671

C_c is the rating factor for buried circuits; C_c is 0.9 for cables buried in the ground, otherwise is 1

C_d is correction factor for depth of buried cable, see Table 4B4.

Assume that the thermosetting cables are laid in standard UK conditions (this would not be appropriate for cables laid in rubble, cinders). See 16.1.

United Kingdom standard conditions:

Soil thermal resistivity	1.2 K.m/W
Ground ambient temperature	15 °C
and $C_a = 1.04$	(Table 4B2)
$C_g = 0.75$	(Table 4C3)
$C_i = 1$	
$C_f = 1$	
$C_s = 1.13$	(Table 4B3)
$C_c = 0.9$	(para 5.1.1 of Appendix 4)
$C_d = 1$	(Table 4B4)

$$\text{Therefore: } I_t \geq \frac{100}{0.75 \times 1.04 \times 1.13 \times 1 \times 1 \times 1 \times 0.9}$$

$$I_t \geq 126.1\,\text{A}$$

(ii) Overload not simultaneous

Assuming:

(a) overload protection is required
(b) simultaneous overload is not possible

then current carrying capacity must be not less than the largest of:

$$I_t \geq \frac{I_b}{C_g C_a C_s C_d C_i C_f C_c}$$ Equation (3) of Appendix 4 of BS 7671

and

$$I_t \geq \frac{1}{C_a C_s C_d C_i}\sqrt{\left[\frac{I_n}{C_c C_f}\right]^2 + 0.48 I_b^2 \left(\frac{1-C_g^2}{C_g^2}\right)}$$ Equation (4) of Appendix 4 of BS 7671

where:

I_b is the design current of the circuit, i.e. the current intended to be carried by the circuit in normal service

I_n is the nominal current or current setting of the device protecting the circuit against overcurrent

That is,

$$I_t \geq \frac{80}{0.75{\times}1.04{\times}1.13{\times}1{\times}1{\times}1{\times}0.9}\ I_t \geq 100.8\text{A}$$

$$\text{and } I_t \geq \frac{1}{1.04{\times}1.13{\times}1{\times}1}\sqrt{\left[\frac{100}{0.9{\times}1}\right]^2 + 0.48 \times 80^2\left(\frac{1-0.75^2}{0.75^2}\right)}\ I_t \geq 103.3\text{A}$$

(iii) No overload

Current carrying capacity I_t must be:

$$I_t \geq \frac{I_b}{C_g C_a C_s C_d C_i C_f}$$ from Equation 5 of Appendix 4 of BS 7671

$$I_t \geq \frac{80}{0.75 \times 1.04 \times 1.13 \times 1 \times 1 \times 1}$$

$I_t \geq 90.8$ A

Summary:
Minimum tabulated current-carrying capacity necessary allowing for:

(i)	Simultaneous overload of all cables	$I_t \geq 126.1$ A
(ii)	Overload but not simultaneous	$I_t \geq 103.3$ A
(iii)	No overload	$I_t \geq 90.8$ A.

18.3 Fault current calculations

In this section we will not be considering the determination of loads or the selection of cable and equipment ratings, as these are considered elsewhere. We will be carrying out specific fault current calculations. Two basic fault current calculations needed to be carried out are as follows:

1. The maximum prospective fault current:

$$I_{sc} = \frac{C_{max}U_o}{Z_x + Z_D}$$

2. The minimum prospective earth fault current:

$$I_{ef} = \frac{C_{min}U_o}{Z_x + Z_D + Z_1 + Z_2 + Z_{PEN}}$$

These two equations are derived in section 6.3.3

Figure 18.1 shows a typical distribution system for which the maximum prospective fault current and maximum earth loop fault impedance are calculated at the busbars and final equipment for the installation. The equipment impedances are given in the tables of this chapter.

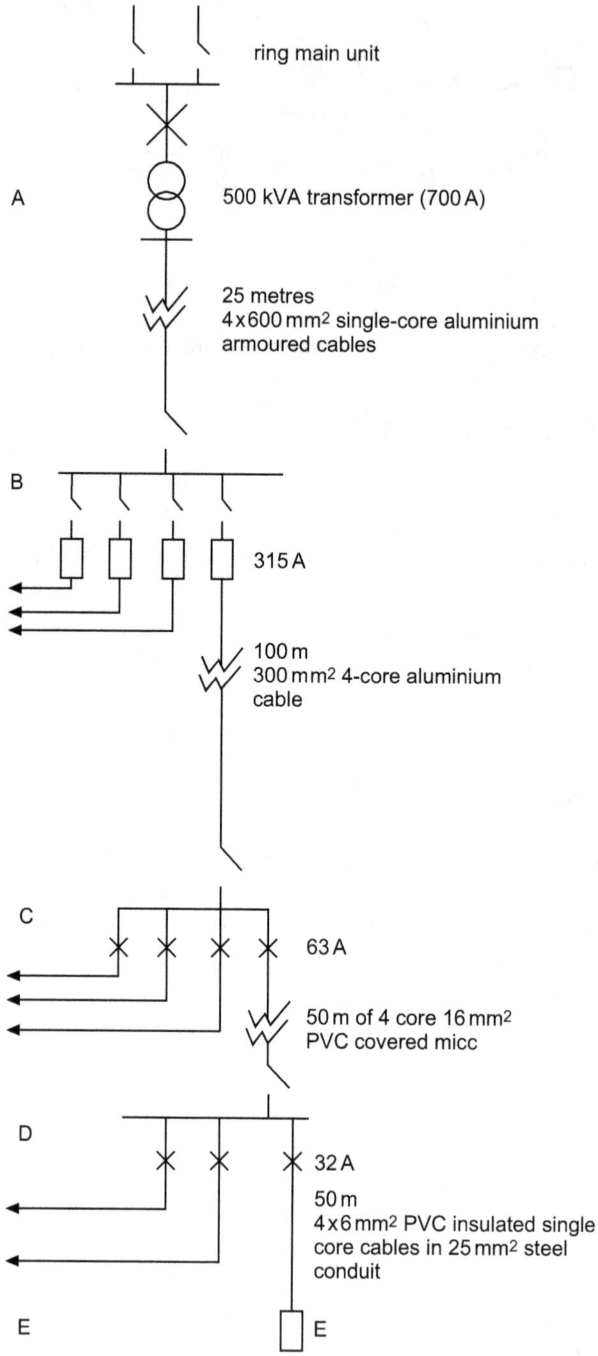

Figure 18.1 Typical distribution system

Calculation of prospective fault currents for a distribution system as Figure 18.1. Impedances are taken from the tables of this chapter.

1. Three-phase to earth current at B

	Impedance (Ω)	
	r	x
500 kVA transformer (Table C5)	0.0051	0.0171
25 metres of 600 mm^2 single-core aluminium armoured (Table C9) phase impedance 25 × (0.0515r + 0.09x)/1000	0.0013	0.0023
Temperature correction (Note 1)	–	–
Total phase impedance at B (Z_{sc})	0.0064	0.0194

$$Z_{sc} = \sqrt{r^2 + x^2} = \sqrt{0.0064^2 + 0.0194^2} = 0.0204\,\Omega$$

$$I_{sc} = \frac{c_{max} U_0}{0.0204} = \frac{1.1 \times 230}{0.0204} = 12404\,A$$

Switchgear at B must have a fault rating exceeding 13 kA. BS 88-2 fuses have a fault rating of 50 kA (see Table 6.9), so would be suitable; circuit-breakers would need an I_{cs} rating of say 20 kA.

2. Three-phase to earth fault current at C

	Impedance (Ω)	
	r	x
Impedance at B from 1 above	0.0064	0.0194
100 metres of 300 mm^2 4-core Al s.w.a. PVC phase impedance Table C7 100 × (0.100r + 0.08x)/1000	0.0100	0.0080
Temperature correction (Note 1)	–	–
Total phase impedance at C (Z_{sc})	0.0164	0.0274

Note 1: Temperature correction is not applied for three-phase to earth faults where the worst condition is a 'cold' installation.

$$Z_{sc} = \sqrt{r^2 + x^2} = \sqrt{0.0164^2 + 0.0274^2} = 0.0319\,\Omega$$

$$I_{sc} = \frac{C_{max} U_0}{0.0319} = \frac{1.1 \times 230}{0.0319} = 7932\text{ A}$$

The breaking capacity of the circuit-breakers at C have to be suitable for a fault level of 7 900 A, say an I_{cs} rating of 10 kA, and must be set to discriminate with the overcurrent device(s) at B.

3. Earth fault loop impedance at C

	Impedance (Ω)	
	r	x
Transformer	0.0051	0.0171
25 m of 600 mm^2 A1 single		
(a) phase impedance at 20 °C	0.0013	0.0023
(b) neutral/earth impedance	0.0013	0.0023
(c) correction to 70 °C $(a + b) \times 0.20$ (Note 2)	0.0005	–
100 metres of 300 mm^2 4-core A1 s.w.a.		
(a) phase impedance at 20 °C Table C7	0.0100	0.0080
(b) armour impedance at 20 °C Table C7		
$100_x(0.52r + 0.3x)/1{,}000\ \Omega$	0.0520	0.030
(c) correction of (a) to 70 °C(× 0.20) (Note 2)	0.0020	–
(d) correction of (b) to 60 °C(× 0.18) (Note 3)	0.0093	–
Total phase impedance at C	0.0815	0.0597

Note 2: Correction factor from 20 °C to70 °C for copper and aluminium conductors is (70 – 20) × 0.004 = 0.20 (see Table C.17). The correction factor is only applied to the resistive component of the impedance.

Note 3: Correction factor from 20 °C to 60 °C for cable armouring (see Table 54B of BS 7671) is (60 – 20) × 0.0045 = 0.18 (see Table C.17).

$$Z_{sc} = \sqrt{r2 + x2} = \sqrt{0.0815^2 + 0.0597^2} = 0.10.11\,\Omega$$

$$I_{ef} = \frac{C_{\min} U_0}{0.1011} = \frac{0.95 \times 230}{0.1011} = 2163\ A$$

From Table 11.9, this fault current is marginally sufficient to operate a 315 A fuse at B within 5 s. It would be appropriate to check the circuit length.

4. Three-phase to earth fault current at D

	Impedance (Ω)	
	r	x
Impedance at C from 2	0.0164	0.0274
50 m of 4-core 16 mm^2 from Table C15		
MICC 50 × 1.16r/1000 at 20°C	0.0580	–
Temperature correction (Note 1)	–	–
Total phase impedance at D (Z_{sc})	0.0744	0.0274

Note 1. Temperature correction is not applied for three-phase to earth faults where the worst condition is a 'cold' installation.

$$Z_{sc} = \sqrt{r^2 + x^2} = \sqrt{0.0744^2 + 0.0274^2} = 0.0793\,\Omega$$

$$I_{sc} = \frac{C_{max} U_0}{0.0793} = \frac{1.1 \times 230}{0.0793} = 3190\,A$$

5. Earth fault loop impedance at D

	Impedance (Ω)	
	r	x
Earth fault loop impedance at C from 3	0.0815	0.0597
50 m of 4-core 16 mm²		
Loop impedance at 70 °C		
50 × (1.4 + 0.604)r/1000 (Table C15) (Note 4)	0.1002	–
Total earth fault loop impedance at D (Z_{ef})	0.1817	0.0597

Note 4: As the coefficient of resistance of MICC sheaths and cores is different, loop impedances are tabulated at full load temperature.

$$Z_{sc} = \sqrt{r^2 + x^2} = \sqrt{0.0815^2 + 0.0597^2} = 0.1913\,\Omega$$

Table 41.3 of BS 7671 indicates that this loop impedance is sufficiently low for the instantaneous operation of 63 A devices Types B and C and operation within 5 seconds for Type D (note: prior to Amd 3 it would not have been suitable for Type D).

6. Earth fault loop impedance at E

	Impedance (Ω)	
	r	x
Loop impedance at D from 5	0.1817	0.0597
50 m of 6 mm² PVC from Table C1		
50 × 3.08r/1000	0.1540	–
Correction to 70 °C 0.154 × 0.20 (Table C17)	0.0308	–
50 m of 25 mm² steel conduit Table C11		
50 × (1.6r + 1.6x)/1000	0.0800	0.0800
Temperature correction of conduit (Note 5)	–	–
Total earth fault loop impedance at E (Z_{ef})	0.4465	0.1397

Note 5: The cross-sectional area and surface area of steel conduit is such that no increase in resistance is presumed.

$$Z_{sc} = \sqrt{r^2 + x^2} = \sqrt{0.4465^2 + 0.1397^2} = 0.4678\,\Omega$$

All 32 A CB types except D operate instantaneously with a loop impedance of 0.47 Ω.

18.4 Motor circuits

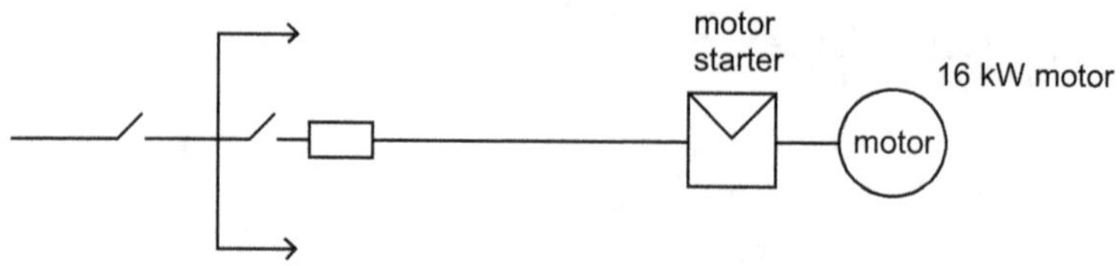

Figure 18.2 A motor sub-circuit

Consider a 16 kW motor at 400 V with a full load power factor of 0.8, see Figure 18.2.

$$I_b = \frac{motor\ kW \times 1000}{\sqrt{3} \times U \times \cos \varnothing}$$

$$I_b = \frac{16 \times 1000}{\sqrt{3} \times 400 \times 0.8} = 28.86\ \text{A}$$

Full-load current is 29 A and a 32M50 motor fuse to BS 88 is selected to prevent fusing on starting.

Circuit-breakers

For circuit-breakers the manufacturer's data must be consulted (see sample table from a Hager catalogue below). A 50 A breaker would be suitable.

Device ratings for 3-phase 400 V assisted starting star-delta

Motor			Circuit-breaker frame type and rating				Fuse
kW	hp	FLC	NC	ND	HD	HN	HRC
3	4	6.3	16	10	–	–	16
4	5	8.2	20	10	–	–	16
5.5	7.5	11.2	32	16	16	–	20
15	20	27	–	40	40	–	32
18.5	25	32	–	50	50	–	40
22	30	38	–	63	63	–	50
30	40	51	–	–	63	–	63
45	60	76	–	–	100	–	80
55	75	91	–	–	125	–	100
75	100	124	–	–	–	200	160

Fuses

Motor circuit fuses usually have a breaking range and utilization category of gM. These fuses have a dual current rating separated by the letter 'M' e.g. g 32M50.

The first lower current rating is the maximum continuous rating, I_n, which also determines the rating and size of equipment to which the fuse is fitted. The higher current rating, I_{ch}, is the time/current characteristic of the fuse, which indicates its ability to withstand the motor starting current. They may allow the use of lower-rated switch and/or fusegear than would be the case using gG fuselinks with a cost saving. Type gG fuselinks however may be used.

Assume that a four-core armoured thermoplastic insulated cable 80 m in length is to be used and it will be clipped to a perforated cable tray. Table 4D4A of BS 7671 indicates that a 4 mm² cable would be suitable. It is proposed to use the cable armour as the circuit protective conductor.

It needs to be confirmed that the overcurrent device selected will protect both the cores and the armour of the cable in the event of a fault.

Let us assume the motor is to be supplied from distribution board C of Figure 18.1.

	Impedance (Ω)	
	r	*x*
Earth loop impedance at C from 18.3 para 3	0.0815	0.0597
80 m of 4 mm² 4-core Cus.w.a.		
(a) phase impedance at 20 °C Col. 2 of Table C7 80 × 4.61*r*/1000	0.3688	–
(b) armour impedance at 20 °C Col. 6 of Table C7 80 × 4.60*r*/1000	0.3680	–
(c) correction of (a) to 70 °C (× 0.20) (Table C17)	0.0738	–
(d) correction of (b) to 60 °C (× 0.18) (Table C17)	0.0662	–
Total loop impedance at motor starter	0.9583	0.0597

$$Z_{sc} = \sqrt{r^2 + x^2} = \sqrt{0.9583^2 + 0.0597^2} = 0.960\,\Omega$$

$$I_{ef} = \frac{C_{\min} U_0}{0.960} = \frac{0.95 \times 230}{0.960} = 227.6\,A$$

32M50 BS 88-2 fuse

From Figure 3A3A of Appendix 3 of BS 7671, the disconnection time t for a fault current of 230 A for a 50 A fuse is 3 s. Using the formula $S \geq \frac{I\sqrt{t}}{k}$ (see section 6.3.6) where k is given by Table 43.1 of BS 7671. For the line conductor

$$S \geq \frac{230\sqrt{3}}{115} = 3.46\,\text{mm}^2.$$

The cable core size is 4 mm^2, so this is satisfactory.

For the armour $S \leq \frac{230\sqrt{3}}{51} = 7.81 \text{ mm}^2$

From Table 11.8, the area of the armouring is 35 mm^2, so this is satisfactory.

Circuit-breaker

For a circuit assuming the 50 A circuit-breaker is operating instantaneously the energy let-through will be small (see Table 11.5).

Using the formula $S \geq \frac{I\sqrt{t}}{k}$ (see section 6.3.6)

where k is given by Table 43.1 of BS 7671.

For the armour $S \geq \frac{230\sqrt{0.1}}{51} = 0.45 \text{ mm}^2$

This is satisfactory. However, a fault close up on the circuit-breaker must be considered as well, then the three-phase-to-earth fault current at C will be as calculated earlier at 7922 A. The I^2t energy let-through of the circuit-breaker at that fault level must be determined from the manufacturer's data e.g. 0.28×10^6 as per Table 11.11 of Chapter 11, to check that the armour csa is sufficient.

From table 11.12 energy let-through for the 50A device is 0.28×10^6 amperes^{2}seconds

And $S^2k^2 \geq 0.28 \times 10^6$
As $k = 51$
$S \geq 10.4 \text{ mm}^2$
Which is satisfactory as S is 35 mm^2 (from Table 11.8 or Table C.7b).

18.5 Final circuits

It is generally impractical to design final circuits on an individual basis (without a software program based on the loop impedances at each distribution board). A more practical approach is to have standard circuits designed for particular supply loop impedances. If the loop impedance at the particular board is below these values, the standard circuits can be used. The limiting factor of a standard circuit is the maximum circuit length that allows all the parameters for circuit design to be met.

In this section on final circuits, the calculations are determined for use in domestic installations, assuming the maximum supply impedances that are quoted by electricity distributors. The circuits can also be used on larger installations where the earth fault loop impedance at the distribution board is below the values assumed for the standard circuit.

18.5.1 Electricity supplies

Enquiries to an electricity company may result in the advice that supply loop impedances will not exceed:

(i) 0.35 Ω for PME supplies; and
(ii) 0.8 Ω for TN-S supplies.

and that the maximum prospective fault current will not exceed 16 000 A. These two parameters are obviously not compatible: a fault level of 16 000 A would indicate an earth loop impedance of 0.015 Ω or similar. However, as discussed earlier in this Commentary, supply parameters might change, and it is advantageous to calculate circuits assuming that the prospective fault current will be high, and the loop impedances low. Then, if the distribution network changes during the life of the installation, the design will remain valid.

18.5.2 Fault rating of switchgear

From a knowledge of the distance of the distribution board or consumer unit from the main distribution, an estimation of the attenuation of the fault level from 16 000 A can be made (see section 6.3.3). However, this is generally not necessary, as consumer units to BS EN 60439-3, Annex ZB, 'Specification for particular requirements of consumer units complete with fuses, MCBs and protective devices', are able to withstand the fault current for prospective fault levels of up to 16 kA when supplied via a Type 2 fuse to BS 88-3 rated at no more than 100 A. This is the standard 100 A cut-out fuse installed by electricity companies. The consumer units to this standard are therefore conditionally rated for a fault level of 16 000 A. This means that, above the fault ratings of the individual overcurrent devices within the consumer unit, the distributor's fuse will be providing disconnection in the event of a fault.

Designers can, of course, use this switchgear for such fault levels, when protected by devices with an equal or lower energy let-through, for fault levels up to 16 000 A.

18.5.3 Selection of overcurrent device and cables

For radial final circuits for which overload protection is to be provided:

$$I_t \geq \frac{I_n}{C_a C_g C_i C_f} \qquad \text{(a)}$$

(see 16.5.1)

for ring final circuits for which overcurrent protection is to be provided (see Regulation 433.1.204):

Current-carrying capacity I_t must satisfy the equation:

$$I_t \geq \frac{20}{C_a C_g C_i C_f} \qquad \text{(b)}$$

where:

I_t is the tabulated current-carrying capacity of a cable found in Appendix 4 of BS 7671.
C_a is the rating factor for ambient temperature, see tables 4B1 and 4B2 of BS 7671
C_g is the rating factor for grouping see tables 4C1 to 4C6 of BS 7671
C_i is rating factor for conductors surrounded by thermal insulation, see Table 52.2 of BS 7671

C_f is a rating factor applied when overload protection is being provided by an overcurrent device with a fusing factor greater than 1.45 (for rewirable fuses)

$$C_f = 1.45 \times \frac{device\ \ rating}{device\ \ fusing\ current} = 1.45\frac{I_n}{I_2}$$

I_2 is the operating current (that is the fusing current) of the device providing protection against overload), see Table 6.2

For socket circuits, it is always presumed that the overcurrent device is providing overload protection. However, there are many circumstances where the overcurrent device will not be providing overload protection, because the load is fixed, as with a shower. In these circumstances the formula for a radial circuit is

$$I_t \geq \frac{I_b}{C_a C_g C_i C_f} \qquad \text{(c)}$$

where I_b is the design current of the circuit.

(See 16.5.2)

18.5.4 Voltage drop

The equation for voltage drop for a radial circuit is given by

$$L \times I_b \times \frac{(mV/A/m)}{1000} \times C_t \leq 5\% \text{ of 230 V for other than lighting}$$

where:

I_b = design current
L = length of cable
C_t = voltage drop correction factor for conductor operating temperature
mV/A/m = voltage drop from Appendix 4 of BS 7671.

Then:

$$L_{VD} = \frac{11.5 \times 1000}{I_b (mV/A/m) \times C_t} \qquad \text{(d)}$$

where L_{VD} is the maximum cable length for other than lighting circuits if the voltage drop limitation is to be met.

For ring circuits:

$$L_{VD} = \frac{4 \times 11.5 \times 1000}{I_b (mV/A/m) \times C_t} \qquad \text{(e)}$$

where:

L_{VD} =	maximum length of cable allowed by voltage drop considerations
I_b =	design current
mV/A/m =	voltage drop from Appendix 4 of BS 7671
C_t =	voltage drop correction factor for conductor operating temperature from Appendix 4 of BS 7671.

$$C_t = \frac{230 + t_p - \left(C_a^2 C_g^2 - \frac{I_b^2}{I_t^2}\right)(t_p - 30)}{230 + t_p} \qquad \text{(f)}$$

18.5.5 Shock protection
Limited disconnection time

The shock protection requirements can be met by meeting the disconnection times of Regulation 411.3.2, by limiting the earth fault loop impedances to those of Tables 41.2 and 41.3 etc. as per Regulation 411.4.5. Where the disconnection time cannot be met supplementary bonding as per Regulation 411.3.2.6 should get a designer out of trouble.

For limited disconnection times, the equations to be met are:

For radial circuits:

$$Z_{41} \geq Z_e + (R_1 + R_2) \times C_r \times L$$

For ring circuits:

$$Z_{41} \geq Z_e + (R_1 + R_2) \times C_r \times \frac{L}{4}$$

The circuit lengths are then limited as follows:

For radial circuits:

$$L_s = \frac{Z_{41} - Z_e}{(R_1 + R_2)C_r} \qquad \text{(g)}$$

where L_s = maximum cable length if the shock protection limitation is to be met.

For ring circuits:

$$L_s = \frac{4(Z_{41} - Z_e)}{(R_1 + R_2)C_r} \qquad \text{(h)}$$

where:

L_s = maximum length of cable to meet the shock protection requirements
Z_{41} = maximum earth fault loop impedance given by the appropriate Tables 41.2 to 41.4
Z_e = earth fault loop impedance – in this section it is assumed to be 0.8 or 0.35 Ω
R_1 = resistance of the line conductor
R_2 = resistance of the protective conductor
C_r = correction factor for temperature – see the notes to Tables 41.2 etc. of BS 7671, or see Table 4.10 of this Commentary.

Values of C_r are given in Table 4.10 of this Commentary. C_r will be taken as 1.20 in these examples; this will correct from 20 °C to a conductor operating temperature of 70 °C, which is appropriate for the thermoplastic (PVC) cables we are presuming.

Adiabatic limit

Where an overcurrent device is not providing protection against overload, or where the protective conductor is of a smaller size than the line conductor, a check that the adiabatic equation:

$$S = \{\sqrt{(I^2 t)}\}/k$$

is met must be made. As discussed in section 11.4.2, it is often easier to check that this requirement is met by carrying out a circuit length calculation, assuming a maximum adiabatic loop impedance Z_a. The equations that then have to be complied with are:

For radial circuits:

$$L_a \leq \frac{Z_a - Z_e}{(R_1 + R_2)C_r} \tag{k}$$

For ring circuits:

$$L_a \leq 4 \times \frac{Z_a - Z_e}{(R_1 + R_2)C_r} \tag{l}$$

where:

Z_a = maximum adiabatic loop impedance, see section 11.4.2
Z_e = external or supply loop impedance
R_1 = resistance of the line conductor
R_2 = resistance of the protective conductor
C_r = correction factor for temperature (see Table 4.10 of this Commentary).

Typical calculations for a range of domestic circuits are tabulated, the notes providing guidance as to the assumptions being made in determining the circuit lengths.

18.5.6 RCDs and short circuit protection

Where a circuit is protected by an RCD as is usual in domestic premises, the RCD will ensure the shock protection requirements are met for a fault between line and protective conductor, but the RCD will not detect a fault between the line and neutral (a short circuit). Consequently when an RCD is installed circuit lengths will not be limited by shock protection but will be by short circuit length limitations (L_{ss}).

Regulation 435.1 states that the overload protective devices listed in Regulation 433.1.201 and having adequate rated short circuit breaking capacity may be considered to provide fault current protection to the downstream cables.

Table 18.1 20A Radial final socket-outlet circuit (Note 7)

Equation or table	Element calculated in table	Device type						
		Fuse BS				c.bs. type		
		88-3	3036	3036	88-2	B	C	D
	I_n (A)	20	20	20	20	20	20	20
	I_b (A) average	16.6	16.6	16.6	16.6	16.6	16.6	16.6
	Required I_t (A)[1]	20	27.60	27.60	20	20	20	20
4D5A(R)	Cable L/PE[2]	2.5/1.5	2.5/1.5	4.0/1.5	2.5/1.5	2.5/1.5	2.5/1.5	2.5/1.5
	Installation method[3]	A	C	100	A	A	A	A
	Cable I_t (A)	20	20	27	20	20	20	20
4D5A (R)	Ct	0.96	0.92	0.92	0.96	0.96	0.96	0.96
	(mV/A/m)	18	18	11	18	18	18	18
	L_{VD} (m)	40.1	42.3	70.1	40.2	40.2	40.2	40.2
	Shock[4]							
41.2, 41.3 (R)	$Z_{41}(\Omega)$	1.93	1.68	1.68	1.68	2.19	1.09	0.55
F1	$R_1 + R_2$(mΩ)/m	19.51	19.51	16.71	19.51	19.51	19.51	19.51
F3	Cr	1.2	1.2	1.2	1.2	1.2	1.2	1.2
	Z_e, 0.8 (Ω)							
	L_s, 0.4 s (m)	48.2	37.6	43.9	37.6	54	12.5	NP
1	Z_e, 0.35 (Ω)							
	L_s, 0.4 s (m)	67.5	56.8	66.3	56.8	73	29.4	7.2
	Adiabatic							
7 A etc.	Z_a (Ω)	3.1	4.1	4.1	2.4			
k	L_a, 0.35 (m)	117	170	310	87	Note 5	Note 5	Note 5
k	L_a, 0.8 (m)	98	208	270	68			
Maximum lengths (no RCD)	L_m 0.8 (m)	40	37	44	37	40	12	NP
Note 6	L_m 0.35 (m)	40	42	70	40	40	29	7.2
Maximum lengths with RCD	L_{m6} 0.8 (m)	40	42	70	40	40	40	40
	L_{m6} 0.35 (m)	40	42	70	40	40	40	40

Note:

1. Overcurrent protection is required.
2. 70 °C thermoplastic flat cable with protective conductor as per Table 4D5A of BS 7671.
3. Circuits have been designed for the installation reference method listed (generally A). The circuit may be used in less onerous conditions. For example a circuit designed for installation reference method A may be used for reference methods A, 100, 102, B and C.
4. Disconnection required in 0.4 seconds for TN installations.
5. See Tables 11.5 and 11.6 for minimum protective conductor sizes.
6. L_m 0.8 signifies the maximum length for external supply impedance $Z_e = 0.8\ \Omega$.
7. Socket circuits will in most circumstances be required to be additionally protected by a 30 mA RCD (Regulation 411.3.3).
8. A rule based selection may require a 4 mm^2 cable, however a designer may exercise judgment and select a 2.5 mm2 cable.
9. Design load current based on 13A at the extremity and the balance to the device rating evenly distributed.

NP Not permitted.

Table 18.2 Ring final circuit supplying socket-outlets (Note 7)

Equation or table	Element calculated in table	Device type					
		Fuse BS			cb type		
		88-3	3036	88-2	B	C	D
	I_n (A)	32	30	32	32	32	32
	I_b (A)[1&8]	25	25	25	26	26	26
433.1.204 (R)	Required I_t (A)[1]	20	20	20	20	20	20
	Cable L/PE[2]	2.5/1.5	2.5/1.5	2.5/1.5	2.5/1.5	2.5/1.5	2.5/1.5
	Installation method	A	A	A	A	A	A
4D5A (R)	I_t (A) $\geq I_z$	20	20	20	20	20	20
f	Ct	0.92	0.92	0.92	0.92	0.92	0.92
4D5A (R)	(mV/A/m)	18	18	18	18	18	18
e	L_{VD} (m)	106	111	106	106	106	106
	Shock[4]						
41.2,.3 (R)	Z_{41}(Ω)	0.91	1.04	0.99	1.37	0.68	0.34
16B(C)	$R_1 + R_2$ (mΩ/m)	19.50	19.50	19.50	19.50	19.50	19.50
4L(C)9C(OSG)	Cr	1.2	1.2	1.2	1.2	1.2	1.2
	Z_e, 0.8 (Ω)						
h	L_s (m)	19	41	32	96	NP	NP
	Z_e, 0.35 (Ω)						
h	L_s (m)	95	117	104	173	56	NP
	Adiabatic						
12 etc. (C)	Z_a (Ω)	1.0	2.40	1.04	note 5	note 5	note 5
1	L_a, 0.35 (m)	111	350	117			
	L_a, 0.8 (m)	34	273	41			
Maximum lengths	L_m, 0.8 (m)	18	41	32	96	NP	NP
(no RCD)[6]	L_m, 0.35 (m)	95	106	106	106	56	NP
Maximum lengths	L_m, 0.8 (m)	106	111	106	106	106	106
with RCD	L_m, 0.35 (m)	106	111	106	106	106	106

Notes:

1. Regulation 433.1.204 requires $I_z > 20$ A. Overcurrent protection is not required; this is significant for rewirable fuses.
2. 70 °C thermoplastic flat cable with protective conductor per Table 4D5A of BS 7671.
3. Circuits have been designed for the installation reference method listed (generally A). The circuit may be used in less onerous conditions. For example a circuit designed for installation reference method A may be used for reference methods A, 100, 102, B and C.
4. Disconnection required in 0.4 seconds for TN installations.
5. See Tables 11.5 and 11.6 for minimum protective conductor sizes.
6. L_m 0.8 signifies the maximum length for installation method A and external supply impedance $Z_e = 0.8$ Ω
7. Socket circuits will in most circumstances be required to be additionally protected by a 30 mA RCD (Regulation 411.3.3).
8. Design load current based on 20 A load at the extremity and the balance to device rating (10 or 12 A) evenly balanced. This equates to 25 A for a 30 A device. 20 A is the maximum current socket-outlets to BS 1363 are required to handle.

NP Not permitted.

Table 18.3 Cooker circuit to be protected by 30/32 A overcurrent device (up to 14.4 kW cooker – Note 7)

Equation or Table	Element calculated in table	Device type						
		Fuse BS				cb. type		
		88-3	3036	3036	88-2	B	C	D
	I_n (A)[1]	32	30	30	32	32	32	32
	I_b (A)	30	30	30	30	30	30	30
a	Req I_t (A) ≥	30	41.30	41.30	30	30	30	30
	Cable L/PE	6/2.5	6/2.5	10/4	6/2.5	6/2.5	6/2.5	6/2.5
	Installation Method	A	C	A	A	A	A	A
4D5A (R)	I_t	32	46	44	32	32	32	32
f	C_t M6	0.98	0.92	0.93	0.98	0.98	0.98	0.98
4D5A (R)	(mV/A/m)	7.30	7.30	4.4	7.30	7.30	7.30	7.30
d	L_{VD} M6 (m)	53.4	57	93	53.4	53.4	53.4	53.4
	Shock							
41.2,41.3 (R)	$Z_{41}(\Omega)$	0.91	1.04	1.04	0.99	1.37	0.68	0.34
4L(C)9C(OSG)	$R_1 + R_2$ (mΩ/m)	10.49	10.49	6.44	10.49	10.49	10.49	10.49
	Cr	1.2	1.2	1.2	1.2	1.2	1.2	1.2
16B(C)	Z_e, 0.8 (Ω) L_s, (m)	8.7	19.7	31.6	15.1	45	NP	NP
g	Z_e, 0.35 (Ω) L_s, (m)	44.5	54.8	89.3	50.8	81	24	NP
	Adiabatic							
7A etc.	$Z_a(\Omega)$	1.6	2.9	3.55	1.82	Note 5	Note 5	Note 5
k	L_a, 0.32 (m)	96	202	413	116			
k	L_a, 0.8 (m)	60	161	355	81	OK	OK	OK
Maximum lengths								
Maximum lengths (no RCD)[6]	L_m, 0.8 (m)	8	19	31	15	45	NP	NP
	L_m, 0.35 (m)	44	54	89	50	53	23	NP
Maximum lengths with RCD	L_m, 0.8 (m)	53	57	93	53	53	53	53
	L_m, 0.35 (m)	53	57	93	53	53	53	53

Notes:

1. Using Table D1 of Appendix D, assume a cooker rated at up to 14.4 kW at 240 V and assuming a socket is incorporated in the cooker control unit, then $I_b = 10 + 0.3(60 - 10) + 5 = 30$ A.
2. 70 °C thermoplastic flat cable with protective conductor per Table 4D5A of BS 7671.
3. Circuits have been designed for the installation reference method listed (generally A). The circuit may be used in less onerous conditions. For example, a circuit designed for installation reference method A may be used for reference methods A, 100, 102, B and C.
4. Disconnection required in 0.4 seconds for TN installations for 30 and 32 A circuits.
5. See Tables 11.5 and 11.6 for minimum protective conductor sizes.
6. L_m 0.8 signifies the maximum length for installation method A and external supply impedance $Z_e = 0.8\ \Omega$.
7. Cooker circuits with a cooker control unit with a socket-outlet will be required to be additionally protected by a 30 mA RCD (Regulation 411.3.3).

NP Not permitted.

Table 18.4 Shower circuits for up to 7.21 kW and 9.6 kW ratings (at 240 V), RCD required – see Note 7

Equation or table	Element calculated in table	Device type											
		Fuse BS						cb type					
		88-3		3036		88-2		B		C		D	
	I_n (A)	32	45	30	45	32	40	32	40	32	40	32	40
	I_b(A)	30	40	30	40	30	40	30	40	30	40	30	40
	rating kW	7.20	9.60	7.20	9.60	7.20	9.60	7.20	9.60	7.20	9.60	7.20	9.60
	Req I_t (A) ≥ 30	30	40	30[1]	40[1]	30	40	30	40	30	40	30	40
a	Cable L/PE[2]	6/2.5	10/4	6/2.5	10/4	6/2.5	10/4	6/2.5	10/4	6/2.5	10/4	6/2.5	10/4
4D5A (R)	Installation method[3]	A	A	A	A	A	A	A	A	A	A	A	A
	cable I_t (A)	32	44	32	44	32	44	32	44	32	44	32	44
f	C_t	0.98	0.98	0.98	0.98	0.98	0.98	0.98	0.98	0.98	0.98	0.98	0.98
4D5B (R)													
d	(mV/A/m)	7.30	4.40	7.30	4.40	7.30	4.40	7.30	4.40	7.30	4.40	7.30	4.40
d	L_{VD} (m)	53.4	66.9	53.4	66.9	53.4	66.9	53.4	66.9	53.4	66.9	53.4	66.9
	Shock[4]												
41.2, 41.3 (R)	Z_{41} (Ω)	0.91	1.04	1.04	1.51	0.94	1.30	1.37	1.09	0.68	0.55	0.34	0.27
16B (C)	$R_1 + R_2$ (Ω)	10.49	6.44	10.49	6.44	10.49	6.44	10.49	6.44	10.49	6.44	10.49	6.44
4L (C)	Cr	1.2	1.2	1.2	1.2	1.2	1.2	1.2	1.2	1.2	1.2	1.2	1.2
	Z_e, 0.8 (Ω)												
g	L_s, (m)	8.7	25	19	91	15	64	39	30	NP	NP	NP	NP
g	Z_e, 0.35(Ω)												
	L_s, (m)	44	84	54	150	50	122	75	89	27	22	NP	NP
Note 3	Adiabatic												
k	Z_a (Ω)	1.64	1.1	3.16	1.92	1.9	1.44						
k	L_a, 0.35(m)	102	97	223	202	124	140	Note 5	Note 5	Note 5	Note 5	Note 5	Note 5
	L_a, 0.8(m)	66	38	187	144	88	82						
Maximum lengths													
Maximum lengths with RCD	L_m, 0.8 (m)	53	66	53	66	53	66	53	66	53	66	53	66
	L_m, 0.35 (m)	53	66	53	66	53	66	53	66	53	66	53	66

Notes

1. Overcurrent protection is not required as shower loads are fixed; this is significant for rewirable fuses.
2. 70 °C thermoplastic flat cable with protective conductor per Table 4D5A of BS 7671.
3. Circuits have been designed for the installation reference method listed (generally A). The circuit may be used in less onerous conditions. For example, a circuit designed for installation reference method A may be used for reference methods A, 100, 102, B and C.
4. Disconnection required in 0.4 seconds for TN installations for 30 and 32 A circuits, 5 seconds for 40 and 45 A.
5. See Tables 11.5 and 11.6 for minimum protective conductor sizes.
6. L_m 0.8 signifies the maximum length for installation method A and external supply impedance Z_e = 0.8 Ω.
7. Shower circuits are required to be additionally protected by a 30 mA RCD (Regulation 701.411.3.3).

NP Not permitted.

Table 18.5 Lighting circuits (take note of device ratings In and Note 1)

Equation or table	Element calculated in table	Device type					
		Fuse BS			cb type		
		88-3	3036	88-2	B	C	D
	I_n (A)[1]	5	5	6	10	6	6
	I_b (A)[1]	5	5	5	5	5	5
a	Required I_t (A) ≥	5	6.90	10	10	5	6
	Cable L/PE	1.5/1.0	1.5/1.0	1.5/1.0	1.5/1.0	1.5/1.0	1.5/1.0
4D5A (R) note 4	Installation Method	103	103	103	103	103	103
	Cable I_t (A)	10	10	10	10	10	10
f	C_t	0.88	0.88	0.88	0.88	0.88	0.88
4D5A (R)	(mV/A/m)	29	29	29	29	29	29
d	L_{VD} M6 (m)	108	108	108	108	108	108
	Shock[4]						
41.2,41.3 (R)	Z_s, 5 s (Ω)	9.93	9.10	7.80	4.37	2.19	1.82
16B (C)	$R_1 + R_2$ (Ω)	30.20	30.20	30.20	30.20	30.20	30.20
4L (C)	C_r	1.2	1.2	1.2	1.2	1.2	1.2
g	Z_e, 0.8 (Ω)						
	L_s (m)	251	229	193	92	38	25
	Z_e, 0.35 (Ω)						
	L_s (m)	264	241	205	104	50	38
	Adiabatic						
	Z_a (Ω)						
7A etc. (C)	L_a 0.35 (m)	20	22	19	Note 5	Note 5	Note 5
g	L_a 0.8 (m)	542	597	502			
g	L_a 0.8 (m)	529	585	514			
Maximum lengths							
Maximum lengths (no RCD)	L_m 0.8 (m)	108	108	108	92	38	32
Note 6	L_m 0.35 (m)	108	108	108	104	50	47
Maximum lengths with RCD	L_m 0.8 (m)	108	108	108	108	108	108
	L_m 0.35 (m)	108	108	108	108	108	108

Notes:

1. Load has been presumed to be 5 A evenly distributed even when 10 A devices are installed (10 A selected for type B cbs to avoid unwanted tripping).
2. 70 °C thermoplastic flat cable with protective conductor per Table 4D5A of BS 7671.
3. Circuits have been designed for the installation reference method listed (generally A). The circuit may be used in less onerous conditions. For example, a circuit designed for installation reference method A may be used for reference methods A, 100, 102, B and C.
4. Disconnection required in 0.4 seconds for TN installations.
5. See Tables 11.5 and 11.6 for minimum protective conductor sizes.
6. L_m 0.8 signifies the maximum length for installation method A and external supply impedance $Z_e = 0.8\ \Omega$

Table 18.6 Immersion heater circuit

Equation or table	Element calculated in table	Device type					
		Fuse BS			cb type		
		88-3	3036	88-2	B	3 or C	D
	I_n (A)[1]	16	15	16	16	16	16
	I_b (A)[1]	12.5	12.5	12.5	12.5	12.5	12.5
a	Required I_t (A) ≥	16	17.24	16	16	16	16
	Cable L/PE	2.5/1.5	2.5/1.5	2.5/1.5	2.5/1.5	2.5/1.5	2.5/1.5
4D5A (R)	Installation method	101	101	101	101	101	101
f	cable I_t (A)	13.5	13.5	13.5	13.5	13.5	13.5
	C_t	0.94	0.94	0.94	0.94	0.94	0.94
4D5A	(mV/A/m)	18	18	18	18	18	18
d	L_{VD} (m)	54	54	54	54	54	54
	Shock[4]						
41.32, 41.3 (R)	Z_{41} (Ω)	2.3	2.43	2.43	2.73	1.37	0.68
	$R_1 + R_2$ (Ω)	19.51	19.51	19.50	19.50	19.50	19.50
16B (C)	C_r	1.2	1.2	1.2	1.2	1.2	1.2
4L (C)	Z_e, 0.8 (Ω)						
g	L_s (m)	64	69	69	76	21	NP
	Z_e, 0.35 (Ω)						
	L_s (m)	83	88	88	96	40	13
	Adiabatic						
12A etc. (C)	Z_a (Ω)	4.6	6.38	4.8	Note 5	Note 5	Note 5
	C_r	1.2	1.2	1.2			
k	L_a, 0.8 (m)	181	257	191			
Maximum lengths (no RCD)	L_m 0.8 (m)	54	54	54	54	21	NP
Note 6	L_m 0.35 (m)	54	54	54	54	40	13
Maximum lengths (no RCD)	L_m 0.8 (m)	54	54	54	54	54	54
Note 6	L_m 0.35 (m)	54	54	54	54	54	54

Notes:

1. Load has been presumed to be 12.5 A (3 kW at 240 V).
2. 70 °C thermoplastic flat cable with protective conductor per Table 4D5A of BS 7671.
3. Circuits have been designed for the installation reference method listed (generally A). The circuit may be used in less onerous conditions. For example, a circuit designed for installation reference method A may be used for reference methods A, 100, 102, B and C.
4. Disconnection required in 0.4 seconds for TN installations.
5. See Tables 11.5 and 11.6 for minimum protective conductor sizes.
6. L_m 0.8 signifies the maximum length for installation method A and external supply impedance Z_e = 0.8 Ω.

NP Not permitted

18.6 Appendix 15

18.6.1 Ring final circuits

The appendix provides guidance on the design of a ring circuit in a household or similar premises such that the load is unlikely to exceed for long periods the minimum 20 A current-carrying capacity of the ring circuit cable.

As noted at the start of the Appendix, calculations still have to be carried out on the ring and radial circuits to confirm that requirements of Chapters 41, 42 and 43 and Section 525 are met.

18.6.2 Spur lengths from rings

The figure in Appendix 15 shows spurs to the ring, but does not provide guidance on how long the spurs can be. For the loop impedance of the spur not to exceed the loop impedance at the farthest point of the ring, spur lengths must not exceed those given in Table 18.7 and Figure 18.3.

Table 18.7 Length of spurs from rings

Position	Position around ring	Spur length	Spur length for 110m ring (m)
A	1/8L	L/8	14
B	1/4L	L/16	7
C	3/8L	L/64	4
D	1/2L	0	0

Where L is the end-to-end cable length.

18.6.3 Accessibility of junction boxes

The guidance given in the appendix is that junction boxes with maintenance-free terminals need not be accessible. Regulation 526.3 allows equipment complying with BS 5733 for a maintenance-free accessory and marked to be inaccessible if installed in accordance with the manufacturer's instructions.

18.6.4 Radial final circuits

Little information is given in the drawings of radial final circuits in Appendix 15 as to how the requirements of Regulation 433.1 are to be met. However, unlike the rings, the circuit protective devices are able to provide overload protection. Appendix 5 of the 15th Edition of the IEE Wiring Regulations provided guidance on the maximum floor area to be served by ring and radial circuits for household installations. An updated version of this guidance is provided in Appendix 8 of the *On-Site Guide* and is reproduced in Table 18.8.

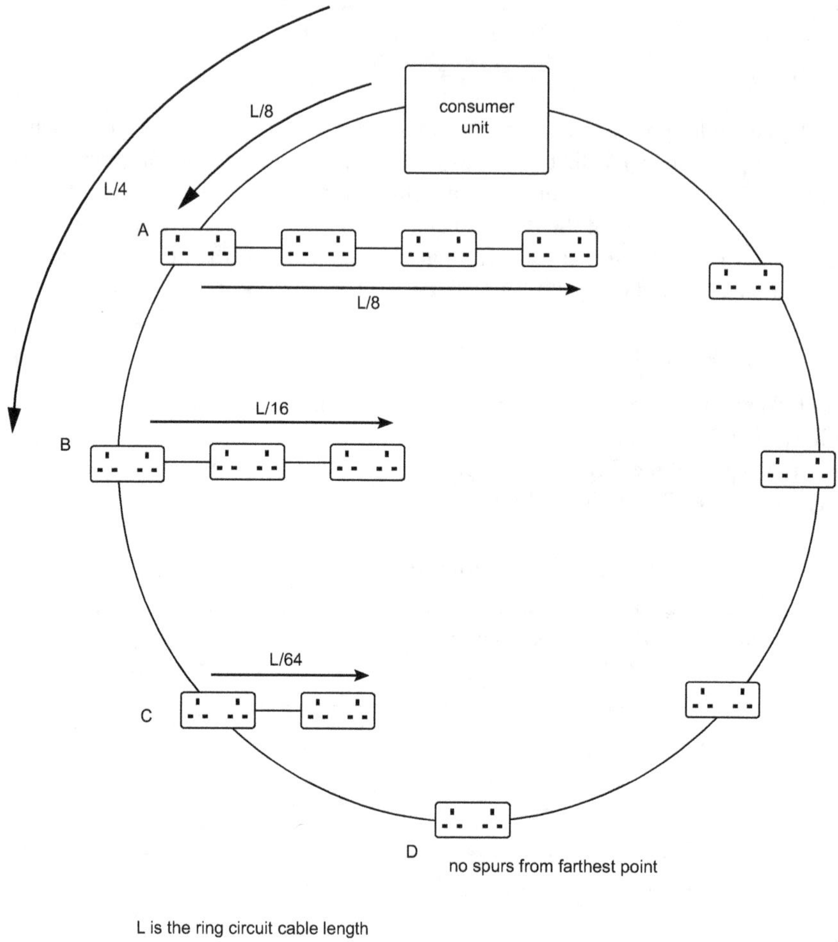

Figure 18.3 Ring circuit spurs

Appendix 15 shows the use of 4 mm^2 cables, with 4 mm^2 spurs. However, it must be noted that it may be very difficult if not impossible to spur a 4 mm^2 cable from a radial, as the accessory terminals are likely to be of insufficient size and the cutting of conductor strands is deprecated.

Table 18.8 Final circuits using BS 1363 socket-outlets and connection units

Type of circuit		Overcurrent protective device rating (A)	Minimum live conductor cross-sectional area* (mm^2)		Maximum floor area served (m^2)
			Copper conductor thermoplastic or thermosetting insulated cables	Copper conductor mineral insulated cables	
1	2	3	4	5	6
A1	Ring	30 or 32	2.5	1.5	100
A2	Radial	30 or 32	4	2.5	75
A3	Radial	20	2.5	1.5	50

* See Section 7 and Table 7.1 of the *On-Site Guide* for the minimum csa for particular installation reference methods. It is permitted to reduce the values of conductor cross-sectional areas for fused spurs.

Source: Table 8A of the *On-Site Guide,* 'Final circuits using BS 1363 socket-outlets and connection units'

Appendix A

The first century of the Wiring Regulations

Brian Bowers, PhD, CEng, MIEE

Abstract: Every edition of the Wiring Regulations has attracted criticism and praise, but no one today questions the IET's right to make the rules. It was not always so. For many years insurance companies produced their own rules, which had to be followed before they would insure buildings with electrical installations. Those responsible for designing and installing wiring and equipment found the multiplicity of rules very irksome, and they pressed for uniformity. The Institution's rules were only gradually accepted as the Institution itself gained acceptance as an authoritative body. By the First World War, however, virtually all insurers accepted the IEE Rules. Now in their 17th Edition, the Rules have been changed as ideas and technology have developed. Many of the early rules sound distinctly strange today: one hundred years ago it was not permitted to have a potential difference of more than 200 V between two points in the same room, but there was no suggestion that anything should be earthed. The language used has always brought complaints, for being too complex and poorly expressed. One early defender of the IEE remarked that the Rules were 'originally drafted in good English but afterwards amended in committee'.

A.1 Introduction

On 11 May 1882, the IEE Council appointed a committee to 'consider and report to the Council upon the rules which they would recommend for adoption for the prevention of fire risks arising from the use of the electric light'. The work took six weeks, and, at another Council meeting, on 21 June 1882, the proposed rules were adopted. Council resolved that the rules should be 'published together with the names of the Committee both in the Society's Journal and in a separate pamphlet form'.

A.2 The need for rules

In 1882 electric lighting was developing rapidly. The new Victoria Embankment in London had been lit with arc lamps since December 1878. Several other parts of London had electric arc lighting in 1881. Private houses first received a supply from a central generating station in the autumn of that year, when the first public electricity supply began at Godalming, and there were numerous private house lighting installations.

Arc lights made use of the brilliant light created when a discharge between two carbon rods brought the tips to white heat. The voltage across an arc lamp was about 40 V, the current typically 10 to 20 A. They were usually operated in series in a constant current system, and so voltages of several hundred or even a few thousand volts were commonplace.

The practical filament lamp, developed in the late 1870s, required a constant voltage supply, and so lamps were nearly always operated in parallel. Currents could be high but voltages were low, usually about 50 V and never more than 100 V in single-circuit systems in the 1880s, although three-wire and five-wire systems would have two times or four times those voltages between the outer conductors.

The risk of shock in a high voltage system had been understood since the eighteenth century, when the American scientist and statesman, Benjamin Franklin, showed that lightning was an electrical phenomenon. Less well known is his Russian contemporary, Professor Richmann, who tried to repeat Franklin's experiments in 1753 and was killed. Franklin himself dreamed of harnessing electricity. In 1749 he wrote that one day there could be an 'electric banquet', with a turkey killed by an electric shock. Franklin did not foresee electric resistance heating, but the turkey was to be roasted over a fire lit by an electric spark.

By 1882 there had been a few fires caused by overheated cables, and the twin risks of fire and shock were appreciated. Shock was the chief risk in arc lamp systems but, as the lamps were usually arranged high in the air, there was little or no danger to the public. There were, however, several fatal accidents among people working on the system. For domestic lighting, filament lamps were used and the main risk was fire. A rather casual attitude to the risk of shock at domestic voltages is illustrated in Crompton's 'Reminiscences'. Referring to one of the electrical exhibitions at the Crystal Palace, Crompton wrote:[1]

'The Prince of Wales, afterward Edward VII, and Princess Alexandra visited our exhibition more than once, and I personally had the honour of conducting her Royal Highness round, to point out and explain to her the electrical phenomena. I assured her that she could touch bare conductors carrying energy at 200 V pressure without danger. She made the experiment, and agreed with me that such a pressure was quite safe to use'.

The words of a man who not only served on the Institution's first Wiring Rules Committee, but twice became President!

Some arc lighting was used domestically, but only in the very largest houses. In his book 'Hatfield House', Lord David Cecil relates how that distinguished scientist and three times Prime Minister, the third Marquess of Salisbury, had electric light installed at Hatfield House in 1881:

'The installation was very dangerous. Apart from the risk of shocks, the naked wires on the Gallery ceiling were apt to break into flame. The family sitting beneath nonchalantly threw up cushions to put the fire out and then went on with their conversations'.

In December 1881 there was a fatal accident at Hatfield.[2] The house had 117 arc lamps supplied from a generator in the estate sawmill. Two wires ran between the house and the mill, so the potential between them must have been around 5000 V. For part of the route the wires ran, unprotected, along a 3 ft (~ 1 m) high wall. A 22 year old labourer, William Dimmock, was helping to lay a telephone wire between the house and the mill, and it seems that he fell and touched the lighting wires. It was stated that 'to avoid similar accidents in future, the wires would all be conveyed either underground or on poles out of reach'. The General Manager of the Brush Electric Light Company found it necessary to write to the press, saying that it was only fair to point out that, although his company had supplied the generator at Hatfield, they had nothing to do with the installation.[3]

The chief risk in filament installations was fire, which could be caused by overloaded cables getting hot, by short-circuits, or by bad joints, again overheating. According to Crompton,[4] there were at least six fires at the 1881 Paris Electrical Exhibition. Consequently, when the 1882 Electrical Exhibition was set up at Crystal Palace, fire insurers charged extra premiums for insuring the electric lighting exhibits. The *Electrician* was at pains to point out that the extra risk, and premium, was not really on account of the electric light.[5] It was because each exhibitor had to install not only a generator and steam engine, but his own boiler and furnace as well. The Institution possesses a splendid volume of photographs of the Paris Exhibition. There is nothing in it about electrical safety, but the photograph of the electric lift suggests the general level of safety consciousness. The cage was open so passengers could lean out. There were no doors with interlocks at the bottom; but a piece of rope was provided to stop people from walking underneath!

A.3 Insurance company rules

The first published rules relating to electrical risks came from insurance companies. The Phoenix Fire Office published its first in February 1882. Those rules were an appendix to a report prepared for the company by Musgrave Heaphy, who was a friend of Lord Salisbury.

Born in London in 1843, Heaphy served with Garibaldi in Italy in 1860-61, then settled down to a career in civil engineering, mainly in connection with railways and docks. He became an engineer and scientific adviser to the Phoenix Fire Office in 1874.[6] He did not become a member of the IEE until 1887, when he was proposed by W.H. Preece 'as being intimately associated with the progress of electric lighting'.[7] That association began with his report to the Phoenix, a document with a splendid long Victorian title: 'A report to the Phoenix Fire Office upon the various systems of electric lighting and the dangers attending them. Also a suggested table of requirements when the light is used in very hazardous risks'.

The very hazardous risks Heaphy had in mind were places 'such as Mills of almost every description, Theatres etc.', but the suggested table of requirements for those places was quickly adopted as the rule for all places that the Phoenix insured.

There were fourteen requirements in Heaphy's suggested rules. The first was that all wires should be so placed that they could be inspected. The conductors had to be capable of carrying safely 50 per cent more current than would ever be required. There were requirements about insulation and mechanical protection. If single-core cables supplied an arc lamp, the two conductors had to be at least 8 inches (0.20 m) apart and spaced 8 inches from all other wires, metal etc. Single-core wires for incandescent light had to be 2.5 inches (0.06 m) apart. All connections and cut-outs had to be placed so that no danger arose if they overheated. Arc lamps had to be enclosed so that particles of hot carbon could not drop off and cause a fire. One rule dealt with fuses – a safety device already in use:

'Whenever a branch wire is led off the main conductor to supply current for one or more incandescent lamps, a short length of lead or other fusible wire must be inserted between the main conductor and one end of the branch; and the lead wire must be of such a section and nature that if the current passing through it exceeds the normal current by 50 per cent, then it will fuse and disconnect the branch'.

Heaphy's rules were widely publicized. Heaphy himself summarized them in a letter to *The Times* in September 1882. The letter was designed to make householders think about their electrical installations. One piece of advice from Heaphy to the public was:

'See that no naked wires are used. Try and arrange that they be covered with a material rendered as incombustible as possible. This you can very easily test for yourself'.

The IEE has never advised householders to test the flammability of their wiring!

A.4 The first IEE rules

When the Council decided, on 11 May 1882, to appoint a Committee to consider rules for the prevention of fire risks, they named fifteen members. The President would have been a member *ex officio,* and two others were added later, Dr John Hopkinson and James Shoolbred, both of whom had practical experience of electric lighting. Thirteen of the eighteen either had been or were to become President. Crompton and Gordon seem to have taken the lead, although no chairman was named. They had a letter published in *The Times* the previous day, which drew favourable attention to Heaphy's Phoenix rules.[8]

The IEE's rules as they first appeared in the Journal are reproduced in Appendix B.

Council directed that a copy be sent to the Fire Offices Committee, to each of the principal newspapers and technical journals, and to each of the electric light companies. The separate pamphlet with the Rules did not appear until April of the following year, and although there is no record of any further discussion in Council or of any further meeting of the Committee, the rules in the pamphlet differ in some respects from those in the Journal.

The first six Rules are instructions about the dynamo machine – a reminder that with almost no public supply every lighting scheme began with a generator.

Eleven rules are about the wires, including junctions, switches and fuses. Rule 9 provides that the size of conductors should be such that no conductor reaches a temperature above 150 °F – rather more precise than the Phoenix rule which required cables to be able to carry 50 per cent more than the designed working current but did not specify how the maximum current-carrying capacity was determined. It was also better than the contemporary New York Board of Fire Underwriters' rule that 'the wires must never be unpleasantly warm'. Rule 16 deals, among other things, with protection from 'the depredations of rats or mice'. Rules 18 and 19 are about lamps and, surprisingly, only arc lamps are mentioned. Lanterns had to be insulated, although there was no concept of earthing. The last two rules were concerned with danger to people. No one should be exposed to more than 60 V, and the potential between two points in the same room should not exceed 200 V.

When the rules appeared in pamphlet form ten months later there were several changes, presumably reflecting comments made at the time. The rule which had required the insulation of the dynamo to be perfect was amended to read 'practically perfect'. The requirements for fixing dynamos on an insulating bed and that switches should have a base of some incombustible substance were dropped. The requirements for fuses at the junction of larger and smaller conductors was amended to permit the use of 'protectors': electromagnetically operated cut-outs which broke the circuit if the current exceeded a certain level. (The changeover from fuses to miniature circuit-breakers began in 1882.) The rule that all indoor wires had to be insulated was relaxed to allow bare wires on insulated supports, and the voltages to which people could be exposed were raised to 100 V mean on a.c. and 200 V on d.c. systems.

Those first rules are sound general advice rather than a wiring specification. They differ in intent, as well as in content, from the 17th Edition. The first rules were expressly 'for the guidance and instruction of those who have...electric lighting apparatus installed in their premises'. The preamble makes the point that the chief dangers of electricity arise from ignorance and the chief element of safety is the employment of skilled and experienced electricians to supervise the work. Hilaire Belloc made the same point, rather more forcibly:

> Lord Finchley tried to mend the electric light
> Himself. It struck him dead – and serve him right!
> For 'tis the duty of the wealthy man
> To give employment to the artisan.

'Skilled and experienced electricians' meant, of course, members of the IEE. The advice was that every electrical installation should be supervised by, in modern terms, a chartered electrical engineer. The actual work was done by Belloc's artisans, the people we now call electricians. According to Crompton, the occupation electrician came in to being in 1882. He had arranged a lighting installation for Octavious Coope, the brewer, at a house he had just built in Essex, Berechurch Hall.

'This was the first time that a new house and premises had been lighted throughout with electric light. It was the occasion, moreover, of a new departure in handicraft. In order to deal with the novel work of wiring and fixing, I took over from the builder of the house a party of his workmen, whose previous trade had been that of bellhangers. Several of these men gradually bettered themselves, and eventually went in to business as master contractors for electrical wiring of houses' (Reference 1, p. 97).

The Rules were criticized on a variety of grounds. One letter in the *Electrician* ran to more than two pages of detailed comment, mainly about alleged obscurities and ambiguities.[10] A writer the following week defended the Institution:[11]

'The English of the rules, although in some cases awkward, is not obscure, and its awkwardness is only what one might expect from rules. . .originally drafted in good English but afterwards amended in committee'.

A.5 Acceptance of the IEE rules

The IEE rules were not universally accepted at first. Fire insurers, and not only the Phoenix, continued to produce their own. So did some consultants. Killingworth Hedges, for example, a member of the IEE and a consulting engineer concerned with house lighting, published a book in 1886 entitled 'Precautions on introducing the electric light'. He included a specimen set of instructions which 'should be hung up in the engine room and should be read over and explained to the man in charge'. It included the injunction 'Never cool an overheated dynamo with water'.

Most rules are built up on practical experience, and early dynamos did overheat frequently. Somewhere in the unrecorded and now forgotten history of electrical engineering must be the pioneer who did cool an overheated dynamo with a bucket of water.

In May 1887 the Wiring Rules Committee was resurrected, under the chairmanship of W.H. Preece, with a view to updating the rules. They sought the cooperation of the fire insurers, but as Preece told the Institution afterwards:[12]

'We found that when we brought these gentlemen connected with the fire offices together it was something like attempting to mix oil and water; it was a signal for a storm, and the result was a general electrical disturbance'.

Preece was seeking Institution rules setting out general principles followed by detailed 'Fire Office Rules', showing how the principles were to be applied in practice. But they could not agree. The second edition of the IEE rules, in April 1888, was just a revised and expanded version of the first. Additions included rules for transformers in systems where distribution was by fairly high voltage alternating current and each house or pair of houses had a step-down transformer. The concern was to make sure that the high voltage could not get into the house if the transformer insulation failed.

Before the third edition was produced, in July 1897, the IEE tried again to persuade the insurance companies to agree on a common set of rules, but with little success.[13]

By the end of the century there were a variety of rules produced by insurers and supply undertakings, with differing requirements. An Associate Member of the Institution, J. Pigg, presented a paper in January 1899 comparing the requirements of various bodies. He found twenty-six sets of rules. The differences may be illustrated by the insulation requirements then in force, which not only varied, but were expressed in different ways:

'The IEE required that an installation should be tested twice, 15 days apart, at twice the working voltage, and the insulation resistance should be not less than 10 megohms divided by the maximum number of amperes taken'.

'The Phoenix required an installation resistance of at least one megohm in a 12-light installation, and correspondingly less in bigger ones'.

'The Board of Trade required that current leakage should not exceed one ten-thousandth of the full load current. (That requirement was written into the 1937 Supply Regulations)'.

'The Liverpool & London & Globe Insurance Company demanded insulation such that leakage was less than one twenty-thousandth part of the maximum load current'.

Chaos reigned! There was clearly a need for one agreed set of rules.

In August 1899 the Secretary of the IEE wrote a circular letter to fire insurance companies and other interested people:[14]

'You are no doubt aware that members of our profession who wire premises for the supply of electrical energy have for some time been complaining that they are harassed by the multiplicity of existing rules. It is generally admitted that the rules framed by a committee of experts representative of the special interests involved, and published by this Institution, have been accepted by a large number of central-station, municipal, and consulting engineers, inspectors of fire insurance offices, and others. The Council of this Institution has appointed a Committee to enquire into the reasons, if there be any, which prevent the Institution rules being accepted by everyone as the standard. The Committee feel that it is very probable that most of the authorities who are using other rules might be induced to accept those of the Institution if these latter were modified in some particulars. As a first step, therefore, towards standardising of rules, the Committee are anxious to receive any reasonable suggestions that may be offered with the above object. I should be glad if you would kindly let me know whether you would, if called upon, be willing to assist the Committee with your advice and criticisms of the existing rules'.

The Fire Officers' Committee, representing the main fire insurers, met on 26th January 1900 to consider the new approach from the IEE. Musgrave Heaphy represented the Phoenix, and the notes he made before the meeting are preserved in the Company's archives.[15] He noted several objections to adopting the IEE rules:

1. First, the Phoenix rules had proved very satisfactory in practice. Why change?
2. The IEE rules were too rigid, and some elasticity was desirable in dealing with insurance cases. 'Hard and fast rules under certain conditions do not always make for safety besides often causing great irritation to the Assured'.
3. If they accepted the IEE rules it would be difficult to make changes quickly when they became necessary.

Heaphy's notes included a list of insurers who were agreed in opposing the IEE rules, and a few rude remarks to be made about the Institution, which 'is largely made up of Contractors, Professors, and Electricians who are desirous of installations being placed as cheaply as possible, in order to increase the consumption current'.

In July 1902 the Institution circulated the draft of the 4th Edition. Two fire office experts had agreed to serve on the Committee, but most insurers still declined to cooperate. The Phoenix pointed out that their rules were in their 31st edition; amply illustrating Heaphy's point that they could change their rules more quickly than could the Institution. When the 4th Edition was published, early in 1903, Heaphy announced that he proposed to take no notice.[16] Opposition was crumbling, however, and several companies did accept the new IEE rules. At the end of 1904 Heaphy retired, which probably made it easier for companies to change their attitude to the Institution. Inside the 7th Edition, published in 1916, is a list of fifty insurers who had adopted the IEE rules. At the time there were about 130 fire insurance companies in Britain, although most of the others were small firms who acted through larger ones, and it seems unlikely that any were still insisting on their own separate wiring rules.

A.6 Development of the IEE rules

Heaphy's assertion that the IEE wanted to encourage cheap installations so as to boost the sales of electricity was not quite fair, but there were certainly economic pressures on the designers of electrical installations. From 1895 the Board of Trade permitted distribution and domestic supplies at voltages up to 250 V, instead of 100 V. The advantage for the supply undertakings was, of course, that the capacity of their mains was more than doubled at no extra cost. But many customers had to rewire their premises to satisfy their insurers, and usually the supply undertakers had to pay.

Test conditions were first specified in the third edition of the rules, in 1897. The first two editions had advocated testing, but said nothing about how tests should be carried out. The 3rd Edition required an insulation test at twice the working voltage. The well known 'Megger' direct-reading ohmmeter was developed for this test.[17] The 1897 rules also called for the earthing of frames of machines operating at above 250 V, although no earthing was required on any equipment operating at domestic voltages.

When the supply industry urges people to use electricity more efficiently, the result can be a fall in demand; that happened around 1907, when tungsten filament lamps began to replace carbon filament ones with a five-fold increase in efficiency. To stimulate demand the industry encouraged people to use electricity for purposes other than lighting, and that added to the pressure for cheaper wiring. One way of reducing costs was to wire up extensions using flexible cables. In 1897 the 3rd Edition banned the practice, although the 4th Edition, in 1903, and the fifth, in 1907, permitted flexible conductors suitably fixed.

The 5th Edition banned the common practice of fitting fuses in wall sockets, ceiling roses and lampholders. During the Second World War the IEE considered permitting that again, but in the event the 12th Edition, in 1950, introduced the fused 13 A plug.

Earthing the metalwork of appliances working at domestic voltages was first demanded in the 8th Edition, in 1924. The fact that an adequate earth is difficult to obtain in some circumstances was recognized by the 9th Edition, in 1930, which introduced a requirement for an earth leakage trip operating at a leakage current not exceeding 30 mA.

The 10th Edition, in 1934, began with 'General Rule 1'. 'Good workmanship is an essential requirement for compliance with these Regulations'. Although the wording has varied that rule remains, although perhaps it is only another way of saying what the first edition said: 'the chief element of safety is the employment of skilled and experienced electricians'.

A.7 Have the rules succeeded?

During the debate, early in the 20th century, about the merits or otherwise of a uniform set of rules, the Phoenix's proud boast was that in no case had they had to pay compensation for an electrically caused fire and therefore 'it is impossible for the Phoenix Rules to be improved upon'.[16]

We have Wiring Rules to ensure that things which could be dangerous are kept safe. The test of the success or otherwise of the Rules must therefore be found in the accident statistics. Over the years, the number of fatalities per unit of consumption has fallen dramatically. In round figures, however, a hundred people are killed in Britain each year by electric shock or by fires caused by electrical faults. Musgrave Heaphy would undoubtedly say they were installations that did not comply with his rules. He and his staff could inspect every installation they insured, but regular inspections of every installation today would be an enormous task. A move to periodic testing was made in the 11th Edition of the IEE rules, 1939. Regulation 6 called for 'an incorrodible white tablet of ivorine, not less than 4 multiplied by 2 in' (0.1×0.05 m) fixed on or near each main distribution board, with the legend:

> **IMPORTANT**
>
> This installation should be periodically inspected and tested, and a report on its condition obtained, as prescribed in the Regulations for the Electrical Equipment of Buildings issued by the Institution of Electrical Engineers.

A test every five years was envisaged, but the rules only required the label to be there. There was no rule that the test must be carried out in order for the installation to comply.

If there is a lesson that emerges from surveying the history of the Wiring Rules it is that the rules can only be a guide to safe practice. The chief element of safety has always been a knowledge of the dangers. We can update the rules from time to time to cover new materials and new techniques, but greater safety can only come from increasing the general electrical awareness of the ordinary man and woman. Preece said:

'There is a danger connected with the burning of a candle, but no insurance office has ever issued a rule that you are not to burn a candle at both ends. . .Nobody would dream of issuing rules to meet self-evident dangers'.

A.8 Acknowledgements

The author is grateful to Mrs E.D.P. Symons, IEE Archivist, Mr Ray Tye, Phoenix Assurance Co. Archivist, and Mrs Rachel Lawrence for information and discussions.

A.9 References

The Council Minutes and copies of all editions of the Wiring Regulations are in the IEE Archives.

1. Crompton, R.E.B., *Reminiscences* (Constable, 1928), p. 100
2. 'Electrical lighting – fatal accident', *Electrician,* 17 December 1881, **8**, p.68
3. Humphreys, J., letter to the editor, *Electrician,* 24 December 1881, **8**, p.89
4. Letter, *The Times,* 10 May 1882
5. 'Fire risk at the Crystal Palace', *Electrician,* 25 March 1882, **8**, p.299
6. Press cuttings in the Phoenix Assurance Company's archives
7. Application form in the IEE Archives
8. *The Times,* 10 May 1882
9. 'Risks attaching to electric light wires', *Electrician,* 18 February 1882, **8**, p.211
10. Moseley, W., letter, *Electrician,* 29 July 1882, **9,** pp. 249–251
11. Letter (signed 'Electric Light'), *Electrician,* 5 August 1882, **9,** pp. 278–279
12. Preece, W.H., 'On fire risks and fire office rules', *J. Soc. Telegraph Engineers,* 1888, **17** (74), p. 479
13. Letter from the IEE to the Phoenix Assurance Company, 2 July 1902
14. 'Wiring Rules', *Electrician,* 18 August 1899, **43**, p.589
15. Heaphy, M., Papers in the Phoenix Assurance Company's Archives
16. 'The Institution Wiring Rules', *Electrician,* 27 March 1903, **50,** pp. 927–8
17. Mellanby, J., *The History of Electric Wiring* (Macdonald, 1957), p. 37

Appendix B
The first IEE rules and regulations

The copy of the rules and regulations below was published in the *Journal of the Society of Telegraph Engineers and of Electricians*, Volume 11 (1882), pp. 361–4, and is dated June 21, 1882. The version printed in the previous edition of the Commentary is not as early and is dated 11 April 1883.

RULES AND REGULATIONS, ETC.

Society of Telegraph Engineers and of Electricians

RULES AND REGULATIONS

FOR THE PREVENTION OF FIRE RISKS ARISING FROM ELECTRIC LIGHTING,

Recommended by the Council in accordance with the Report of the Committee appointed by them on May 11, 1882, to consider the subject.

MEMBERS OF THE COMMITTEE.

Professor W G. Adams, F.R.S., *Vice-President.*
Sir Charles T. Bright.
T. Russell Crampton.
R. E. Crompton.
W Crookes, F.R.S.
Warren De la Rue, D.C.L., F.R.S.
Professor G. C. Foster, F.R.S., *Past President.*
Edward Graves.
J. E. H. Gordon.
Dr. J. Hopkinson, F.R.S.
Professor D. E. Hughes, F.R.S., *Vice-President.*
W H. Preece, F.R.S., *Past President.*
Alexander Siemens.
C. E. Spagnoletti, *Vice-President.*
James N. Shoolbred.
Augustus Stroh.
Sir William Thomson, F.R.S., *Past President.*
Lieut-Colonel C. E. Webber, R.E., *President.*

These rules and regulations are drawn up not only for the guidance and instruction of those who have electric lighting apparatus installed on their premises, but for the reduction to a minimum of those risks of fire which are inherent to every system of artificial illumination.

The chief dangers of every new application of electricity arise mainly from ignorance and inexperience on the part of those who supply and fit up the requisite plant.

The difficulties that beset the electrical engineer are chiefly internal and invisible, and they can only be effectually guarded against by 'testing', or probing with electric currents. They depend chiefly on leakage, undue resistance in the conductor, and bad joints, which lead to waste of energy and the production of heat. These defects can only be detected by measuring, by means of special apparatus, the currents that are either ordinarily or for the purpose of testing, passed through the circuit. Bare or exposed conductors should always be within visual inspection, since the accidental falling on to, or the thoughtless placing of other conducting bodies upon such conductors might lead to 'short circuiting', or the sudden generation of heat due to a powerful current of electricity in conductors too small to carry it.

It cannot be too strongly urged that amongst the chief enemies to be guarded against, are the presence of moisture and the use of 'earth' as part of the circuit. Moisture leads to loss of current and to the destruction of the conductor by electrolytic corrosion, and the injudicious use of 'earth' as a part of the circuit tends to magnify every other source of difficulty and danger.

The chief element of safety is the employment of skilled and experienced electricians to supervise the work.

I. The Dynamo Machine.

1. The dynamo machine should be fixed in a dry place.
2. It should not be exposed to dust or flyings.
3. It should be kept perfectly clean and its bearings well oiled.
4. The insulation of its coils and conductors should be perfect.
5. It is better, when practicable, to fix it on an insulating bed.
6. All conductors in the Dynamo Room should be firmly supported, well insulated, conveniently arranged for inspection, and marked or numbered.

II. The Wires.

7. Every switch or commutator used for turning the current on or off should be constructed so that when it is moved and left to itself it cannot permit of a permanent arc or of heating, and its stand should be made of slate, stoneware, or some other incombustible substance.

8. There should be in connection with the main circuit a safety fuse constructed of easily fusible metal which would be melted if the current attain any undue magnitude, and would thus cause the circuit to be broken.

9. Every part of the circuit should be so determined, that the gauge of wire to be used is properly proportioned to the currents it will have to carry, and changes of circuit from a larger to a smaller conductor, should be sufficiently protected with suitable safety fuses so that no portion of the conductor should ever be allowed to attain a temperature exceeding 150 °F.

N.B.—These fuses are of the very essence of safety. They should always be enclosed in incombustible cases. Even if wires become perceptibly warmed by the ordinary current, it is a proof that they are too small for the work they have to do, and that they ought to be replaced by larger wires.

10. Under ordinary circumstances complete metallic circuits should be used, and the employment of gas or water pipes as conductors for the purpose of completing the circuit, should in no case he allowed.

11. Where bare wire out of doors rests on insulating supports it should be coated with insulating material, such as india-rubber tape or tube, for at least two feet on each side of the support.

12. Bare wires passing over the tops of houses should never be less than seven feet clear of any part of the roof, and they should invariably be high enough, when crossing thoroughfares, to allow fire escapes to pass under them.

13. It is most essential that the joints should be electrically and mechanically perfect. One of the best joints is that shown in the annexed sketches. The joint is whipped around with small wire, and the whole mechanically united by solder.

14. The position of wires when underground should be efficiently indicated, and they should be laid down so as to be easily inspected and repaired.

15. All wires used for indoor purposes should be efficiently insulated.

16. When these wires pass through roofs, floors, walls, or partitions, or where they cross or are liable to touch metallic masses, like iron girders or pipes, they should be thoroughly protected from abrasion with each other, or with the metallic masses, by suitable additional covering; and where they are liable to abrasion from any cause, or to the depredations of rats or mice, they should be efficiently encased in some hard material.

17. Where wires are put out of sight, as beneath flooring, they should be thoroughly protected from mechanical injury, and their position should be indicated.

N.B.—The value of frequently testing the wires cannot be too strongly urged. It is an operation, skill in which is easily acquired and applied. The escape of electricity cannot be detected by the sense of smell, as can gas, but it can be detected by apparatus far more certain and delicate. Leakage not only means waste, but in the presence of moisture it means destruction of the conductor and its insulating covering, by electric action.

III. Lamps.

18. Arc lamps should always be guarded by proper lanterns to prevent danger from falling incandescent pieces of carbon, and from ascending sparks. Their globes should be protected with wire netting.

19. The lanterns, and all parts which are to be handled, should be insulated from the circuit.

IV. Danger to Person.

20. To secure persons from danger inside buildings, it is essential so to arrange the conductors and fittings, that no one can be exposed to the shocks of alternating currents exceeding 60 volts; and that there should never be a difference of potential of more than 200 volts between any two points in the same room.

21. If the difference of potential within any house exceeds 200 volts, whether the source of electricity be external or internal, the house should be provided outside with a 'switch', so arranged that the supply of electricity can be at once cut off.

By Order of the Council.

F. H. WEBB, *Secretary.*

Offices of the Society,
4, The Sanctuary, Westminster,
June 21, 1882.

Appendix C
Tables

Table C.1 Values of resistance/metre for copper and aluminium conductors at 20 °C in milliohms/metre (see section 4.11)

Cross-sectional area (mm^2)		Resistance/metre or $(R_1 + R_2)$/m ($m\Omega$/m)	
Line conductor	Protective conductor	Copper	Aluminium
1	–	18.1	
1	1	36.2	
1.5	–	12.1	
1.5	1	30.2	
1.5	1.5	24.2	
2.5	–	7.41	
2.5	1	25.51	
2.5	1.5	19.51	
2.5	2.5	14.82	
4	–	4.61	
4	1.5	16.71	
4	2.5	12.02	
4	4	9.22	
6	–	3.08	
6	2.5	10.49	
6	4	7.69	
6	6	6.16	
10	–	1.83	
10	4	6.44	
10	6	4.91	
10	10	3.66	
16	–	1.15	1.91
16	6	4.23	–
16	10	2.98	–
16	16	2.30	3.82
25	–	0.727	1.20
25	10	2.557	–
25	16	1.877	–
25	25	1.454	2.40
35	–	0.524	0.87
35	16	1.674	2.78

Continues

Table C.1 Continued

Cross-sectional area (mm^2)		Resistance/metre or $(R_1 + R_2)$/m (mΩ/m)	
Line conductor	Protective conductor	Copper	Aluminium
35	25	1.251	2.07
35	35	1.048	1.74
50	–	0.387	0.64
50	25	1.114	1.84
50	35	0.911	1.51
50	50	0.774	1.28

Notes:

1. From BS 6360, Table 2.
2. For larger sizes – see Table C7.

Table C.2 Ambient temperature multipliers to be applied to Table 1.1 resistances to convert resistances at 20 °C to other ambient temperatures (see section 4.11 and Table 4.13)

Expected ambient temperature (°C)	Correction factor, C_{9B}
5	0.94
10	0.96
15	0.98
20	1.00
25	1.02
30	1.04

Note:

The correction factor is given by:

$$\{1 + 0.004(\text{ambient temp} - 20\ °\text{C})\}$$

where 0.004 is the simplified resistance coefficient per °C at 20 °C given by BS EN 60228 for copper and aluminium conductors.

Table C.3 Conductor temperature multiplier C_r, to convert conductor resistance at 20 °C to conductor resistance at conductor maximum operating temperature (Table 9C, On-Site Guide) (see section 4.11 and Table 4.10)

Conductor installation	Conductor operating temperature and conductor insulation type		
	70 °C thermoplastic (PVC)	85 °C thermosetting (Note 4)	90 °C thermosetting (Note 4)
Protective conductor not incorporated in the cable and not bunched with the cable (Notes 1, 3)	1.04	1.04	1.04
Conductors incorporated in a cable or bunched (Notes 2, 3)	1.20	1.26	1.28

Notes:

1. See Table 54.2 of BS 7671. These factors apply when protective conductor is not incorporated or bunched with cables, or for bare protective conductors in contact with cable covering. They correct from 20 °C to 30 °C.
2. See Table 54.3 of BS 7671. These factors apply when the protective conductor is a core in a cable or is bunched with cables.
3. The factors are given by $F = \{1 + 0.004$ (conductor operating temperature – 20 °C)$\}$, where 0.004 is the simplified resistance coefficient per °C at 20 °C given in BS 6360 for copper and aluminium conductors.
4. If cable loading is such that the maximum operating temperature is 70 °C, thermoplastic (70 °C) factors are appropriate.

Table C.4 Coefficients of resistance for conductors (see section 4.11 and Table 4.9)

Material	Coefficient of resistance α at 20 °C
Annealed copper	0.00393*
Hard drawn copper	0.00381
Aluminium	0.00403*
Lead	0.00400
Steel	0.0045

* An average value of 0.004 is often used for copper and aluminium.

Table C.5 Impedance of distribution transformers

Impedance of distribution transformers referred to 415, 480 or 240 V systems as appropriate

Transformer		Resistance per phase (Ω)	Reactance per phase (Ω)
Type	Rating (kVA)		
Single-phase two-wire in 240V system	5	0.4300	0.3620
	10	0.1910	0.2060
	15	0.1180	0.1460
	16	0.1080	0.1390
	25	0.0612	0.0944
	25*	0.0570	0.0920
	50	0.0266	0.0496
	50*	0.0270	0.0497
Single-phase three-wire in 240V system	25	0.0853	0.0943
	50	0.0393	0.0513
	100	0.0165	0.0255
Single-phase three-wire in 480V system	25	0.2330	0.3650
	50	0.1090	0.1950
	100	0.0445	0.1020
Three-phase in 415V system and 240V system	25	0.20800	0.2660
	50	0.08760	0.1440
	100	0.03710	0.0810
	200	0.01580	0.0406
	300	0.00948	0.0281
	315	0.00901	0.0268
	500	0.00509	0.0171
	750	0.00313	0.0115
	800	0.00291	0.0107
	1000	0.00219	0.0086

*Three-wire transformer with links arranged for two-wire output.

Source: Electricity Association Engineering Recommendation P28, Table D6

Table C.6 Maximum resistance and reactance values at 20 °C for PVC insulated single-core copper cables

Nominal area of conductors (mm^2)	Resistance r (mΩ/m)	Reactance x (mΩ/m)
35	0.5240	0.095
50	0.3870	0.094
70	0.2680	0.090
95	0.1930	0.090
120	0.1530	0.085
150	0.1240	0.085
185	0.0991	0.085
240	0.0754	0.083
300	0.0601	0.082

Notes:

1. The values of reactance given above apply only when two cables are installed touching throughout or when three cables are installed in trefoil formation touching throughout.
   Resistance values from BS 6360. Reactance values from BICC.
2. For other cable configurations resistance and reactance values for phase conductors can be derived from Table 4D1B of the Regulations at 70 °C.
   The resistance component of voltage drop in the three-phase column is divided by ($\sqrt{3} \times 1.20$) and the reactive component divided by $\sqrt{3}$.

For example:

Conductor area (mm^2)	Reference Methods A and B enclosed in conduit etc.				Reference Methods C and F in trefoil			
	Volt drop from 4D1B at 70 °C (V)		Impedance at 20 °C(Ω)		Volt drop from 4D1B at 70 °C(V)		Impedance at 20 °C(Ω)	
	r	x	r	x	r	x	r	x
95	0.42	0.24	0.20	0.14	0.41	0.155	0.20	0.089

Table C.7(a) *PVC cables: Impedance of conductor and armour for two-core, three-core and four-core cables having steel-wire armour*

Nominal cross-sectional area of conductor (mm^2)	Impedance in ohms per kilometre (Ω/km) of cable at 20 °C																	
	Copper conductor		Aluminium conductor		Steel-wire armour													
					Cables with stranded copper conductors								Cables with solid aluminium conductors					
					Two-core 600/1000 V		Three-core 600/1000 V		Four core (equal) 600/1000 V		Four-core (reduced neutral) 600/1000 V		Two-core 600/1000 V		Three-core 600/1000 V		Four-core 600/1000 V	
1	2		3		4		5		6		7		8		9		10	
	r	*x*	*r*	*x*	*r*	*x*	*r*	*x*	*r*	*x*	*r*	*x*	*r*	*x*	*r*	*x*	*r*	*x*
1.5*	12.100	–	–	–	10.70	–	10.2	–	9.50	–	–	–	–	–	–	–	–	
2.5*	7.410	–	–	–	9.10	–	8.8	–	7.90	–	–	–	–	–	–	–	–	–
4	4.610	–	–	–	7.50	–	7.0	–	4.60	–	–	–	–	–	–	–	–	–
6	3.080	–	–	–	6.80	–	4.6	–	4.10	–	–	–	–	–	–	–	–	–
10	1.830	–	–	–	3.90	–	3.7	–	3.40	–	–	–	–	–	–	–	–	–
16	1.150	0.09	1.910	0.09	3.50	–	3.2	–	2.20	–	–	–	–	–	–	–	–	–
25	0.727	0.09	1.200	0.09	2.60	–	2.4	–	2.10	–	–	–	3.7	–	3.40	–	2.40	–
35	0.524	0.08	0.868	0.08	2.40	–	2.1	–	1.90	–	2.10	–	2.9	–	2.50	–	2.30	–
50	0.387	0.08	0.641	0.08	2.10	0.3	1.9	0.3	1.30	0.3	1.90	–	2.7	–	2.30	–	2.00	–
70	0.268	0.08	0.443	0.08	1.90	0.3	1.4	0.3	1.20	0.3	1.70	0.3	2.4	0.3	2.00	0.3	1.40	–
95	0.193	0.08	0.320	0.08	1.30	0.3	1.2	0.3	0.98	0.3	1.20	0.3	2.1	0.3	1.40	0.3	1.30	0.3
120	0.153	0.08	0.253	0.08	1.20	0.3	1.1	0.3	0.71	0.3	1.00	0.3	1.5	0.3	1.30	0.3	1.10	0.3
150	0.124	0.08	0.206	0.08	1.10	0.3	0.74	0.3	0.65	0.3	0.73	0.3	–	–	1.20	0.3	0.78	0.3
185	0.0991	0.08	0.164	0.08	0.78	0.3	0.68	0.3	0.59	0.3	0.67	0.3	–	–	0.82	0.3	0.71	0.3
240	0.0754	0.08	0.125	0.08	0.69	0.3	0.60	0.3	0.52	0.3	0.60	0.3	–	–	0.73	0.3	0.64	0.3
300	0.0601	0.08	0.100	0.08	0.63	0.3	0.54	0.3	0.47	0.3	0.54	0.3	–	–	0.65	0.3	0.52	0.3
300	0.0601	0.08	0.100	0.08	–	–	–	–	–	–	0.49 (150 mm^2)	0.3	–	–	0.59	0.3	0.52	0.3
400	0.0470	0.08	–	0.08	0.56	0.3	0.49	0.3	0.34	0.3	0.47 (185 mm^2)	0.3	–	–	–	–	–	0.3
											0.35	0.3	–	–	–	–	–	–

*The values apply to cables with either solid or stranded conductors.

Source: BS 6346,Table 31

Table C.7(b) *Gross cross-sectional area of steel-wire armour for two-core, three-core and four-core 600/1000 V PVC insulated cables to BS 6346*

Nominal area of conductor (mm^2)	Cross-sectional area of round armour wires (mm^2)						
	Cables with stranded copper conductors				Cables with solid aluminium conductors		
	Two-core	Three-core	Four-core	Four-core (reduced neutral)	Two-core	Three-core	Four-core
1.5*	15	16	17	–	–	–	–
2.5*	17	19	20	–	–	–	–
4*	21	23	35	–	–	–	–
6*	24	36	40	–	–	–	–
10*	41	44	49	–	–	–	–
16*	46	50	72	–	42	46	66
25	60	66	76	76	54	62	70
35	66	74	84	82	58	68	78
50	74	84	122	94	66	78	113
70	84	119	138	135	74	113	128
95	122	138	160	157	109	128	147
120	131	150	220	215	–	138	201
150	144	211	240	235	–	191	220
185	201	230	265	260	–	215	245
240	225	260	299	289	–	240	274
300	250	289	333	323	–	265	304
400	279	319	467	452	–	–	–

*Circular conductors.

Source: BS 6346, Table 36

Table C.8(a) *Thermosetting cables: Impedance of conductor and armour for two-core, three-core and four-core cables having steel-wire armour*

Nominal cross-sectional area of conductor (mm^2)	Impedance in ohms per kilometre (Ω/km) of cable at 20 °C																	
	Copper conductor		Aluminium conductor		Steel-wire armour													
					Cables with stranded copper conductors								Cables with solid aluminium conductors					
					Two-core 600/1000 V		Three-core 600/1000 V		Four core (equal) 600/1000 V		Four-core (reduced neutral) 600/1000 V		Two-core 600/1000 V		Three-core 600/1000 V		Four-core 600/1000 V	
1	2		3		4		5		6		7		8		9		10	
	r	*x*	*r*	*x*	*r*	*x*	*r*	*x*	*r*	*x*	*r*	*x*	*r*	*x*	*r*	*x*	*r*	*x*
1.5*	12.100	–	–	–	9.40			–	8.50	–		–	–	–	–	–		–
2.5*	7.410	–	–	–	8.80	–	9.10	–	7.70	–	–	–	–	–	–	–	–	–
4	4.610	–	–	–	7.90	–	8.20	–	6.80	–	–	–	–	–	–	–	–	–
6	3.080	–	–	–	7.00	–	7.50	–	4.30	–	–	–	–	–	–	–	–	–
10	1.830	–	–	–	6.00	–	6.60	–	3.70	–	–	–	–	–	–	–	–	–
16	1.150	0.09	1.910	0.09	3.80	–	4.00	–	3.20	–	–	–	3.7	–	3.9	–	–	–
25	0.727	0.09	1.200	0.09	3.70	–	3.60	–	2.30	–	–	–	2.9	–	3.1	–	3.40	–
35	0.524	0.08	0.868	0.08	2.50	–	2.50	–	2.00	–	2.30	–	2.7	–	2.9	–	2.40	–
50	0.387	0.08	0.641	0.08	2.30	–	2.30	–	1.80	–	2.10	–	2.4	0.3	2.6	0.3	2.20	0.3
70	0.268	0.08	0.443	0.08	2.00	–	2.00	–	1.20	–	1.90	–	2.1	0.3	2.3	0.3	1.90	0.3
95	0.193	0.08	0.320	0.08	1.40	–	1.80	–	1.10	–	1.30	–	1.5	0.3	1.6	0.3	1.30	0.3
120	0.153	0.08	0.253	0.08	1.30	–	1.30	–	0.76	–	1.10	–	–	–	–	–	1.20	0.3
150	0.124	0.08	0.206	0.08	1.20	–	1.20	–	0.68	–	0.96	–	–	–	–	–	0.82	0.3
185	0.0991	0.08	0.164	0.08	0.82	–	0.78	–	0.61	–	0.71	–	–	–	–	–	0.74	0.3
240	0.754	0.08	0.125	0.08	0.73	–	0.71	–	0.54	–	0.63	–	–	–	–	–	0.67	0.3
300	0.0601	0.08	0.100	0.08	0.67	–	0.63	–	0.49	–	0.56	–	–	–	–	–	0.59	0.3
300	0.0601	0.08	0.100	0.08	–	–	0.58	–	–	–	0.52 (150 mm^2)	–	–	–	–	–	0.54	–
400	0.0470	0.08	–	0.08	–	–	–	–	0.35	–	0.49 (185 mm^2)	–	–	–	–	–	–	–
						–	0.52				0.46						–	

*The values given are for plain annealed copper conductors. For tinned conductors, reference should be made to BS 6360.
Source: BS 5467, Table 28

Table C.8(b) *Gross cross-sectional area of steel-wire armour for two-core, three-core and four-core 600/1000 V cables with thermosetting insulation to BS 5467*

Nominal area of conductor (mm^2)	Cross-sectional area of round armour wires (mm^2)						
	Cables with stranded copper conductors				Cables with solid aluminium conductors		
	Two-core	Three-core	Four-core	Four-core (reduced neutral)	Two-core	Three-core	Four-core
1.5*	16	17	18	–	–	–	–
2.5*	17	19	20	–	–	–	–
4*	19	21	23	–	–	–	–
6*	22	23	36	–	–	–	–
10*	26	39	43	–	–	–	–
16*	41	44	49	–	40	42	46
25	42	62	70	70	38	58	66
35	62	70	80	76	54	64	72
50	68	78	90	86	60	72	82
70	80	90	131	128	70	84	122
95	113	128	147	144	100	119	135
120	125	141	206	163	–	131	191
150	138	201	230	220	–	181	211
185	191	220	255	250	–	206	235
240	215	250	289	279	–	230	265
300	235	269	319	304	–	250	289
400	265	304	452	343	–	–	–

* Circular conductors.

Source: BS 5407, Table 33

Table C.9 Impedance values at 20 °C for PVC single-core cables having solid aluminium conductors and aluminium strip armour

Nominal area of conductor (mm^2)	Conductor resistance (mΩ/m) r	Conductor reactance (mΩ/m) x	Strip armour resistance	
			600/1000V	1900/3300V
600	0.0515	0.09	Not applicable, not used as earth conductor	

Table C.10 Cross-sectional areas of heavy gauge steel conduit and trunking to BS 4568 Pt 1

Nominal diameter (mm)	Minimum steel cross-sectional area (mm^2)
16	64.4
20	82.6
25	105.4
32	137.3

Steel surface trunking BS 4678-1 (sample sizes)

Nominal size (mm × mm)	Minimum steel cross-sectional area without lid (mm^2)
50 × 50	135
75 × 75	243
100 × 50	216
100 × 100	324
150 × 100	378

Steel underfloor trunking BS 4678-2 (sample sizes)

Nominal size (mm × mm)	Minimum steel cross-sectional area without lid (mm^2)
75 × 25	118
100 × 50	142
100 × 100	213
150 × 100	284

Table C.11 Impedance of steel conduit at 20 °C

Nominal conduit size (mm)	Typical impedance (mΩ/m) (Note 3)	
	Resistance r (Note 1)	Reactance x (Note 1)
Heavy gauge		
16	3.3	3.3
20	2.4	2.4
25	1.6	1.6
32	1.4	1.4
Light gauge		
16	4.5	4.7
20	3.7	3.7
25	2.1	2.1
32	1.5	1.5

Notes:

1. Typical values are taken at fault currents greater than 100 A. When I_f is less than 100 A, the tabulated impedances should be doubled.
2. When touch voltages on the conduit are being determined, the product of current and resistance (r) gives a good approximation. The inductance of the conduit is reflected into the enclosed cables and is not effective on the conduit itself.
3. The above values are at 20 °C but may be assumed to be independent of temperature and are used for design and verification.

Table C.12 Typical impedance at 20 °C of steel trunking to BS 4678 (Note 2)

Size (mm × mm)	Resistance r (mΩ/m)	Reactance x (mΩ/m)
50 × 37	2.96	2.96
50 × 50	2.44	2.44
75 × 50	1.75	1.75
75 × 75	1.37	1.37
100 × 50	1.52	1.52
100 × 75	1.21	1.21
100 × 100	0.87	0.87
150 × 50	1.05	1.05
150 × 75	0.87	0.87
150 × 100	0.81	0.81
150 × 150	0.52	0.52

Notes:

1. When determining touch voltages on the trunking, the product of current and resistance (r) gives a good approximation.
2. The above values may be assumed to be independent of temperature and are used for design and verification.

Table C.13 Typical impedance at 20 °C of steel underfloor trunking to BS 4678 Part 2 (Note 2)

Size (mm × mm)	Resistance r (mΩ/m)	Reactance x (mΩ/m)
75 × 25	1.28	1.28
75 × 37.5	1.16	1.16
100 × 25	1.08	1.08
100 × 37.5	0.99	0.99
150 × 25	0.74	0.74
150 × 37.5	0.69	0.69
225 × 25	0.52	0.52
225 × 37.5	0.49	0.49

Notes:

1. When determining touch voltages on the ducting, the product of current and resistance (r) gives a good approximation.
2. The above values are at 20 °C but may be assumed to be independent of temperature and are used for design and verification.

Table C.14 Resistance values for 500 V (light duty) mineral insulated multicore cables exposed to touch or PVC covered – copper

	20 °C		Exposed to touch 70 °C sheath		Not exposed to touch 105 °C sheath	
Cable ref.	R_1 Conductor resistance (Ω/km)	R_2 Sheath resistance (Ω/km)	R_1 Conductor resistance (Ω/km)	R_2 Sheath resistance (Ω/km)	R_1 Conductor resistance (Ω/km)	R_2 Sheath resistance (Ω/km)
2L1	18.10	3.95	21.87	4.47	24.50	4.84
2L1.5	12.10	3.35	14.62	3.79	16.38	4.10
2L2.5	7.41	2.53	8.95	2.87	10.03	3.10
2L4	4.61	1.96	5.57	2.22	6.24	2.40
3L1	18.10	3.15	21.87	3.57	24.50	3.86
3L1.5	12.10	2.67	14.62	3.02	16.38	3.27
3L2.5	7.41	2.23	8.95	2.53	10.03	2.73
4L1	18.10	2.71	21.87	3.07	24.50	3.32
4L1.5	12.10	2.33	14.62	2.64	16.38	2.85
4L2.5	7.41	1.85	8.95	2.10	10.03	2.27
7L1	18.10	2.06	21.87	2.33	24.50	2.52
7L1.5	12.10	1.78	14.62	2.02	16.38	2.18
7L2.5	7.41	1.36	8.95	1.54	10.03	1.67

Table C.15 Resistance values for 750 V (heavy duty) mineral insulated cables – copper

	20 °C		Exposed to touch 70 °C sheath		Not exposed to touch 105 °C sheath	
Cable ref.	R_1 Conductor resistance (Ω/km)	R_2 Sheath resistance (Ω/km)	R_1 Conductor resistance (Ω/km)	R_2 Sheath resistance (Ω/km)	R_1 Conductor resistance (Ω/km)	R_2 Sheath resistance (Ω/km)
1H10	1.83	2.23	2.21	2.53	2.48	2.73
1H16	1.16	1.81	1.40	2.05	1.57	2.22
1H25	0.727	1.40	0.878	1.59	0.984	1.72
1H35	0.524	1.17	0.633	1.33	0.709	1.43
1H50	0.387	0.959	0.468	1.09	0.524	1.18
1H70	0.268	0.767	0.324	0.869	0.363	0.94
1H95	0.193	0.646	0.233	0.732	0.261	0.792
1H120	0.153	0.556	0.185	0.63	0.207	0.681
1H150	0.124	0.479	0.15	0.542	0.168	0.587
1H185	0.101	0.412	0.122	0.467	0.137	0.505
1H240	0.0775	0.341	0.0936	0.386	0.105	0.418
2H1.5	12.10	1.90	14.62	2.15	16.38	2.33
2H2.5	7.41	1.63	8.95	1.85	10.03	2
2H4	4.61	1.35	5.57	1.53	6.24	1.65
2H6	3.08	1.13	3.72	1.28	4.17	1.38
2H10	1.83	0.887	2.21	1.005	2.48	1.09
2H16	1.16	0.695	1.40	0.787	1.57	0.852
2H25	0.727	0.546	0.878	0.618	0.984	0.669
3H1.5	12.10	1.75	14.62	1.98	16.38	2.14
3H2.5	7.41	1.47	8.95	1.66	10.03	1.8
3H4	4.61	1.23	5.57	1.39	6.24	1.51
3H6	3.08	1.03	3.72	1.17	4.17	1.26
3H10	1.83	0.783	2.21	0.887	2.48	0.959
3H16	1.16	0.622	1.40	0.704	1.57	0.762
3H25	0.727	0.50	0.878	0.566	0.984	0.613
4H1.5	12.10	1.51	14.62	1.71	16.38	1.85
4H2.5	7.41	1.29	8.95	1.46	10.03	1.58
4H4	4.61	1.04	5.57	1.18	6.24	1.27
4H6	3.08	0.887	3.72	1	4.17	1.09
4H10	1.83	0.69	2.21	0.781	2.48	0.845
4H16	1.16	0.533	1.40	0.604	1.57	0.653
4H25	0.727	0.423	0.878	0.479	0.984	0.518

Table C.15 Continued

	20 °C		Exposed to touch 70 °C sheath		Not exposed to touch 105 °C sheath	
Cable ref.	R_1 Conductor resistance (Ω/km)	R_2 Sheath resistance (Ω/km)	R_1 Conductor resistance (Ω/km)	R_2 Sheath resistance (Ω/km)	R_1 Conductor resistance (Ω/km)	R_2 Sheath resistance (Ω/km)
7H1.5	12.10	1.15	14.62	1.3	16.38	1.41
7H2.5	7.41	0.959	8.95	1.09	10.03	1.18
12H1.5	12.10	0.744	14.62	0.843	16.38	0.912
12H2.5	7.41	0.63	8.95	0.713	10.03	0.772
19H1.5	12.10	0.57	14.62	0.646	16.38	0.698

Note: The calculation of $R_1 + R_2$ for mineral insulated cables differs from other cables, in that the loaded conductor temperature is not tabulated. Table 4G1A of Appendix 4 of BS 7671 gives normal full load sheath operating temperatures of 70 °C for PVC sheathed types and 105 °C for bare cables not in contact with combustible materials, in a 30 °C ambient. Magnesium oxide is a relatively good thermal conductor, and, being in a thin layer, it is found that conductor temperatures are usually only some 3 °C higher than sheath temperatures.

The sheath is of copper to a different material standard to that of the conductors and the coefficient of resistance 0.004 does not apply. A coefficient of 0.00275 at 20 °C can be used for sheath resistance change calculations. Table C.15 gives calculated values of R_1 and R_2 at a standard 20 °C and at standard sheath operating temperatures, and these can be used directly for calculations at full load temperatures for devices in Appendix 3 of BS 7671.

Table C.16 Resistance values for 1000 V (heavy duty) mineral insulated single-core cables – copper conductors

Number and cross-sectional area of conductors	Effective sheath area*	Resistance at 20 °C		Cables exposed to touch: bare and PVC covered						Cables NOT exposed to touch: bare cables					
				Loop resistance at full load ($R_1 + R_2$)			Loop resistance during earth fault ($R_1 + R_2$)			Loop resistance at full load ($R_1 + R_2$)			Loop resistance during earth fault ($R_1 + R_2$)		
		Conductor	Sheath	Single-phase	Three-phase		Single-phase	Three-phase		Single-phase	Three-phase		Single-phase	Three-phase	
				2-wire	3-wire	4-wire	2-wire	3-wire	4-wire	2-wire	3-wire	4-wire	2-wire	3-wire	4-wire
(No. × mm^2)	(mm^2)	(mΩ/m)	(mΩ/m)	(mΩ/m)	(mΩ/m)	(mΩ/m)	(mΩ/m)	(mΩ/m)	(mΩ/m)	(mΩ/m)	(mΩ/m)	(mΩ/m)	(mΩ/m)	(mΩ/m)	(mΩ/m)
1 × 6	7.8	3.0800	2.20	4.70	4.30	4.10	5.70	5.30	5.10	5.00	4.60	4.40	5.90	5.50	5.20
1 × 10	9.5	1.8300	1.80	3.10	2.80	2.60	3.70	3.40	3.20	3.30	2.90	2.80	3.90	3.50	3.30
1 × 16	12	1.1500	1.50	2.10	1.90	1.70	2.60	2.30	2.10	2.30	2.00	1.80	2.60	2.30	2.20
1 × 25	15	0.7270	1.10	1.50	1.30	1.20	1.80	1.50	1.40	1.60	1.30	1.20	1.80	1.60	1.40
1 × 35	18	0.5240	0.97	1.10	1.00	0.90	1.40	1.20	1.00	1.20	1.00	0.90	1.40	1.20	1.10
1 × 50	22	0.3870	0.79	0.90	0.70	0.64	1.00	0.85	0.77	0.90	0.76	0.68	1.00	0.88	0.79
1 × 70	27	0.2680	0.64	0.70	0.54	0.48	0.75	0.65	0.57	0.70	0.57	0.51	0.77	0.66	0.59
1 × 95	32	0.1930	0.53	0.52	0.42	0.37	0.59	0.51	0.44	0.55	0.44	0.39	0.60	0.52	0.46
1 × 120	37	0.1530	0.46	0.43	0.35	0.30	0.49	0.41	0.36	0.46	0.37	0.32	0.50	0.42	0.37
1 × 150	44	0.1240	0.39	0.36	0.29	0.25	0.40	0.34	0.30	0.38	0.30	0.26	0.42	0.34	0.31
1 × 185	54	0.1010	0.32	0.29	0.23	0.20	0.33	0.27	0.24	0.31	0.25	0.21	0.34	0.28	0.25
1 × 240	70	0.0775	0.25	0.23	0.18	0.16	0.25	0.21	0.19	0.24	0.19	0.17	0.26	0.22	0.19

* The term *effective sheath area* is used because the conductivity of the copper used for the sheath is lower than that used for the conductors. The value shown is calculated as if the materials were the same and enables a direct comparison to be made between conductor and sheath cross-sectional areas.

Note:

When using single-core cables the protective conductor is made up of: 2 sheaths in parallel for single-phase circuits; 3 sheaths in parallel for three-phase, three-wire circuits and 4 sheaths in parallel for three-phase, four-wire circuits.

For cables with phase conductors greater than 35 mm^2 the inductance may need to be considered; however, for single-core cables the inductance will vary with the method of installation and the manufacturer should be contacted for further information.

Table C.17 Correction factors for change of protective conductor resistance with temperature

The product of these correction factors and the conductor resistance at 20 °C gives the increase in resistance.

Conductor type	Conductor material	Insulation of protective conductor or cable covering							
		70 °C PVC		90 °C PVC		85 °C rubber		95 °C thermosetting	
		full load	fault	full load	fault	full load	fault	full load	fault
Table 54.2: insulated protective conductor not incorporated in a cable and not bunched with cables, or for separate bare protective conductor in contact with cable covering but not bunched with cables.	Copper	0.040	0.300	0.040	0.300	0.040	0.420	0.040	0.480
	Aluminium	0.040	0.300	0.040	0.300	0.040	0.420	0.040	0.480
	Steel	0.045	0.338	0.045	0.338	0.045	0.450	0.045	0.540
	Assumed initial temperature	20 °C	20 °C	20 °C	20 °C	20 °C	20 °C	20 °C	20 °C
	Final temperature	30 °C	95 °C	30 °C	95 °C	30 °C	125 °C	30 °C	140 °C
Table 54.3: conductor incorporated in a cable or bunched with cables	Copper	0.20	0.38	0.28	0.42	0.26	0.53	0.28	0.60
	Aluminium	0.20	0.38	0.28	0.42	0.26	0.53	0.28	0.60
	Assumed initial temperature	20 °C	20 °C	20 °C	20 °C	20 °C	20 °C	20 °C	20 °C
	Final temperature	70 °C	115 °C	99 °C	125 °C	85 °C	152 °C	90 °C	170 °C
Table 54.4: protective conductor as a sheath or armour of a cable	Aluminium	0.16	0.44	0.24	0.48	0.22	0.51	0.24	0.48
	Steel	0.18	0.50	0.27	0.54	0.25	0.58	0.27	0.54
	Lead	0.16	0.44	0.24	0.48	0.22	0.51	0.24	0.48
	Assumed initial temperature	20 °C	20 °C	20 °C	20 °C	20 °C	20 °C	20 °C	20 °C
	Final temperature	60 °C	130 °C	80 °C	140 °C	75 °C	148 °C	80 °C	140 °C

Appendix D

Maximum demand and diversity

(Originally taken from Appendix 4 of the 15th Edition of the Wiring Regulations and amended as necessary.)

(See also section 3.3 of this Commentary.)

This appendix gives some information on the determination of the maximum demand for an installation and includes the current demand to be assumed for commonly used equipment. It also includes some notes on the application of allowances for diversity.

The information and values given in this appendix are intended only for guidance because it is impossible to specify the appropriate allowances for diversity for every type of installation and such allowances call for special knowledge and experience. The figures given in Table D.2, therefore, may be increased or decreased as decided by the competent person responsible for the design of the installation concerned. For blocks of residential dwellings, large hotels, industrial and large commercial premises, the allowances should be assessed by a competent person.

The current demand of a final circuit is determined by adding the current demands of all points of utilization and equipment in the circuit and, where appropriate, making an allowance for diversity. Typical current demands to be used for this summation are given in Table D.1.

The current demand of a distribution circuit supplying a number of final circuits may be assessed by using the allowances for diversity given in Table D.2, which are applied to the total current demand of all the equipment supplied by that circuit and not by summing the current demands of the individual final circuits obtained as outlined above. In Table D.2 the allowances are expressed either as percentages of the current demand or, where followed by the letters f.l., as percentages of the rated full load current of the current-using equipment. The current demand for any final circuit which is a standard circuit arrangement complying with Appendix 8 (see Note at the end of this Appendix) is the rated current of the overcurrent protective device of that circuit.

An alternative method of assessing the current demand of a circuit supplying a number of final circuits is to add the diversified current demands of the individual circuits then apply a further allowance for diversity but with this method the allowances given in Table D.2 should not be used, the values to be chosen being the responsibility of the installation designer.

The use of other methods of determining maximum demand is not precluded where specified by a competent person. After the design currents for all the circuits have been determined, enabling the conductor sizes to be chosen, it is necessary to check that the design complies with the requirements of Part 4 of BS 7671 and that the limitation on voltage drop is met.

Table D.1 Current demand to be assumed for points of utilization and current-using equipment

Point of utilization or current-using equipment	Current demand to be assumed
Socket-outlets other than 2 A socket-outlets and 13 A socket-outlets	Rated current
2 A socket-outlets	At least 0.5 A
Lighting outlet*	Current equivalent to the connected load, with a minimum of 100 W per lampholder
Electric clock, shaver supply unit (complying with BS EN 61558-2-5), shaver socket-outlet (complying with BS 4573), bell transformer, and current-using equipment of a rating not greater than 5 VA	May be neglected
Household cooking appliance	The first 10 A of the rated current plus 30% of the remainder of the rated current plus 5 A if a socket-outlet is incorporated in the control unit
All other stationary equipment	British Standard rated current, or normal current

* Final circuits for discharge lighting must be arranged so as to be capable of carrying the total steady current, viz. that of the lamp(s) and any associated controlgear and also their harmonic currents. Where more exact information is not available, the demand in volt-amperes is taken as the rated lamp watts multiplied by not less than 1.8. This multiplier is based on the assumption that the circuit is corrected to a power factor of not less than 0.85 lagging, and takes into account controlgear losses and harmonic current.

Table D.2 Allowances for diversity

Purpose of final circuit fed from conductors or switchgear to which diversity applies	Type of premises		
	Individual household installations, including individual dwellings of a block	Small shops, stores, offices and business premises	Small hotels, boarding houses, guest houses etc.
1. Lighting	66% of total current demand	90% of total current demand	75% of total current demand
2. Heating and power (but see 3 to 8 below)	100% of total current demand up to 10 A + 50% of any current demand in excess of 10 A	100% f.l. of largest appliance + 75% f.l. of remaining appliances	100% f.l. of largest appliance + 80% f.l. of second largest appliance + 60% f.l. of remaining appliances
3. Cooking appliances	10 A + 30% f.l. of connected cooking appliances in excess of 10 A + 5 A if socket-outlet incorporated in control unit	100% f.l. of largest appliance + 80% f.l. of second largest appliance + 60% f.l. of remaining appliances	100% f.l. of largest appliance + 80% f.l. of second largest appliance + 60% f.l. of remaining appliances
4. Motors (other than lift motors, which are subject to special consideration)	Not applicable	100% f.l. of largest motor + 80% f.l. of second largest motor 60% f.l. of remaining motors	100% f.l. of largest motor + 50% f.l. of remaining motors
5. Water heater (instantaneous type)*	100% f.l. of largest appliance + 100% f.l. of second largest appliance + 25% f.l. of remaining appliances	100% f.l. of largest appliance + 100% f.l. of second largest appliance + 25% f.l. of remaining appliances	100% f.l. of largest appliance + 100% f.l. of second largest appliance + 25% f.l. of remaining appliances
6. Water heaters (thermostatically controlled)		No diversity allowable†	
7. Floor-warming installations		No diversity allowable†	

Continues

Table D.2 Continued

Purpose of final circuit fed from conductors or switchgear to which diversity applies	Type of premises		
	Individual household installations, including individual dwellings of a block	Small shops, stores, offices and business premises	Small hotels, boarding houses, guest houses etc.
8. Thermal-storage space-heating installations		No diversity allowable†	
9. Standard arrangements of final circuits in accordance with Appendix H (see Note)	100% of current demand of largest circuit + 40% of current demand of every other circuit	100% of current demand of largest circuit + 50% of current demand of every other circuit	
10. Socket-outlets other than those included in 9 above and stationary equipment other than those listed above	100% of current demand of largest point of utilization + 40% of current demand of every other point of utilization	100% of current demand of largest point of utilization + 75% of current demand of every other point of utilization	100% of current demand of largest point of utilization + 75% of current demand of every point in main rooms (dining rooms, etc.) + 40% of current demand of every other point of utilization

* For the purpose of this table an instantaneous water heater is deemed to be a water heater of any loading that heats water only while the tap is turned on and therefore uses electricity intermittently.

† It is important to ensure that the distribution boards and consumer units are of sufficient rating to take the total load connected to them without the application of any diversity.

- For water heaters, floor warming and storage heating installations, it is important to ensure that the distribution boards and consumer units are of sufficient rating to take the total load connected to them without the application of any diversity. For mixed loads, the DB/CU can be rated for the diversified load, allowing for potential expansion if required. For further information see *section 10.6.2* of the IET Commentary on rated diversity factor (RDF) for continuously and simultaneously loaded outgoing circuits.
- The current demand may be that estimated for example in accordance with Table D.1. Where the circuit is a standard circuit for household or similar installations the current demand is the rated current of the overcurrent protective device of the circuit.

Note: Appendix 8 of the 15th Edition is not reproduced in the 17th Edition but is found in the IET's *On-Site Guide* as Appendix H and in Guidance Note 1 as Appendix H.

Bibliography

Chapter 1

CENELEC Operational Procedures

IEC Operational Procedures[1]

CEN/CENELEC internal regulations

Chapter 2

Building Regulations 2000, Statutory Instrument No. 2531

Building (Scotland) Act 2003 (2003 asp 8)

Building Regulations (Northern Ireland) 2000 (as amended)

Construction (Design and Management) Regulations 2015 (CDM 2015)

(Contains public sector information licensed under the Open Government Licence v3.0.)

Construction Products Regulations (Statutory Instrument 1991 No. 1620)

Electrical Equipment (Safety) Regulations 1994

Electricity Safety, Quality and Continuity Regulations 2002 (as amended) and DTI guidance

(Contains public sector information licensed under the Open Government Licence v3.0.)

Electricity at Work Regulations 1989

Electromagnetic Compatibility (EMC) Regulations 2006 (SI 2006 No. 3418)

Health and Safety at Work etc. Act 1974 (OPSI)

[1] The author thanks the International Electrotechnical Commission (IEC) for permission to reproduce Information from its International Standards. All such extracts are copyright of IEC, Geneva, Switzerland. All rights reserved. Further information on the IEC is available from www.iec.ch. IEC has no responsibility for the placement and context in which the extracts and contents are reproduced by the author, nor is IEC in any way responsible for the other content or accuracy therein.

Memorandum of guidance on the Electricity at Work Regulations 1989 (HSE)

Supply of Machinery (Safety) Regulations 1992 (as amended)

Technical Standards for Compliance with Building Standards (Scotland) Regulations (OPSI)

Approved Document B: Fire safety (OPSI)

Approved Document J: Heat producing appliances (OPSI)

Approved Document L: Conservation of fuel and power (OPSI)

Chapter 3

Building Regulations (2010)
(Contains public sector information licensed under the Open Government Licence v3.0.)

Construction (Design and Management) Regulations 2015
(Contains public sector information licensed under the Open Government Licence v3.0.)

Electricity Safety, Quality and Continuity Regulations 2002 (as amended) and DTI guidance
(Contains public sector information licensed under the Open Government Licence v3.0.)

Technical Report No. 113: *Notes of guidance for the protection of private generating sets up to 5 MW for operation in parallel with electricity boards' distribution networks* (Energy Networks Association)

Engineering Recommendation G.59/3: *Recommendations for the connection of embedded generating plant to the regional electricity companies' distribution system* (Energy Networks Association)

IEC 64 (Secretariat) 254: *Estimation of maximum demand* (BSI)

Public Affairs Board Report: *The Possible Harmful Biological Effects of Low-level Electromagnetic Fields of Frequencies up to 300 GHz* (IET Position Statement, May 2006)

Chapter 4

CENELEC Technical Report PD CLC/TR50480

IEC TS 60479-1:2005/COR2:2013 *Effects of current on human beings and livestock - Part 1: General aspects*

IEC 71: *Insulation co-ordination* (BSI)

IEC 71A: *Supplement to Publication 71: Recommendations for insulation coordination*(BSI)

IEC 664: *Insulation co-ordination within low voltage systems including clearances and creepage distances for equipment* (BSI)

IEC 61200-413:1996 *Electrical Installation Guide, Clause 413: Explanatory notes to measures of protection against indirect contact by automatic disconnection of supply* (withdrawn)

Chapter 5

BS EN 60332 series *Tests on electrical and optical fibre cables*

Guidance Note 4: *Protection Against Fire* (London, IET)

Fire Statistics United Kingdom (Government Statistical Services)

(Contains public sector information licensed under the Open Government Licence v3.0.)

Safe Hot Water and Surface Temperatures (NHS Estates)

BS EN ISO 13732-1:2006 *Ergonomics of the thermal environment. Methods for the assessment of human responses to contact with surfaces. Hot surfaces*

Chapter 6

BS EN 60898-1 *Specification for Circuit Breakers for Overcurrent Protection for Household and Similar Installations*

CENELEC technical report CLC/TR 50480: Determination of cross-sectional area of conductors and selection of protective devices

CLC TR 50480 *Determination of cross-sectional area of conductors and selection of protective devices*

IEC TR 60943:1989+AMD1:2008 *Guidance concerning the permissible temperature rise for parts of electrical equipment, in particular for terminals*

IEC 949 (BS 7454): *Calculation of thermally permissible short-circuit currents, taking into account non-adiabatic heating effect* (BSI)

Engineering Recommendation P.25: *The short-circuit characteristics of electricity boards low voltage distribution networks and the co-ordination of overcurrent protective devices on 240 V single-phase supplies up to 100 A* (Energy Networks Association)

Engineering Recommendation P.26: *The estimation of the maximum prospective short-circuit current for three-phase 415 V supplies* (Energy Networks Association)

Electricity Safety, Quality and Continuity Regulations 2002 (SI 2002 No. 2665, OPSI)

(Contains public sector information licensed under the Open Government Licence v3.0.)

IEC 60287 *Electric cables. Calculation of the current rating. Current rating equations (100% load factor) and calculation of losses.* Series

Electra: CIGRE's Bilingual Bimonthly Journal for Power System Professionals, http://www.cigre.org/Electra

Chapter 7

BS EN 50522:2010 *Earthing of power installations exceeding 1 kV a.c*

IEC 61936-1: *Power installations exceeding 1 kV a.c. - Part 1: Common rules*

BS 6701:2010 *Code of practice for installation of apparatus intended for connection to certain telecommunication systems*

DD IEC/TS 60479-1:2005 *Effects of current on human beings and livestock. General aspects*

Central Networks Earthing Manual, Section E5, 'Standard Distribution Earthing Arrangements'

Distribution Code

Guide to the Distribution Code of Licensed Distribution Network Operators of Great Britain

Electricity Safety, Quality and Continuity Regulations 2002 (SI 2002 No. 2665, OPSI)

(Contains public sector information licensed under the Open Government Licence v3.0.)

Engineering Recommendation S36-1 *Procedure to identify and record hot substations*

Engineering Recommendation 41-24 *Framework for design and planning*

Engineering Recommendation P28 *Planning limits for voltage fluctuations caused by industrial, commercial and domestic equipment in the UK*

Chapter 8

BS 6701 Part 1:1980 *Code of practice for the installation of apparatus intended for connection to certain telecommunication systems*

HD 472: *Nominal voltages for low voltage public electricity supply systems*(BSI)

IEC 724: *Guide to the short-circuit temperature limits of electric cables with a rated voltage not exceeding 0.6/1.0 kV* (BSI)

IEC 949: *Calculation of thermally permissible short-circuit currents, taking into account non-adiabatic heating effects* (BSI)

Engineering Recommendation P.28: *Planning limits for voltage fluctuations caused by industrial, commercial, and domestic equipment in the United Kingdom* (Energy Networks Association)

IEC 364-4-444: *Protection against voltage disturbances and electromagnetic disturbances*(BSI)

IEC 364-5-548: *Earthing arrangements and equipotential bonding for information technology installations* (BSI)

Public Affairs Study Group Report: *Electromagnetic Interference* (IET)

Chapter 9

Guidelines on the Positioning and Colour Coding of Underground Utilities, Volume 1: Apparatus (National Joint Utilities Group)

Electricity Safety, Quality and Continuity Regulations 2002 (SI 2002 No. 2665, OPSI)

(Contains public sector information licensed under the Open Government Licence v3.0.)

Guidance on the Electricity Safety, Quality and Continuity Regulations 2002 (DTI, URN 02/1544)

(Contains public sector information licensed under the Open Government Licence v3.0.)

Memorandum of guidance on the Electricity at Work Regulations 1989 (HSE)

BS EN 62305 *Protection against lightning*

BS 6004 *Electric cables. PVC insulated, non-armoured cables for voltages up to and including 450/750 V, for electric power, lighting and internal wiring*

BS 6701:2004 *Telecommunications equipment and telecommunications cabling*

BS 6891:2005 *Installation of low-pressure gas pipework in domestic premises*

Sheath Voltages in Single Point Bonded Systems (ERA Report)

Chapter 10

Electricity Safety, Quality and Continuity Regulations 2002 (as amended) and DTI guidance

(Contains public sector information licensed under the Open Government Licence v3.0.)

IEC 64 (Austria) 50: *Proposal for a new standard for electricity installations in buildings* (BSI)

IEC 1200-413: *Explanatory notes to measures of protection against indirect contact by automatic disconnection of supply* (BSI)

Supply of Machinery (Safety) Regulations 2008 (SI 2008 No. 1597)
(Contains public sector information licensed under the Open Government Licence v3.0.)

BS 88 *Cartridge fuses for voltages up to and including 1000 V a.c. and 1500 V d.c.*

BS 1361 *Specification for cartridge fuses for a.c. circuits in domestic and similar premises*

BS 3036 *Specification. Semi-enclosed electric fuses (ratings up to 100 A and 240 V to earth)*

BS EN 60947 *Specification for low voltage switchgear and controlgear*

BS 5467 *Specification for cables with thermosetting insulation for electricity supply for rated voltages up to and including 600/1000 V and up to and including 1000/3000 V*

BS EN 60204 *Safety of Machinery*

BS EN 60664-1 *Insulation coordination for equipment within low-voltage systems*

BS EN 61439-1 *Low-voltage switchgear and controlgear assemblies. General rules*

BS EN 62305 *Protection against lightning*

BS 6724 *Specification for armoured cables for electricity supply having thermosetting insulation with low emission of smoke and corrosive gases when affected by fire*

Chapter 11

CENELEC technical report CLC TR 50480 *Determination of cross-sectional area of conductors and selection of protective devices*

R. Parr and E. Leary, *Earth fault impedance of steel-wire armour cable* (ERA Report No. 84-0067)

Temperature rise of cables passing through short lengths of thermal insulation (ERA Report No. 85-0111)

Fault current temperatures of flexible cords (ERA Report No. 89-0561)

Guidance for the design, installation, testing and maintenance of main earthing systems in substations (EATS 41-24, Energy Networks Association)

Chapter 12

Engineering Recommendation G83: *Recommendations for the connection of small scale embedded generators (up to 16 amperes per phase) in parallel with public low-voltage distribution networks* (Energy Networks Association)

Electricity Safety, Quality and Continuity Regulations 2002 (SI 2002 No. 2665, OPSI)
(Contains public sector information licensed under the Open Government Licence v3.0.)

Guidance on the Electricity Safety, Quality and Continuity Regulations 2002 (DTI, URN 02/1544)
(Contains public sector information licensed under the Open Government Licence v3.0.)

Technical Report No. 113: *Notes of guidance for the protection of private generating sets up to 5 MW for operation in parallel with electricity boards' distribution networks* (Energy Networks Association) replaced by G59/3

Engineering Recommendation G.59/3: *Recommendations for the connection of embedded generating plant to the regional electricity companies' distribution system*(Energy Networks Association)

Engineering Recommendation G83 *Recommendations for the Connection of Type Tested Small-scale Embedded Generators (Up to 16A per Phase) in Parallel with Low-Voltage Distribution Systems*

BS EN 50438 *Requirements for the connection of micro-generators in parallel with public low voltage distribution networks*

Chapter 13

BS 5266-1 *Emergency lighting*

BS 5839-1 *Fire detection and fire alarm systems*

Chapter 14

Guidance Note 3: *Inspection & Testing* (London, IET)

On-Site Guide (London, IET)

Chapter 16

The economical application of ambient and group rating factors (ERA Report No. 85-0027, by kind permission of the British Cable Makers' Confederation)

R. Parr, 'Derivation of coefficient C_a' (ERA paper)

R. Parr, 'Shock voltages in single-point bonded systems' (ERA paper)

Appendix C

The earth fault loop impedances of steel trunking (ERA Report No. 84-0066, by kind permission of the British Electrical Association)

Impedance of steel conduit (ERA Report No. 82-62)

Resources

Copies of publications are available from the following organisations.

BSI British Standards
Customer Services
389 Chiswick High Road
London W4 4AL
Tel: +44 (0)20 8996 9001
Fax: +44 (0)20 8996 7001
Email: orders@bsi-global.com

Energy Networks Association Ltd
6th Floor
Dean Bradley House
52 Horseferry Road
London SW1P 2AF
Tel: +44 (0)20 7706 5100
Fax: +44 (0)20 7706 5101
Email: richard.legros@energynetworks.org

ERA Technology Ltd
Cleeve Road
Leatherhead
Surrey KT22 7SA
Tel: +44 (0)1372 367000
Fax: +44 (0)1372 367099
Email: info@era.co.uk

HSE Books
PO Box 1999
Sudbury
Suffolk CO10 2WA
Tel: +44 (0)1787 881165
Email: hsebooks@prolog.uk.com
Web: www.hsebooks.com

Stationery Office Ltd
St Crispins
Duke Street
Norwich NR3 1PD
Tel: +44 (0)870 6005522
Email: customer.services@tso.co.uk
Web: www.tsoshop.co.uk

For other offices and locations see http://www.tso.co.uk/contact-us

Regulation index

Subject index

Note: page numbers in *italics* refer to figures; page numbers in **bold** refer to tables.

www.ingramcontent.com/pod-product-compliance
Lightning Source LLC
LaVergne TN
LVHW011858200926
848484LV00002B/476